Development of Visual Pathways in Mammals

Neurology and Neurobiology

Volume 1 • **Cytochemical Methods in Neuroanatomy,** Victoria Chan-Palay and Sanford L. Palay, *Editors*

Volume 2 • **Basic Mechanisms of Neuronal Hyperexcitability,** Herbert H. Jasper and Nico M. van Gelder, *Editors*

Volume 3 • **Anorexia Nervosa: Recent Developments in Research,** Padraig L. Darby, Paul E. Garfinkel, David M. Garner, and Donald V. Coscina, *Editors*

Volume 4 • **Clinical and Biological Aspects of Peripheral Nerve Diseases,** Leontino Battistin, George A. Hashim, and Abel Lajtha, *Editors*

Volume 5 • **The Physiology of Excitable Cells,** Alan D. Grinnell and William J. Moody, Jr., *Editors*

Volume 6 • **Developing and Regenerating Vertebrate Nervous Systems,** Penelope W. Coates, Roger R. Markwald, and Alexander D. Kenny, *Editors*

Volume 7 • **Glutamine, Glutamate, and GABA in the Central Nervous System,** Leif Hertz, Elling Kvamme, Edith G. McGeer, and Arne Schousboe, *Editors*

Volume 8 • **Catecholamines,** Earl Usdin, Arvid Carlsson, Annica Dahlström, and Jörgen Engel, *Editors*
Published in three volumes: *Part A: Basic and Peripheral Mechanisms; Part B: Neuropharmacology and Central Nervous System—Theoretical Aspects; Part C: Neuropharmacology and Central Nervous System—Therapeutic Aspects*

Volume 9 • **Development of Visual Pathways in Mammals,** Jonathan Stone, Bogdan Dreher, and David H. Rapaport, *Editors*

Volume 10 • **Monoamine Innervation of Cerebral Cortex,** Laurent Descarries, Tomas R. Reader, and Herbert H. Jasper, *Editors*

Volume 11 • **The Neurobiology of Zinc,** C.J. Frederickson, G.A. Howell, and E.J. Kasarskis, *Editors; Part A: Physiochemistry, Anatomy, and Techniques; Part B: Deficiency, Toxicity, and Pathology*

Development of Visual Pathways in Mammals

Proceedings of a Satellite Symposium of the XXIX International Congress of the Union of Physiological Sciences held in Sydney, Australia, August 24–27, 1983

EDITORS

Jonathan Stone
School of Anatomy
University of New South Wales
Kensington, Australia

Bogdan Dreher
Department of Anatomy
University of Sydney
Sydney, Australia

David H. Rapaport
School of Anatomy
University of New South Wales
Kensington, Australia

Alan R. Liss, Inc., New York

Address all inquiries to the Publisher
Alan R. Liss, Inc., 150 Fifth Avenue, New York, NY 10011

Printed in the United States of America.

Library of Congress Cataloging in Publication Data

Main entry under title:

Development of visual pathways in mammals.

(Neurology and neurobiology ; v. 9)
Includes bibliographies and index.
1. Mammals—Physiology—Congresses. 2. Mammals—Development—Congresses. 3. Visual pathways—Congresses.
4. Vision—Congresses. 5. Developmental neurology—Congresses. I. Stone, Jonathan, 1942- . II. Dreher, Bogdan. III. Rapaport, David H. IV. International Congress of Physiological Sciences (39th : 1983 : Sydney, Australia) V. Series.
QL739.2.D48 1984 599′.0332 84-10022
ISBN 0-8451-2710-1

Contents

Session Chairs

Lyn D. Beazley, Neurobiology Laboratory, Department of Psychology, University of Western Australia, Nedlands, Western Australia 6009, Australia

Jean Bullier, Laboratory of Experimental Neuropsychology, INSERM. Unité 94, 69500 Bron, France

Brian G. Cleland, Department of Physiology, University of Sydney, Sydney, New South Wales 2006, Australia

Brian Freeman, School of Anatomy, University of New South Wales, Kensington, New South Wales 2033, Australia

Geoff H. Henry, Department of Physiology; Australian National University, Canberra, A.C.T. 2601, Australia

William R. Levick, Department of Physiology, Australian National University, Canberra, A.C.T. 2601, Australia

Guy A. Orban, Laboratorium voor Neuro- en Psychofysiologie, Faculteit der Geneeskunde, Katholieke Universiteit Leuven, B-3000 Leuven, Belgium

Tauba Pasik, Department of Neurology; Mount Sinai School of Medicine of the City University of New York, New York, NY 10029

Ann J. Sefton, Department of Physiology, University of Sydney, Sydney, New South Wales 2006, Australia

Contributors and Speakers

Lyn D. Beazley, Neurobiology Laboratory, Department of Psychology, University of Western Australia, Nedlands, Western Australia 6009, Australia, **[75]**

Maxwell R. Bennett, Neurobiology Research Centre, University of Sydney, Sydney, New South Wales 2006, Australia **[39]**

Colin Blakemore,* University Laboratory of Physiology, University of Oxford, Oxford OX1 3PT, England **[463]**

Leo M. Chalupa,* Department of Psychology, University of California, Davis, CA 95616 **[89]**

David P. Crewther,* National Vision Research Institute, Carlton, Victoria 3053, and School of Optometry, University of New South Wales, Kensington, New South Wales 2033, Australia **[363]**

S.G. Crewther, National Vision Research Institute, Carlton, Victoria 3053, and School of Optometry, University of New South Wales, Kensington, New South Wales 2033, Australia **[363]**

Timothy J. Cunningham,* Department of Anatomy, The Medical College of Pennsylvania/EPPI Division, Philadelphia, PA 19129 **[315]**

Bogdan Dreher,* Department of Anatomy, University of Sydney, Sydney, New South Wales 2006, Australia **[xiii,39]**

Sarah A. Dunlop,* Neurobiology Laboratory, Department of Psychology, University of Western Australia, Nedlands, Western Australia 6009, Australia **[75]**

Stephen E. Fish, Neurobiology Program, Northeastern Ohio Universities College of Medicine, Rootstown, OH 44272 **[231]**

Michael J. Friedlander,* Department of Physiology and Biophysics, University of Alabama in Birmingham, Birmingham, AL 35294 **[155]**

Luiz de Gonzaga Gawryszewski, Instituto de Biofísica da UFRJ, Centro de Ciências da Saúde, Cidade Universitária, Rio de Janeiro 21941, Brazil **[179]**

Pierre Godement,* Laboratoire de Neurobiologie du Développement, Université Paris-Sud, 91405 Orsay, France **[127]**

Alan R. Harvey,* Unit of Human Physiology, Centre for Neuroscience, Flinders University of South Australia, Bedford Park, South Australia 5042, Australia **[303]**

Forrest A. Haun, Department of Anatomy, The Medical College of Pennsylvania/EPPI Division, Philadelphia, PA 19129 **[315]**

Helmut V.B. Hirsch,* Neurobiology Research Center, State University of New York at Albany, Albany, NY 12222 **[425]**

*Denotes a speaker at the symposium.

A boldface number in brackets is the opening page number of a contributor's article.

Klaus-Peter Hoffmann,* Abteilung für Vergleichende Neurobiologie, Universität Ulm, D-7900 Ulm, Federal Republic of Germany

Hisako Ikeda,* Vision Research Unit of Sherrington School, The Rayne Institute, St. Thomas' Hospital, London SE1 7EH, England **[115]**

Giorgio M. Innocenti,* Institut d'Anatomie, Université de Lausanne, CH-1011 Lausanne-CHUV, Switzerland **[243]**

Jon H. Kaas, Department of Psychology, Vanderbilt University, Nashville, TN 37240 **[289]**

Ronald E. Kalil,* Department of Ophthalmology, University of Wisconsin, Madison, WI 53706 **[257]**

Janusz J. Kulikowski,* Visual Sciences Laboratory, Department of Ophthalmic Optics, UMIST, Manchester M60 10D, England

Audie G. Leventhal,* Department of Anatomy, University of Utah School of Medicine, Salt Lake City, UT 84132 **[347]**

Raymond D. Lund, Department of Cell Biology and Anatomy, School of Medicine, University of Pittsburgh, Pittsburgh, PA 15261 **[303]**

Rafael Malach, School of Optometry, University of California, Berkeley, CA 94720; present address: Department of Psychology, Massachusetts Institute of Technology, Cambridge, MA 02139 **[393]**

Juliani Maslim, School of Anatomy, University of New South Wales, Kensington, New South Wales 2033, Australia **[3]**

Rosalia Méndez-Otero, Instituto de Biofísica da UFRJ, Centro de Ciências da Saúde, Cidade Universitária, Rio de Janeiro 21941, Brazil **[179]**

Donald E. Mitchell,* Psychology Department, Dalhousie University, Halifax, Nova Scotia B3H 4J1, Canada **[381]**

Richard D. Mooney, Department of Anatomy, University of Medicine and Dentistry of New Jersey–School of Osteopathic Medicine and Rutgers Medical School, Piscataway, NJ 08854 **[231]**

Dean M. Murakami, Department of Psychology, University of California, Riverside, CA 92521 **[331]**

E. Hazel Murphy,* Department of Anatomy, The Medical College of Pennsylvania/EPPI Division, Philadelphia, PA 19129 **[439]**

Kathryn M. Murphy, Psychology Department, Dalhousie University, Halifax, Nova Scotia B3H 4J1, Canada **[381]**

Jeanine Nguyen-Legros,* INSERM Unité 154, Hôpital Saint-Vincent de Paul, 75674 Paris 14, France **[103]**

Solomon Y.K. Ni, Department of Anatomy, University of Sydney, Sydney, New South Wales 2006, Australia **[39]**

John D. Parsons, Department of Psychology, University of California, Riverside, CA 92521 **[331]**

Pedro Pasik,* Department of Neurology, Mount Sinai School of Medicine of the City University of New York, New York, NY 10029 **[175]**

Tauba Pasik, Department of Neurology, Mount Sinai School of Medicine of the City University of New York, New York, NY 10029 **[175]**

V. Hugh Perry,* Department of Experimental Psychology, University of Oxford, Oxford OX1 3UD, England **[59]**

John D. Pettigrew,* Department of Physiology and Pharmacology, University of Queensland, St. Lucia, Queensland 4067, Australia

Rebecca A. Potts, Department of Anatomy, University of Sydney, Sydney, New South Wales 2006, Australia **[39]**

Pasko Rakic,* Section of Neurotomy, School of Medicine, Yale University, New Haven, CT 06510

Ary Sabatino Ramôa, Instituto de Biofísica da UFRJ, Centro de Ciências da Saúde, Cidade Universitária, Rio de Janeiro 21941, Brazil **[179]**

David H. Rapaport,* School of Anatomy, University of New South Wales, Kensington, New South Wales 2033, Australia **[xiii,3,23]**

Robert W. Rhoades,* Department of Anatomy, University of Medicine and Dentistry of New Jersey–School of Osteopathic Medicine and Rutgers Medical School, Piscataway, NJ 08854 **[231]**

Jonathan Robbins, Vision Research Unit of Sherrington School, The Rayne Institute, St. Thomas' Hospital, London SE1 7EH, England **[115]**

Stephen R. Robinson, School of Anatomy, University of New South Wales, Kensington, New South Wales 2033, Australia **[23]**

Carlos Eduardo Rocha-Miranda,* Instituto de Biofísica da UFRJ, Centro de Ciências da Saúde, Cidade Universitária, Rio de Janeiro 21941, Brazil **[179]**

Ken J. Sanderson,* School of Biological Sciences, Flinders University of South Australia, Bedford Park, South Australia 5042, Australia **[145]**

Hiromichi Sato, Department of Physiology, Kanazawa University Medical School, Takaramachi, Kanazawa 920, Japan **[217]**

Peter D. Spear,* Department of Psychology, University of Wisconsin, Madison, WI 53706 **[275]**

Barry E. Stein,* Department of Physiology and Biophysics, Medical College of Virginia, Richmond, VA 23298 **[199]**

Jonathan Stone,* School of Anatomy, University of New South Wales, Kensington, New South Wales 2033, Australia **[xiii,3,23]**

Michael P. Stryker,* Department of Physiology, University of California, San Francisco, CA 94143

Kouhei Suda, Department of Physiology, Kanazawa University Medical School, Takaramachi, Kanazawa 920, Japan **[217]**

Brian Timney,* Department of Psychology, University of Western Ontario, London, Ontario N6A 5C2, Canada **[405]**

Tadaharu Tsumoto,* Department of Physiology, Kanazawa University Medical School, Takaramachi, Kanazawa 920, Japan **[217]**

Richard C. Van Sluyters,* School of Optometry, University of California, Berkeley, CA 94720 **[393]**

Claudine Versaux-Botteri, INSERM Unité 154, Hôpital Saint-Vincent de Paul, 75674 Paris 14, France **[103]**

Annette Vigny, Institut de Biologie Physico-Chimique, 75005 Paris, France **[103]**

François Vital-Durand, Laboratoire de Neuropsychologie Expérimentale, INSERM Unité 94, 69500 Bron, France **[463]**

Eliane Volchan, Instituto de Biofísica da UFRJ, Centro de Ciências da Saúde, Cidade Universitária, Rio de Janeiro 21941, Brazil **[179]**

Rosalyn E. Weller,* Department of Psychology, Vanderbilt University, Nashville, TN 37240 **[289]**

Robert W. Williams, Department of Psychology, University of California, Davis, CA 95616; present address: Section of Neuroanatomy, Yale University School of Medicine, New Haven, CT 06510 **[89]**

Paul D. Wilson,* Department of Psychology, University of California, Riverside, CA 92521 **[331]**

Foreword

This symposium represents a landmark for visual neurobiology in Australia. There has been a strong and broad tradition of work on the adult visual system in Australia since the 1950's, when P.O. Bishop established laboratories in the University of Sydney and subsequently at the Australian National University. Although several important developmental studies of the mammalian visual pathways were published from Australian laboratories before 1970, laboratories concentrating on mammalian visual development were not established until the late 1970's. Since 1975, laboratories in all states have focused on this area and their work has provided the scientific background for the organisation in Sydney of the Symposium on the Development of Visual Pathways in Mammals, whose proceedings are collected in this volume. As far as we are aware, this symposium was the first held in Australia on mammalian visual development, and we were delighted at the number of distinguished neurobiologists from many countries who were able to attend and contribute. Although its developmental thrust was new to Australian visual science, the meeting was influenced by earlier IUPS satellite meetings on vision, in particular the highly successful Satellite Meeting on the Structure and Function of the Visual Pathways held at the same location in 1974, and organised by R.W. Rodieck.

Scientifically, the papers covered a broad spectrum of topics, with five themes attracting a number of contributions:

Development of retinal topography. The papers by (in order of presentation) Stone, Rapaport, Dreher, Perry, and Dunlop, centered on the emergence of the regional specialisations of adult retina from the undifferentiated foetal or early postnatal state. Experimental work and discussion dealt with problems of the role of cell death, the interaction of synaptogenesis and cytogenesis, the possible transformation of ganglion cells into amacrine cells and the formation of ganglion cell mosaics. Perry's presentation explored the role of dendritic competition in determining the eventual distribution of ganglion cells. The theme was strengthened by the presentations of Chalupa and Rakic on the overproduction of optic nerve axons and ganglion cells during the development of cat and monkey; by the work of Nguyen-Legros on the development of a population of transmitter-specific amacrine cells; and by Ikeda's description of the development of transmitter-specific responses of cat ganglion cells.

Development of neuronal structure. Friedlander and Pasik addressed this problem, with very distinct techniques (intracellularly applied HRP, and quantitative EM). Their work provided a new level of detail in the study of developing visual pathways, allowing correlations to be drawn, for instance, between the differentiation of dendrites and the invasion of afferents. These studies also provide new evidence of differential rates of maturation of the different functional classes of relay cells (W-, X-, and Y-cells), and of the over-provision and elimination of axonal arborisations within the LGN and visual cortex.

Development of visual connections. The development of connections between visual centres was the subject of presentations by Godement, Sanderson, Rocha-Miranda, Hoffmann, Stein, Tsumoto, Stryker, Rakic, Chalupa, Rhoades, and Innocenti. A recurring theme of these studies, which employed both physiological and anatomical techniques, was evidence of broad patterns of visual projections, contracting during maturation to the specific and differentiated patterns found in the adult. Chalupa, Rakic, Stryker, and Rhoades all described experimental manipulations which modified or prevented the normal differentiation and contraction of projections, and thereby gave further index of the plasticity of developing pathways. The work of Rhoades and Innocenti concerned the development of callosal connections of visual cortex. Hoffmann described the development of the nucleus of the optic tract and its control of optokinetic nystagmus. Tsumoto described the development and differentiation of the corticotectal pathway, and Stein described the impact of the developing corticotectal pathway on the receptive fields of cells in the superior colliculus.

Reorganisation of visual pathways following damage or transplantation. Presentations by Harvey, Cunningham, Kalil, Spear, and Weller explored the response of developing pathways to damage or transplants. They described both degenerative and reorganisational changes that demonstrate considerable plasticity in the pathways, but also some clear limits to that plasticity. The latter three presentations all dealt with the potential of axon branches to form and expand on their normal connections, following damage to another branch.

Environmental modification. Presentations by Murphy, Kulikowski, Pettigrew, Wilson, Leventhal, Hirsch, Timney, Mitchell, Van Sluyters,Crewther, and Blakemore, provided a rich summary of current work on this topic. The papers explored new parameters of environmental impact, for example on the W-cell relay in the lateral geniculate nucleus (Wilson), on the orientation of dendritic fields of cortical cells (Hirsch), and on areas 18 and 19 (Leventhal). The potential for recovery from environmental modifications was powerfully explored (Crewther, Timney, Mitchell, Van Sluyters), and extensive new work on the rabbit (Murphy) and primate (Blakemore) was summarised.

We thank contributors for their care, promptness, and patience in preparation of the manuscripts; they have contributed to making this volume a substantial record of a successful meeting. Alan R. Liss, Inc. deserves our thanks for their promptness in joining and completing the publication of these proceedings.

Jonathan Stone
Bogdan Dreher
David H. Rapaport

Organising Committee
November 1983

Development of Retina and Optic Nerve

Development of Visual Pathways in Mammals, pages 3–21

THE DEVELOPMENT OF THE TOPOGRAPHICAL ORGANISATION OF THE CAT'S RETINA

Jonathan Stone, Juliani Maslim and David Rapaport

School of Anatomy
University of New South Wales
Sydney, N.S.W., Australia

INTRODUCTION

Topography of the Adult Retina

In many mammals the retina has developed striking regional variations in its structure, and in associated visual functions. In particular, the variation in ganglion cell density between central and peripheral regions is very great in some mammals (60-100 fold in the cat, several hundred-fold in some primates), and there is concurrent variation in the classes of ganglion cell present and in the morphology and circuitry of individual ganglion cells. There are also sharp gradients in the concentrations of cones and rods (Polyak 1957; Holländer, Stone 1972; Steinberg, Reid, Lacy 1973), and in horizontal cells (Wässle, Peichl, Boycott 1978). By contrast, no corresponding variation in cell density, morphology or circuitry has been described between regions of the lateral geniculate nucleus, superior colliculus or visual cortex which represent central and peripheral regions of retina.

The regional specialisations of the retina are of major importance in visual function. In species like the cat, monkey and human, the area or fovea centralis is the region of maximal spatial and chromatic acuity and, although it comprises 1% or less of the total area of the retina, loss of function of this particular region is highly disabling. Most eye movements seem specialised to keep images of interest on the foveal or areal region, and the representation of the region occupies a disproportionate part of the visual nuclei of the brain. As a consequence, accounts of the function and topography of the visual pathway in primates and the cat necessarily begin with reference to this central specialisation. In other species, such as the rabbit, the visual streak is the most prominent regional specialisation of the retina, and is correspondingly important functionally.

Retinal Genesis

The adult retina develops from neuroectodermal cells at the tip of the optic stalk. The tip invaginates after reaching the overlying ectoderm, and the cells lining the concave surface thus formed become the neural retina. The genesis of the retina from the optic stalk follows the basic pattern of the genesis of other regions of the central nervous system from the neural tube. Mitotic cells are found at the convex outer surface of the developing retina; this is the surface close to the optic vesicle and corresponds to the ventricular surface of other regions of the neural tube. The layer of dividing cells thus corresponds to the ventricular zone specified by Angevine et al. (1970). The repetitive 'interkinetic migration' of the nuclei of dividing cells, first proposed by Sauer (1936), also occurs in the retina (Sidman 1961); in this movement the nuclei of cells mitosing at the ventricular surface move away from the surface during G_1 interphase, remain away during S-phase and return during G_2 interphase. Moreover, as retinal cells leave the mitotic cycle, they are believed to move radially from the ventricular surface to take up their final positions, just as cells in, for example, the primate telencephalon (Sidman, Rakic 1973). The first cells to leave the cycle appear to be the ganglion cells, followed in overlapping sequence by amacrine, bipolar, horizontal and receptor cells (Sidman 1961; Blanks, Bok, 1977; Johns, Rusoff, Dubin 1979; Carter-Dawson, LaVail 1979).

However, the pattern of retinal genesis is distinct in two substantial ways. First, in most regions of the central nervous system, cells which leave the mitotic cycle migrate well away from the ventricular surface (where they are born), before forming the 'cortical plate', which becomes the adult nervous system (Angevine et al. 1970). The adult retina is no more than 200um thick, however, and many cells settle and form synaptic connections, in the path of the interkinetic migration of nuclei during the mitotic cycle. There is both reason and evidence to suspect, therefore, that the differentiation of cell classes in the retina and the formation of stable synapses between them will interfere with and eventually stop the cell movements characteristic of dividing cells. Second, there is, at least in some species, a long delay between central and peripheral regions of retina in their maturation. The different regions pass through very similar developmental sequences, but with the peripheral regions delayed (in the cat) by up to several weeks. These two features of the formation of the cortical plate of the optic stalk - i.e. of the neural retina - may underlie some characteristic features of the adult retina.

Genesis of the Area Centralis and Visual Streak

Given their functional importance in the adult it is of interest to determine the mechanisms which produce the regional specialisations of the adult retina, and attention was given initially to the most marked specialisation of the cat retina, the high density of ganglion cells found at the area centralis. Our initial working hypothesis, that the concentration of ganglion cells at the area centralis is produced by a 'hot-spot' of ganglion cell generation, did not appear tenable. No evidence emerged that ganglion cells are generated in high numbers at the area centralis; indeed their distribution appeared to be uniform until about 15 days before birth, after which it was 'shaped' by some non-uniform process. Four possible processes have been considered:

(a) A <u>trans-retinal migration</u> of ganglion cells to form the concentration at the area centralis.

(b) <u>Differential growth</u> of the retina, in which the retina expands more at its edge that at the area centralis region, so that peripherally located ganglion cells are more widely spread (Mastronarde, Thibeault, Dubin 1980).

(c) <u>Cell death occurring preferentially</u> among ganglion cells in peripheral retina.

(d) <u>Transformation of a high proportion of ganglion cell precursors in peripheral retina</u> into amacrine cells (Hinds, Hinds 1978, 1983).

These mechanisms are not exclusive. To test them we have made observations on the numbers and distribution of cells in the ganglion cell layer; on the occurrence and determinants of cell death in that layer; and on the timing and topography of synaptogenesis in the plexiform layers, and of cytogenesis in the ventricular zone.

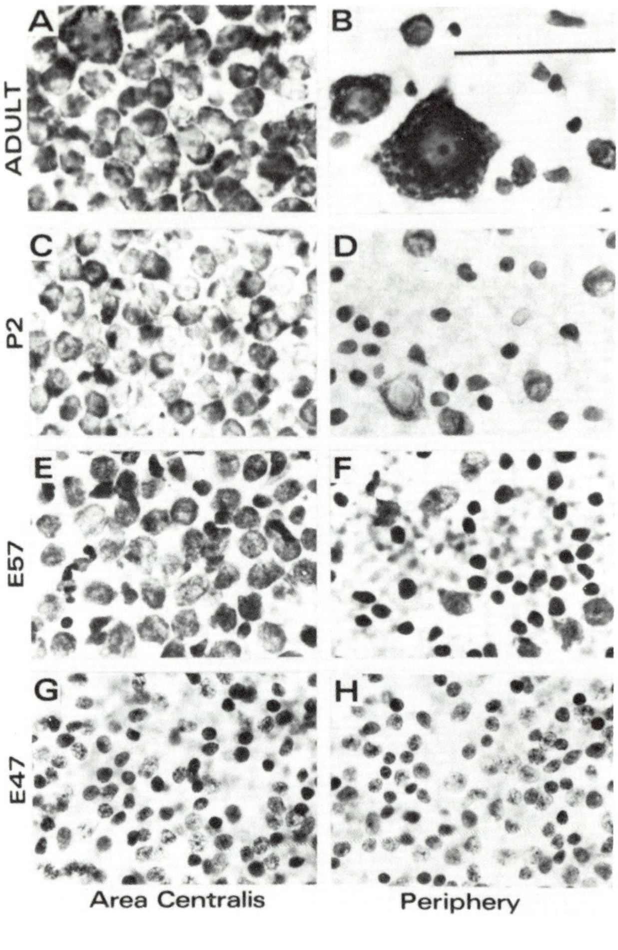

Figure 1. Areas of the ganglion cell layer from central and peripheral regions of Nissl-stained whole mounts of retinas from animals between E47 and adulthood. The calibration bar represents 50um. From Stone, Rapaport, Williams, Chalupa (1982).

THE MATURATION OF THE GANGLION CELL LAYER

Until E50: Uniformity of Precursor Distribution

At E47, i.e. the 47th day of the cat's gestational period of 65 days, the retina comprises two cell layers, the outer 'neuroblast' layer and an inner layer of cells in the position of the adult ganglion cell layer. Among these cells are the precursors of ganglion cells, which were generated 11-26 days earlier (Kliot, Shatz 1982; Walsh, Polley, Hickey, Guillery 1983). Mitotic cells concentrate at the ventricular surface and, as already noted, many of the nuclei in the neuroblast layer are mitotic cells in interphase or S-phase. Four features of the ganglion cell layer at this age are of interest (Stone, Rapaport, Williams, Chalupa 1982): (a) the precursor cells appear immature, with small (4-7um diameter) somas, little cytoplasm and scattered chromatin granules in their nuclei; (b) the density of precursors appears to be uniform over the retina (Fig. 1G,H); (c) the density of precursors appears higher than the highest density of ganglion cells found in the adult; and (d) the total number of precursors appears to be very high (over 800,000, thus several-fold higher than the adult number of ganglion cells).

The identification of precursor cells was based on their appearance in Nissl-stained material (Stone, Rapaport, Williams, Chalupa 1982), which does not demonstrate that the cells counted send axons to the optic nerve. Counts of optic nerve axons at this age, however, demonstrate that the number of axons is several times higher than found in the adult (Ng, Stone 1982; Williams, Bastiani, Chalupa 1983), supporting the view that at least a large proportion of the immature cells in the ganglion cell layer are potential ganglion cells.

This apparently uniform retinal distribution of precursors has been observed as early as E33 (Rapaport, Stone 1983b); it indicates that the area centralis is not produced by a 'hot-spot' of cell generation, but by factors which operate after the precursors are formed. The high density of the precursors makes it unlikely that they migrate across the retina to concentrate at the area centralis. The high number of precursors indicates that the area centralis and other non-uniformities of ganglion cell distribution are 'sculpted' from an initial excess of precursors which undergo either death or transformation into another cell type.

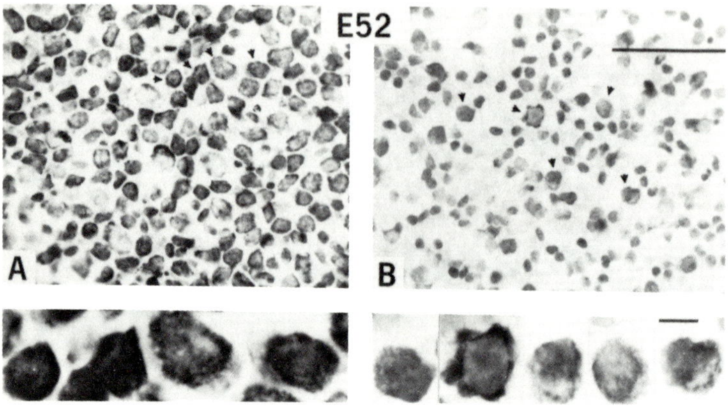

Figure 2: Areas of central (A) and peripheral (B) retina of an E52 retina. This is the earliest age at which we have observed a centro-peripheral gradient in ganglion cell density. The montages at bottom compare neurone-like cells at the area centralis (left) and in the periphery. At this early-stage, the area centralis cells are slightly larger. The calibration bars represent 100um (upper) and 5um (lower). From Rapaport, Stone (1983b).

E47-E52: Central Beginning of Ganglion Cell Maturation

Between E47 and E52, ganglion cell precursors begin to accumulate the granular cytoplasm which characterises adult neurones. Individual cells grow in soma diameter, and a centro-peripheral gradient in ganglion cell density appears. As Figure 2 shows, the growth and differentiation of somas is initially most advanced at the area centralis; in peripheral retina the differentiating somas are relatively small, and many apparently undifferentiated nuclei remain.

E57- Maturation

By E57, however, individual ganglion cells in peripheral retina seem to have 'caught up' with area centralis cells in their growth and accumulation of cytoplasm, and the centro-peripheral gradient in ganglion cell density is very obvious (Fig. 1E,F). The ganglion cell density ratio between area centralis and periphery is approximately 18:1 (Stone, Rapaport, Williams, Chalupa 1982), about 1/3rd of the adult

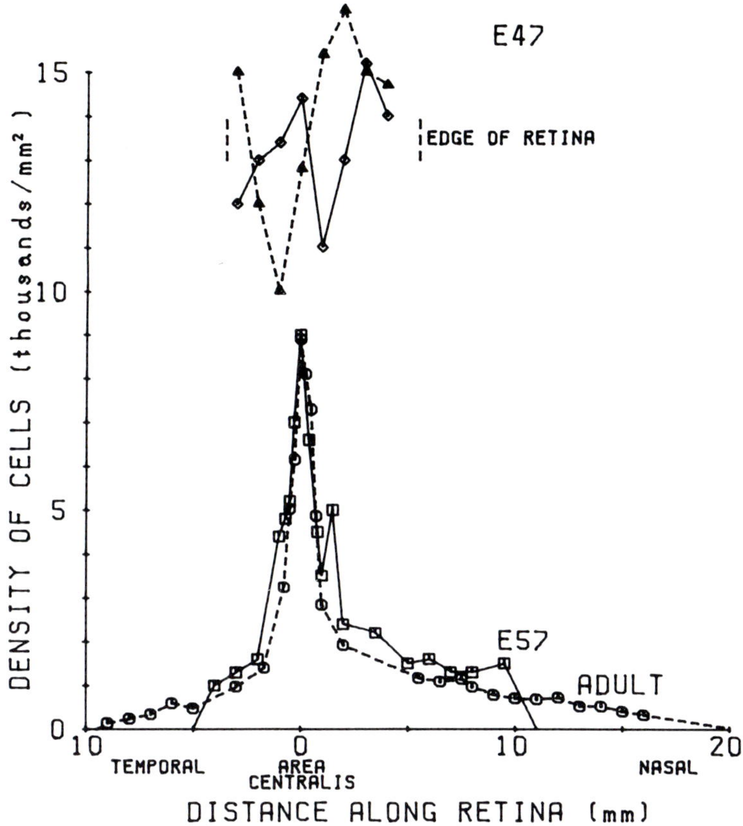

Figure 3: Gradients of ganglion cells or presumed precursor cells along horizontal transects of the retina across the area centralis. At E47 (two transects shown) the retina is small and the density of precursor cells in high and without a centro-peripheral gradient. A clear gradient is present at E57. By adulthood the retina has grown considerably;, reducing cell densities in the periphery of the retina, but not at the area centralis.

ratio (Figure 3). The growth in retinal area between E47 and E57 is between 2- and 3-fold, from 58mm^2 to 126mm^2. If this growth occurred mostly in peripheral retina, it would contribute to the centro-peripheral gradient in ganglion cell density (Mastronarde, Thibeault, Dubin 1980), but it could account for only a minor part of the 18:1 gradient observed (Stone, Rapaport, Williams, Chalupa 1982).

The growth of the retina continues to be substantial after E57. Overall, it grows about 3-fold in area between E47 ($58mm^2$) and birth ($187mm^2$), and a further three-fold by adulthood (Rapaport, Stone 1983b). Growth is differential in the postnatal period (Mastronarde, Thibeault, Dubin 1980), and the centro-peripheral ganglion cell density gradient increases correspondingly, from approximately 20:1 at birth (Stone, Rapaport, Williams, Chalupa 1982) to more than 60:1 in the adult.

It is of interest to note that the appearance of a centro-peripheral gradient occurs concurrently with the morphological differentiation of individual cells. Their somas accumulate granular cytoplasm, grow and begin to show the size groupings characteristic of the adult ganglion cell population (Rapaport, Stone 1983b). The development of the density gradient thus seems to be part of a more general maturational process.

The Role of Cell Death

Considerable numbers of cells in the ganglion cell layer die during the development of retinal topography. There is a large reduction during this period in the number of axons in the optic nerve (Ng, Stone 1982), and pyknotic cells are present in the ganglion cell layer from E33 to early postnatal life. However, two observations go against the view that the death of ganglion cell precursors contributes to the formation of the area centralis. First, pyknotic cells appear evenly distributed over the retina. At E52 for example (Figure 4), pyknotic cells are present at a density of approximately $100/mm^2$, but without any tendency to concentrate in peripheral retina. We have studied the distribution of pyknotic cells in the ganglion cell layer from when they first appear, in the early E30's, to P30 when their density has reduced to very low levels. At no stage did their distribution show significant concentrations in particular areas of the retina. Second, removal of one eye at E42 produces a small elevation in the ganglion cell population of the surviving eye (see also Chalupa, Williams, this volume). This extra complement of cells presumably survives because the cells are relieved of competition from axons of the other eye for synaptic space in the superior colliculus and lateral geniculate nucleus. However, the surviving cells do not concentrate in peripheral retina; ganglion cell density appears up to 20%

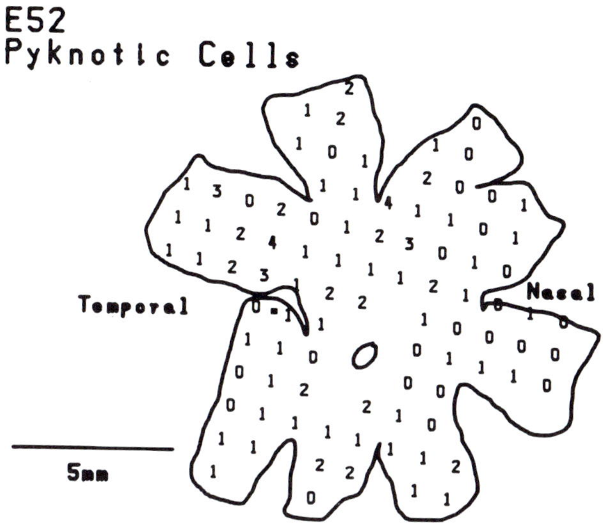

Figure 4: Map of the distribution of pyknotic cells in the ganglion cell layer of an E52 retina. The numbers represent pyknotic cells/ 0.01mm^2. The dot marks the area centralis.

higher in the experimental eye than in a littermate control, at all eccentricities (Stone, Rapaport, in preparation).

In summary, ganglion cell precursors appear to be produced in high numbers and to be present at high, uniform density until approximately E50. In explaining the subsequent development of the area centralis, two hypotheses - that the area centralis is produced by a 'hot-spot' of cytogenesis, or by trans-retinal migration of cells - do not seem useful. Cell death does occur during this period among cells of the ganglion cell layer, but not in a pattern which would contribute to adult topography. The retina grows considerably and at least some of that growth is differential (greater at the retinal edge) and contributes to the formation of the area centralis. Growth between E47 and E57 is insufficient, however, to account for the sharp cell gradients which develop in that period. It seems necessary to consider other mechanisms.

FORMATION OF THE OUTER PLEXIFORM LAYER

The observation that the central connections formed by ganglion cells may not determine the topography of the ganglion cell layer led us to examine events in more peripheral layers of the retina, as a first step at the formation of the outer plexiform layer (OPL).

Histogenesis

Seen in the light microscope, the OPL first appears at about E50. It forms in the middle of the neuroblast layer, separating it into the inner and outer nuclear layers found in the adult. Since the nuclei of cells dividing at the ventricular surface move across the neuroblast layer during their interkinetic migration (Sidman 1961; Rapaport, Robinson, Stone, following paper), the OPL forms directly in their path.

At E52 (Figure 5), the OPL has formed over a small patch at the site of the developing area centralis. Over subsequent weeks the layer extends across the entire retina, reaching the edge at P10, over three weeks after its initial appearance (Rapaport, Stone 1982). At its expanding edge, the layer is discontinuous with columns of cells appearing to span or cross it at intervals.

Thus the formation of the OPL, like the differentiation of ganglion cells, commences at the area centralis. The slow spread of the OPL is of interest. We suggest below that it is related to mechanisms of retinal growth.

Synaptogenesis

Synapses appear in the OPL 6-9 days after the layer can be detected by light microscopy. At E56 the layer is formed only close to the area centralis, where it is characterised by large extracellular spaces, processes of dying cells and growth processes extending from cells on either side. At E59, synaptic ribbons, characteristic of receptor terminals in the adult were apparent, but were few in number and small. Over the next several weeks these OPL synapses increased in number and differentiation. At E59, for example (Figure 6), each synaptic ribbon in a receptor terminal was initially apposed to only 2 postsynaptic processes. The third, central process of the adult 'triad' postsynaptic formation appeared only after birth. The point is of interest because the central process is considered to be a dendrite of an invaginating bipolar cell (Boycott, Kolb 1973), indicating that bipolar cells differentiate somewhat

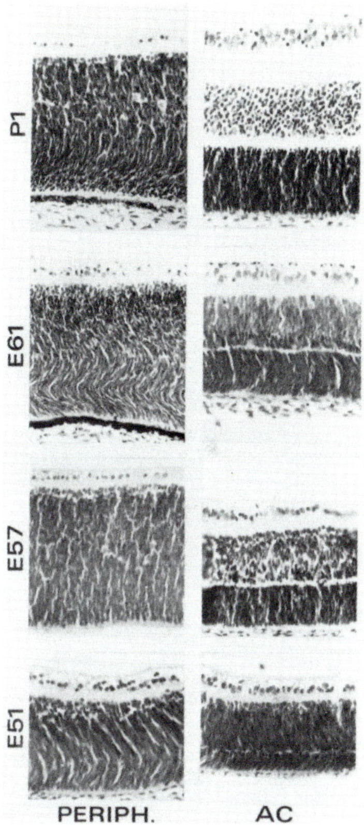

Figure 5: Sections of retina from the periphery and area centralis of foetal and neonatal retina. The OPL appears first at the area centralis at E52, and then matures steadily. At birth the OPL still has not formed in the periphery. From Rapaport, Stone (1982).

later than horizontal cells, which form the lateral processes of the triad. The OPL did not appear fully differentiated until after P20.

A similar temporal sequence of synapse development in the OPL was seen at each point in the retina examined, but delayed by up to three weeks, in accordance with the light microscopic observations described above.

Implications

These observations have two substantial implications for

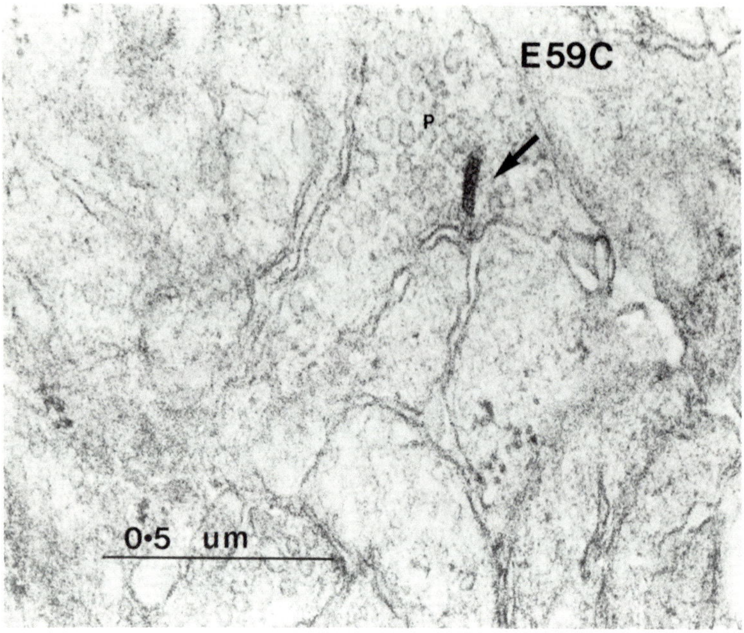

Figure 6: Ribbon synapse in the OPL at E59, the earliest age at which such synapses were observed. The material is from the region of the area centralis.

the genesis of retinal topography. First, because OPL formation begins at the same time and place as the differentiation of the ganglion cell population, it is possible that the two processes are triggered by a single mechanism. Second, the formation of the OPL seems likely to interfere with, and perhaps terminate, the genesis of cells at the ventricular surface of the retina. This implication led us to examine events, particularly cytogenesis, in the most peripheral layer of the developing retina, the ventricular surface.

CYTOGENESIS IN THE VENTRICULAR ZONE

The distribution of mitosing cells in the ventricular zone of the cat's retina was mapped in cresyl-violet stained retinas, prepared as whole mounts with their ventricular surface uppermost. The dividing cells concentrate at this surface, and are readily viewed and mapped (Rapaport, Stone 1983a).

Tempo and Topography of Cytogenesis

From as early as E21 until E50, mitotic cells are evenly distributed over the retina, at a density of 1,000-3,000/mm^2. By E52, mitosis at the ventricular surface has ceased over a small region at the site of the developing area centralis. Thus (Rapaport, Stone 1983a), the area centralis becomes apparent at the ventricular surface as a 'cold spot' of zero cytogenesis, surrounded by a 'cool zone' of reduced density of mitotic cells. The two zones expand steadily until, at birth, mitotic activity at the ventricular surface has ceased over the central 50% of the retina, and mitotic cell density over the peripheral 50% is falling. At P10, mitotic activity at the ventricular surface has ceased, even at the peripheral margin of the retina. These results are consistent with the view that cytogenesis is limited by the formation of the OPL.

Implications

These observations also have two substantial implications for the genesis of retinal topography. First, the cessation of cytogenesis begins at the same time and place as the differentiation of the ganglion cell layer and the formation of the OPL, reinforcing the idea that these processes may be linked and determined by a common mechanism. By analogy with Rakic's (1981) discussion of the developing primate telencephalon, information in the ventricular zone of the developing retina may determine the tempo and topography of developmental events in all the layers of the developing retina.

Second, the delay between central and peripheral regions of the retina in their cyto- and synaptogenesis is longer than is known to occur within other functionally homogeneous regions of the central nervous system. The delay could simply be a mechanism of retinal growth, enabling the retina to extend as the eyeball grows. We suggest, however, that the delay has a more subtle impact on retinal topography; that it may be the mechanism by which the relatively uniform distributions of cells in the inner and outer nuclear layers are established. The cells generated at the retinal periphery after cytogenesis has ceased in central retina are destined for these layers, the ganglion cells having been born considerably earlier. As the peripheral regions of the inner and outer nuclear layers are laid down, therefore, the ganglion cells are spread increasingly thinly over them (Figure 3).

FORMATION OF THE INNER PLEXIFORM LAYER

We have already noted that differential growth of the retina contributes to the development of the centro-peripheral gradient in ganglion cell density, but is insufficient to account for the appearance of a sharp centro-peripheral gradient between E47 and E57. Cell death is also an inadequate explanation for the gradient, for it is non-topographic. One remaining option is that the gradient is formed by the transformation of ganglion cell precursors into amacrine cells, and the movement of these cells back across the IPL, to the inner margin of the INL (Hinds, Hinds 1983). Cell movement across the OPL appears to be limited by the formation of synapses. When do IPL synapses form?

Tempo and Topography of IPL Formation

Seen in the light microscope, the IPL can be distinguished in the mid-E30's, when the ganglion cell layer first forms separately from the neuroblast layer. At this age, only a few synapse-like structures were observed, with apposed membrane densities and no associated vesicles (Figure 7). At E48, the number of synapse-like structures was still very limited, although a few clear-cut conventional synapses were observed. The presynaptic element contained vesicles which seemed to be associated with a thickening of a presynaptic membrane, a synaptic cleft and a thickening of the postsynaptic membrane. No synaptic ribbons, characteristic of bipolar cell synapses, were seen however. At E57 and E62 (Figure 8), there was an increase in the frequency of conventional synapses, but ribbon synapses were not seen until P2. Even then they were few in number and the ribbons small in size, confirming Morrison (1982). Only at P7 were ribbon synapses seen in substantial numbers.

Despite the early appearance of the IPL, therefore, the genesis of many of its synapses is delayed until after the formation of the OPL. Further, the delay between peripheral and central regions of retina in the formation of IPL synapses was only a few days, much less than the delay in OPL synaptogenesis. In retrospect, the long delay in the synaptogenesis of the IPL is not surprising; the synapses

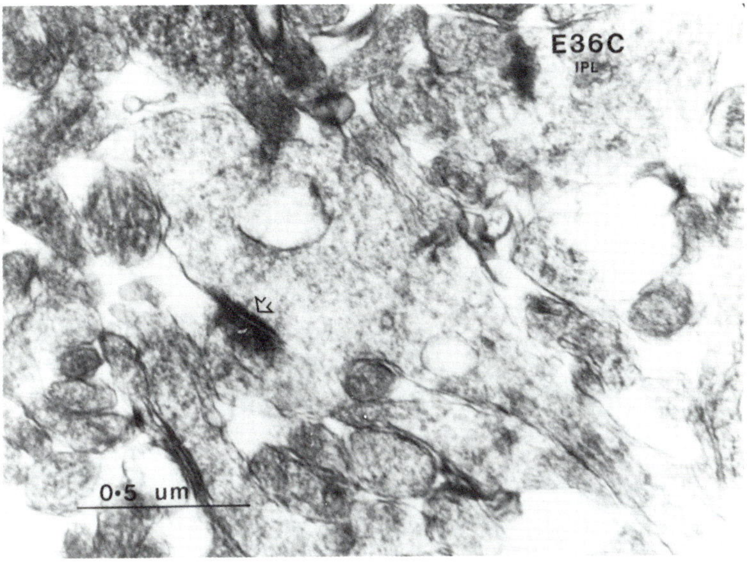

Figure 7: Area of the IPL in an E36 retina. Few of the processes appear differentiated. The arrow indicates a synapse-like structure characterised by opposed membrane densities; vesicles were not apparent.

should not be expected to form until after pre- as well as postsynaptic cells are formed, and the generation of INL cells continues until at least P10 (Johns, Rusoff, Dubin 1979). What is surprising is that IPL synaptogenesis is less topographic than similar events in the OPL. As a result there is a considerable delay between the formation of OPL and IPL synapses at the area centralis.

One feature of IPL synaptogenesis deserves further comment. Ribbon synapses in the IPL are considered to be formed by the axons of bipolar cells; they did not appear in our material until just after birth. The central process of the postsynaptic triads in the OPL are believed to be formed by the dendrites of bipolar cells; these also did not appear until after birth. It would seem therefore that bipolar cell synapses are among the last to form in a particular retinal region.

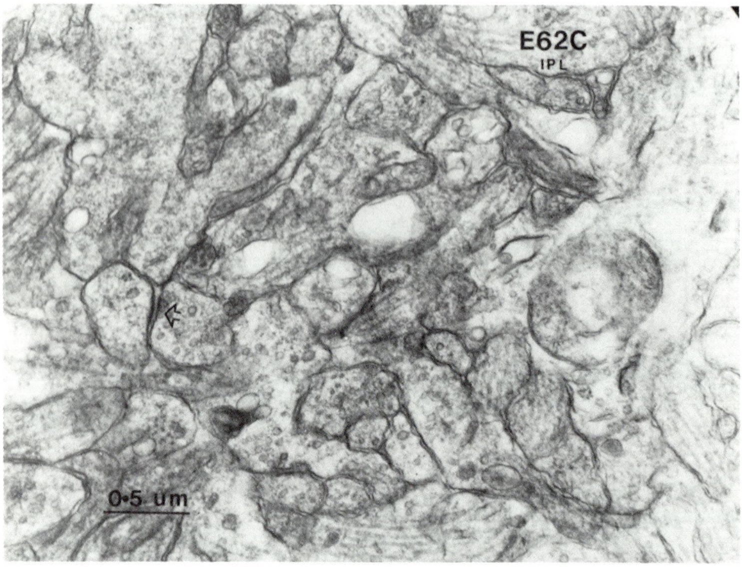

Figure 8: Area of the IPL at E62. The arrow indicates a synapse, with apposed membrane densities, and synaptic vesicle in the presynaptic process.

Implications

Two features of IPL synaptogenesis seem of importance for the genesis of retinal topography. First, a major component of IPL synaptogenesis (the formation of ribbon synapses) follows OPL synaptogenesis. It is possible therefore that the full genesis of IPL synapses is determined by the more peripheral events, occurring in the OPL or at the ventricular surface. This in turn supports the idea that the maturational processes of all layers of the retina are determined by events in the ventricular layer, rather than by trophic influences from the brain. Second, the delay in IPL synaptogenesis may be required to allow the somas of ganglion cells which are transforming into amacrine cells to move outward from the ganglion cell layer to the inner margin of the INL.

CONCLUSIONS

Six conclusions seem available from these data concerning the development of the area centralis:

The hypotheses that the area centralis is generated by a localised high rate of cell generation, or by the transretinal migration of ganglion cell precursors towards the area centralis, were not supported.

The growth of the retina may be differential, and may therefore contribute to the formation of the area centralis. However, growth in the period E47 to E57 seems too little to account for the gradient of ganglion cell density which develops during this period.

The continued generation of cells of the inner and outer nuclear layers at the periphery of the retina, for a substantial period after the generation of ganglion cells is complete, may account for the relative uniformity of cell distribution in these layers.

Cell death does occur during the period of area centralis formation, and is partly determined by competition for synapses in the visual centres of the brain. However, cell death appears to occur diffusely over the retina and may, therefore, make no contribution to the formation of regional specialisations such as the area centralis.

The simultaneity of developmental events in the ventricular, neuroblast and ganglion cell layers suggests that their maturation may be determined by a single mechanism. This idea is explored in the following paper.

The hypothesis that the concentration of ganglion cells at the area centralis is formed in part by the transformation of a high proportion of the ganglion cell precursors in peripheral retina into another cell type, such as amacrine cells (Hinds, Hinds 1978, 1983), deserves examination.

ACKNOWLEDGEMENTS

We are indebted to Mss. Jenni Duncan, Mignon Wegner and Kathy Hickson, who managed the breeding colony and organised the foetal surgery. Mr. Paul Halasz developed the computer programs used for several aspects of the study, and in particular Figures 3 and 4. Mr. Collin Yeo and Ms. Martina Wong provided skilled assistance with the electron microscopy. The work was supported by a grant from the N.H. and M.R.C. of Australia. D.H.R. was supported by NIH

Postdoctoral Fellowship EY05433.

REFERENCES

Angevine JB, Bodian D, Coulombre AJ, Eddes MV, Hamburger V, Jacobson M, Lyser KM, Prestige MC, Sidman RL, Varon S and Weiss P (1970) Embryonic vertebrate nervous system: revised terminology. Anat Rec 166: 257.

Blanks JC, Bok D (1977) An autoradiographic analysis of postnatal cell proliferation in the normal and degenerative mouse retina. J Comp Neurol 174:317.

Boycott BB, Kolb H (1973). The connections between bipolar cells and photoreceptors in the retina of the domestic cat. J Comp Neurol 148:91.

Carter-Dawson LD, Lavail MM (1979). Rods and cones in the mouse retina II: autoradiographic analysis of cell generation using tritiated thymidine. J Comp Neurol 188:263.

Chalupa LM, Williams RW (1984) Prenatal development and reorganization in the visual system of the cat. In Stone J, Dreher B, Rapaport DH (eds.): "Development of Visual Pathways in Mammals", New York: Alan R. Liss (this volume)

Hinds JW, Hinds PL (1978). Early development of amacrine cells in the mouse retina: an electron microscopic, serial section analysis. J Comp Neurol 179:277.

Hinds JW, Hinds PL (1983). Development of retinal amacrine cells in the mouse embryo: evidence for two modes of formation. J Comp Neurol 213:1.

Holländer H, Stone J (1972). Receptor pedicle density in the cat's retina. Brain Res 42:497.

Johns P, Rusoff AC, Dubin MW (1979). Postnatal neurogenesis in the kitten retina. J Comp Neurol 187:545.

Kliot M, Shatz CJ (1982). Genesis of different ganglion cell types in the cat. Soc Neurosci Abstr 8:815.

Mastronarde DN, Thibeault MA, Dubin MW (1980). How ganglion cells redistribute during postnatal growth of the cat's retina. ARVO Abstracts 70

Morrison JD, (1982) Postnatal development of the area centralis of the kitten retina: an electron microscopic study. J Anat 135:255.

Ng AYK, Stone J (1982). The optic nerve of the cat: appearance and loss of axons during normal development. Dev Brain Res 5:263.

Polyak S (1957 The Vertebrate Visual System. Chicago

University Press, Chicago

Rakic P (1981). Neuronal-glial interaction during brain development. Trends Neurosci 4:184.

Rapaport DH, Stone J (1982). The site of commencement of maturation in mammalian retina: observations in the cat. Dev Brain Res 5:273.

Rapaport DH, Stone J (1983a). The topography of cytogenesis in the developing retina of the cat. J Neurosci 3:1824.

Rapaport DH, Stone J (1983b). Time course of the morphological differentiation of cat retinal ganglion cells: influences on cell size. J Comp Neurol 221:42.

Sauer FC (1935). Mitosis in the neural tube J Comp Neurol 62:377.

Sidman RL (1961). Histogenesis of mouse retina studied with thymidine-H^3. In Smelser GK (ed): "Structure of the Eye", New York: Academic Press, p 487.

Sidman RL, Rakic P (1973). Neuronal migration, with special reference to developing human brain: a review. Brain Res 62:1.

Steinberg RH, Reid M, Lacy PL (1973). The distribution of rods and cones in the retina of the cat (Felis domesticus). J Comp Neurol 148:229.

Stone J, Rapaport DH, Williams RW, Chalupa L (1982). Uniformity of cell distribution in the ganglion cell layer of prenatal cat retina: implications for mechanisms of retinal development. Dev Brain Res 2:231.

Walsh C, Polley EH, Hickey TL, Guillery RW (1983). Generation of cat retinal ganglion cells in relation to central pathways. Nature 302:611.

Wässle H, Peichl L, Boycott BB (1978). Topography of horizontal cells in the retina of the domestic cat. Proc R Soc London B, 203:269.

Williams RW, Bastiani MJ, Chalupa LM (1983) Loss of axons in the cat optic nerve following fetal unilateral enucleation: an electron microscopic analysis. J Neurosci 3:133.

Development of Visual Pathways in Mammals, pages 23–38

CELL MOVEMENT AND BIRTH IN THE DEVELOPING CAT RETINA

David H. Rapaport, Stephen R. Robinson and Jonathan Stone

School of Anatomy
University of New South Wales
Sydney, New South Wales 2033
AUSTRALIA

The cat's retina matures rapidly during the last two weeks of gestation and the first few weeks of postnatal life. Prior to this period (i.e. prior to about the 50th embryonic day (E50) in the gestational period of 65 days), mitotic activity is high throughout the retina and there is no evidence of the regional variations in retinal structure (area centralis, visual streak) which characterize the adult. Although the inner layers of the retina (nerve fiber, ganglion cell and inner plexiform layers) are apparent by mid-gestation, the outer layers (inner nuclear, outer plexiform, outer nuclear and receptor layers) cannot be distinguished before E50. Instead the outer half of the neural retina consists of the 'neuroblast' layer, which contains neuroblasts and probably glioblasts in active mitosis. By three weeks after birth, mitotic activity has ceased, the numbers and classes of retinal neurons have stabilized, the regional variations of retinal structure are well developed and the laminar structure of the retina is fully established.

Recent studies have identified several components of the process of retinal maturation, and have described their time course and spatial patterns. These component processes include the maturation of the ganglion cell layer (Stone et al. 1982; Rapaport, Stone 1983b; Stone, Maslim, Rapaport 1984), the development of the outer plexiform layer (OPL) (Greiner, Weidman 1980; Rapaport, Stone 1982); the cessation of cytogenesis (Rapaport, Stone 1983a); and the genesis of synaptic contacts (Greiner, Weidman 1980; Maslim, Stone 1983; Stone, Maslim, Rapaport 1984). These several processes begin at the same place - the area centralis - and at approximately the same time, the 50th embryonic day. Their close synchrony suggests that they are precipitated and controlled by the same developmental mechanisms.

To investigate these mechanisms, we have examined the relationship between two of the component processes of retinal maturation: the formation of the OPL, which divides the neuroblast layer into the inner and outer nuclear layers (INL and ONL), and the cessation of cytogenesis, in particular the genesis of cells by mitosis at the ventricular surface of the retina. The results raise the possibility that cytogenesis at the ventricular surface does not continue after the OPL forms because that layer interferes with the migration of the nuclei of mitotic cells. However, the results also provide evidence of a previously unrecognized process of retinal cytogenesis: a layer of mitotic cells was identified in the inner part of the neuroblast layer and, after the formation of the OPL, in the INL. Cytogenesis among these cells has a distinct time course and spatial pattern - it is not limited by the formation of the OPL for example - and presumably makes a distinct contribution to retinal development.

SIMULTANEITY OF OPL DEVELOPMENT AND THE CESSATION OF CYTOGENESIS

In previous papers, the development of the OPL and the cessation of cytogenesis had been described from different retinas. As a first step in this study, we used a computer-aided system for reconstruction of the retinal surface from serial sections to demonstrate these processes in the same retina, at each of several ages. Maps of the retinas of three prenatal animals are shown in Figure 1.

At embryonic day 42 (E42) the density of mitotic profiles at the ventricular surface is high and uniform across the retina, and there is no sign of the OPL. At E52, mitotic activity at the ventricular surface continues at a comparable density over most of the retina, but has ceased over a small region at the site of the developing area centralis. The OPL has also begun to form, and the region over which it has formed coincides closely with the "cold spot" of zero mitotic activity. By E57, the "cold spot" and the region of OPL formation have both expanded considerably, and they remain closely coincident. At later ages, both regions expand steadily, reaching the edge of the retina at about P(postnatal day)10 (Rapaport, Stone 1982, 1983a).

At both E52 and E57, the "cold spot" is surrounded by an

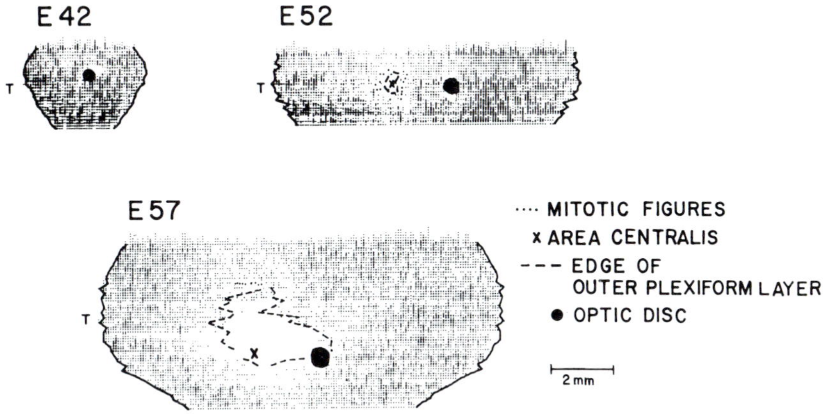

Figure 1. Maps of the distribution of dividing cells at the ventricular surface, and of the extent of formation of the OPL. The maps were constructed from serial horizontal sections with the aid of a computer graphics system. At E42 dividing cells are numerous and uniformly distributed, and the OPL has not formed. At E52, and to a greater extent at E57, mitosis has ceased over a small region at the area centralis, where the OPL has also formed. "T" denotes the temporal edge of the retina.

annular "cool zone", in which mitotic cells are present but at a lower density than in more peripheral levels. The "cool zone" lies outside the region over which the OPL has formed, indicating that mitotic activity begins to decrease before the OPL has formed a continuous layer. Once the OPL has formed, however, mitotic activity at the ventricular surface ceases.

MOVEMENT OF THE NUCLEI OF MITOTIC CELLS

In the retina, as in other regions of the developing central nervous system, cells mitosing at the ventricular surface change their shape in a characteristic and quite dramatic way with each mitotic cycle (Hinds, Hinds, 1974). After division, each daughter cell elongates, extending a process to the vitreal surface of the retina. The cell remains attached to the ventricular surface, but the nucleus moves some distance away from it. The cell then enters the S (synthetic) phase of the cycle, during which DNA is synthesized and the cell doubles its complement of chromosomes preparatory to the next division. When S-phase is complete, the nucleus moves back to the ventricular surface and the cell withdraws its vitreal process, becoming rounded before mitosing. This to-and-fro movement of the nuclei of mitotic cells is the 'interkinetic nuclear migration' proposed by Sauer (1935).

Interkinetic nuclear migration has been demonstrated in the mouse retina by Sidman (1961) and Hinds and Hinds (1974), but has not previously been examined in the cat. As a first step we followed previous workers (e.g. Sidman, 1961) in using thymidine autoradiography to establish the location of nuclei in S-phase. Thymidine is incorporated by cells in S-phase into the DNA of their nuclei; S-phase nuclei can then be detected by their radioactivity in tissue from animals sacrificed soon after an injection of ^{3}H-thymidine. Given that the nuclei divide at the ventricular surface, knowledge of the location of nuclei in S-phase would establish the amplitude of interkinetic migration.

Tritiated thymidine was administered in one of two ways: either, 50-70µCi was injected into the vitreous chamber of the eye, or 5µCi per gram body weight was injected intracardially. The latter method proved more satisfactory. Although less direct, it provided a more even distribution of thymidine over the retina, and the short period over which vascularly administered thymidine is available to tissues (Nowakowski, Rakic 1974) simplified the interpretation of results. The tissue was embedded in paraffin with the eyes in the orbits, sectioned, and processed for autoradiography using conventional methods (Sidman 1970; Edwards, Hendrickson, 1981). The injections were made in postnatal kittens, aged 1, 6 and 10 days.

In the retinas of kittens sacrificed 1 or 2 hours after an injection, some radioactive nuclei were found in the ganglion cell layer. They were fairly evenly distributed over the retina, and most seemed to be endothelial cells, associated with blood vessels, which are growing actively at this time (Halasz, Stone 1983). They will not be considered any further in this paper.

In peripheral regions of the retina, where the OPL was not formed, the radioactive cells were mostly located at the inner margin of the neuroblast layer (Figure 2). Thus, in the cat, the interkinetic migration of nuclei spans the thickness of the neuroblast layer, as Sidman (1961) had observed in the mouse retina.

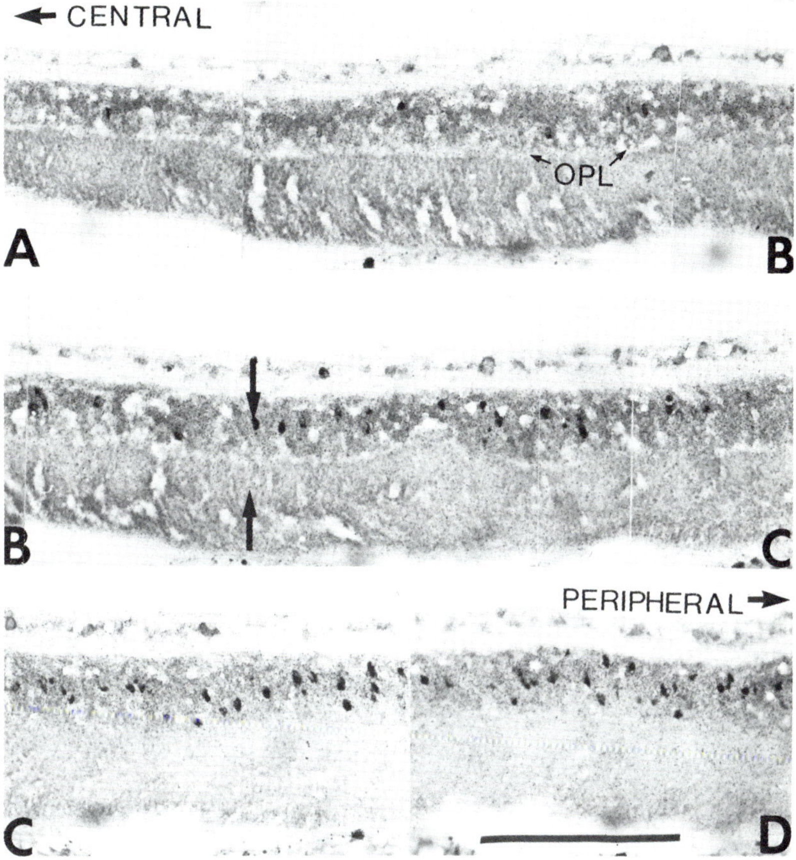

Figure 2. Radioactively labelled cells in the retina of a two day old kitten sacrificed two hours after an intraocular injection of ^{3}H-thymidine. The photomicrographs are adjacent panels from a strip of retina and are continuous at B and C. The density of labelled cells is low in the more central regions of the retina where the OPL is present (top panel), and higher in retina peripheral to the edge of the developing OPL (arrowed in middle panel).

This demonstration of the amplitude of nuclear movement during cytogenesis has a striking implication. The path of nuclear migration crosses the future site of the OPL. The formation of the OPL may prevent the genesis of cells in

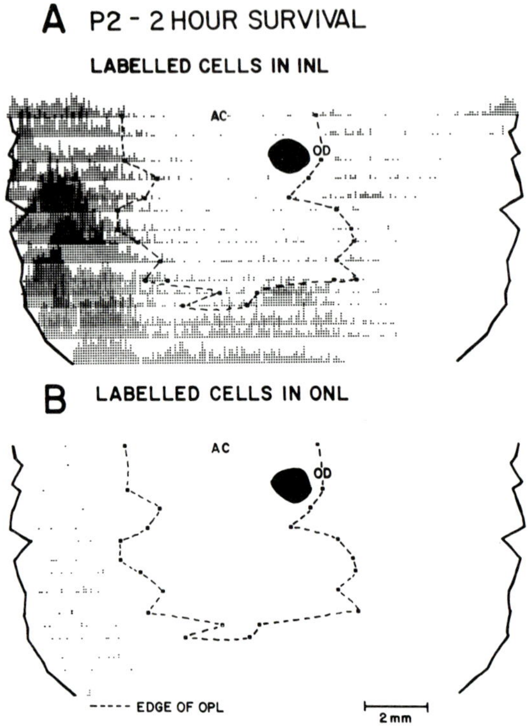

Figure 3. Maps of the distribution of radioactively labelled nuclei in a P2 kitten sacrificed two hours after an intraocular injection of ^{3}H-thymidine. The lack of labelled cells in nasal retina is probably due to poor diffusion of thymidine from the injection site at the temporal edge of the eye.

Labelled cells in the inner nuclear layer (INL) form a dense ring peripheral to the edge of the outer plexiform layer (OPL). Some labelled cells were found in the INL in more central regions of retina, where the OPL is well formed. Only a few labelled cells are apparent in the outer nuclear layer (ONL), and all of these occur between the edge of the retina and the edge of the OPL.

Abbreviations: AC- area centralis; OD- optic disc.

the neuroblast layer by physically blocking interkinetic nuclear migration, and may thus synchronize the formation of the OPL with the cessation of cytogenesis. This synchrony of a synaptogenic event with the end of cytogenesis is not a feature of other regions of the developing neural tube, probably because in other regions the wall of the tube is considerably thicker than the retina, allowing the 'cortical plate' of maturing cells to form well clear of the ventricular zone.

THE TOPOGRAPHY OF CELLS IN S-PHASE

In addition to allowing the study of nuclear movement, thymidine autoradiography provided a second way of examining the relationship between cytogenesis and the formation of the OPL. For example, Figure 3 shows the distribution of S-phase nuclei in a reconstruction of the retina of a P2 kitten which had survived the injection by 2 hours. S-phase nuclei were found in all regions of the retina, but over a large central region their density was very low. Most were restricted to an annular region peripheral to the central region over which the OPL was formed (Figures 2,3). For a limited distance peripheral to the edge of the developing OPL, the density of S-phase nuclei was graded. Thus the distribution of nuclei in S-phase matches closely the distribution of mitotic cells at the ventricular surface (Figure 1). These observations confirm earlier conclusions that a considerable proportion of cells in the neuroblast layer are involved in cytogenesis at the ventricular layer and the associated pattern of nuclear migration.

Two groups of S-phase cells in Figure 3 deserve further comment. First, a small population of radioactive nuclei in the peripheral region of retina was found in the neuroblast layer, but some distance from the inner plexiform layer (IPL). For reasons developed in the next section we assume that these nuclei entered G_2 interphase soon after the injection of ^{3}H-thymidine, and had begun to move back to the ventricular layer before the animal was sacrificed. Second, a small but consistent population of S-phase cells was present in the central region of retina where the OPL is formed and cytogenesis at the ventricular surface has ceased (Figure 3A). These cells were restricted to the INL. For reasons discussed below we believe they represent a previously unrecognized population of cells which divide within the INL.

A similar pattern was apparent in the retina of a 10 day old kitten which survived an intraocular injection of

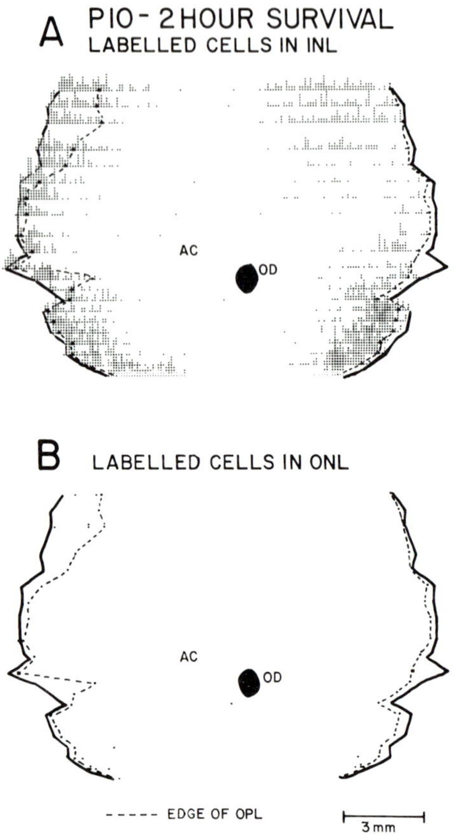

Figure 4. Maps of the distribution of radioactively labelled nuclei in the retina of a P10 kitten sacrificed two hours after an intraocular injection of ^{3}H-thymidine. The OPL extends across almost the entire retinal surface and labelled cells in the INL are restricted to a narrow annulus at the edge of the retina. Some labelled cells are present in the INL in central regions, where the OPL is well formed. Only a few labelled cells were found in the ONL, most of them near the edge of the retina.

Abbreviations: AC- area centralis; OD- optic disc.

thymidine by 2 hours (Figure 4). Most S-phase cells were found around the edge of the retina, where the OPL still had not fully formed. They concentrated at the inner margin of the neuroblast layer. Scattered radioactive nuclei were also found in the INL in the central region of retina, where the OPL had formed.

THE VELOCITY OF INTERKINETIC MIGRATION

Thymidine autoradiography with longer survival times allowed study of the movement of nuclei from their location during S-phase at the inner margin of the neuroblast layer towards the ventricular layer. This movement, which occurs during G_2 interphase, was studied in kittens of three ages (P1, P2, P10), with survival times of up to 24 hours. In these animals, the ^{3}H-thymidine was injected intracardially.

Movement was evident as an increase in the distance of radioactive nuclei from the IPL with increasing survival time. Not all nuclei had moved equally. Our analysis assumes that (a) all nuclei in a particular region of retina move at the same rate, and (b) the nuclei furthest from the IPL are those which were near the end of S-phase at the time of injection, and began to move with minimal delay.

In each eye studied the maximal distance of radioactive nuclei from the IPL was measured, as a function of position along the section. From that distance, and the survival time, the velocity with which these nuclei had moved could be estimated. Typical velocities in regions in which the OPL had not formed were 10 to 15µm/hr, close to values reported in other areas of the mammalian central nervous system (e.g. Atlas, Bond 1965).

The velocity of nuclear movement varied markedly with position relative to the border of the developing OPL. Figure 5 shows the variation in velocity with position along a horizontal transect of the retina through the area centralis. In a P1 retina (Figure 5A) the velocity was close to zero throughout the central region over which the OPL had formed, but was as high as 9-10µm/hr at the retinal edge. For a short distance peripheral to the edge of the OPL, velocity was graded. Figure 5B shows a similar analysis of a P2 retina. Again velocity of nuclear movement was close to zero across the central region in which the OPL is formed, while values of 10-15µm/hour were obtained at positions within 2 millimeters of the edge of the retina. At intermediate positions velocity varied systematically with position.

At these ages the variation in retinal structure between regions in which the OPL has formed and regions in which it has not, can also be viewed as a developmental sequence

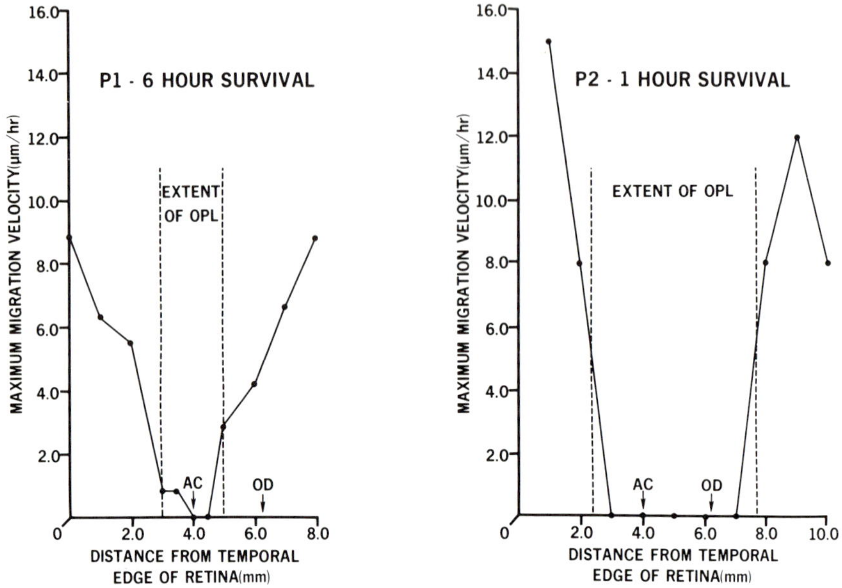

FIGURE 5. Rate of interkinetic nuclear migration as a function of retinal eccentricity. Migration rates are zero in central regions of retina where the OPL has formed, and maximal at the retinal edge.

A: P1 kitten sacrificed 6 hours after an intraocular injection of ^{3}H-thymidine.

B: P2 kitten sacrificed one hour after an intraocular injection of ^{3}H-thymidine.

Abbreviations: AC- area centralis; OD- optic disc.

through which each region of retina passes. Viewed in this way, our results indicate that the velocity of movement of the nuclei of cells in the G_2 interphase of the mitotic cycle is approximately 10µm/hour before the OPL forms. As the OPL begins to form velocity drops, and is effectively zero after the OPL is fully formed. This deceleration may be an aspect of the lengthening of the mitotic cycle with maturation, described in other areas of the CNS (reviewed in Chapter 2 of Jacobson, 1978).

The observation that velocity of migration is zero in regions of retina in which S-phase nuclei are still present, raises several intriguing possibilities. It suggests that some mitotic cells will not reach the ventricular surface to

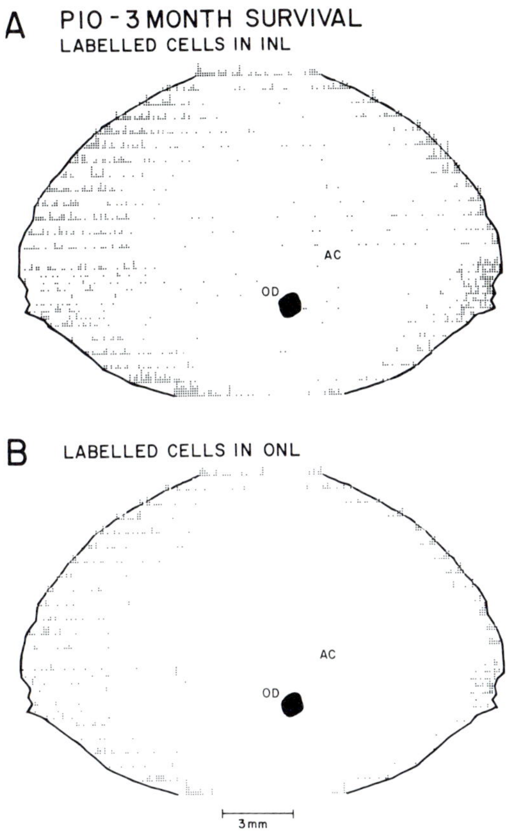

Figure 6. Maps of the distribution of radioactively labelled nuclei in the retina of a kitten injected intraocularly with ^{3}H-thymidine at P10, and sacrificed when three months of age. Labelled cells are concentrated around the edge of the retina in both the INL and ONL. In central retina, labelled cells are only found in the INL.

mitose. Do they then die, or do they mitose in the INL regardless? If the latter, is this an adaptation to the presence of the OPL, or do cells divide in this region independent of the formation of the OPL? If they normally divide here, do these cells represent a particular population of cells in the adult retina? Answers to these questions are approached in the following two sections.

LAMINAR POSITION AND DISTRIBUTION OF CELLS BORN POSTNATALLY

Thymidine autoradiography with long survival times allows identification of cells which left the mitotic cycle at the first or second cell division after the injection. Figure 6 shows the distribution of radioactive cells in the retina of a kitten which was injected intracardially at P10 and survived to P90. By this age, cytogenesis and cell migration are believed to be completed.

At the edge of the retina, radioactive nuclei were found in both the INL and the ONL. These are the cells born from the concentration of S-phase nuclei around the edge of the P10 retina, illustrated in Figure 4. Those in the ONL are presumably the nuclei of photoreceptors; since they are some distance from the outer limiting membrane, they are probably the nuclei of rods. Those in the INL may be the nuclei of bipolar, Müller, horizontal or amacrine cells; discrimination between these cell types requires additional analysis.

In the central region of the retina in Figure 6, radioactive nuclei are found in the INL. These are presumably the progeny of the S-phase nuclei found in the INL in the central region of a P10 retina (Figure 4). This observation indicates that some of these S-phase nuclei subsequently divide and produce viable daughter cells, despite their immobility.

THE OCCURRENCE OF MITOTIC ACTIVITY AWAY FROM THE VENTRICULAR SURFACE

The thymidine autoradiography of developing retina described above pointed strongly to the mitosis of cells in the INL, something not previously reported in any mammalian retina. Mitotic profiles at the ventricular surface were easily seen in sections stained with cresyl violet or hematoxylin and eosin, but we were never convinced we could distinguish mitotic profiles within the densely packed neuroblast layer. We were able to distinguish them, however, in sections stained with a combination of nuclear fast red (Kerchentrot) and crystal violet in a protocol developed by Fraser (1982) to demonstrate mitotic profiles in the developing brain.

The mitotic profiles found in the INL (Figure 7) resemble in morphology those found at the ventricular surface, and in other parts of the developing brain. Dividing cells were also found in a similar laminar position

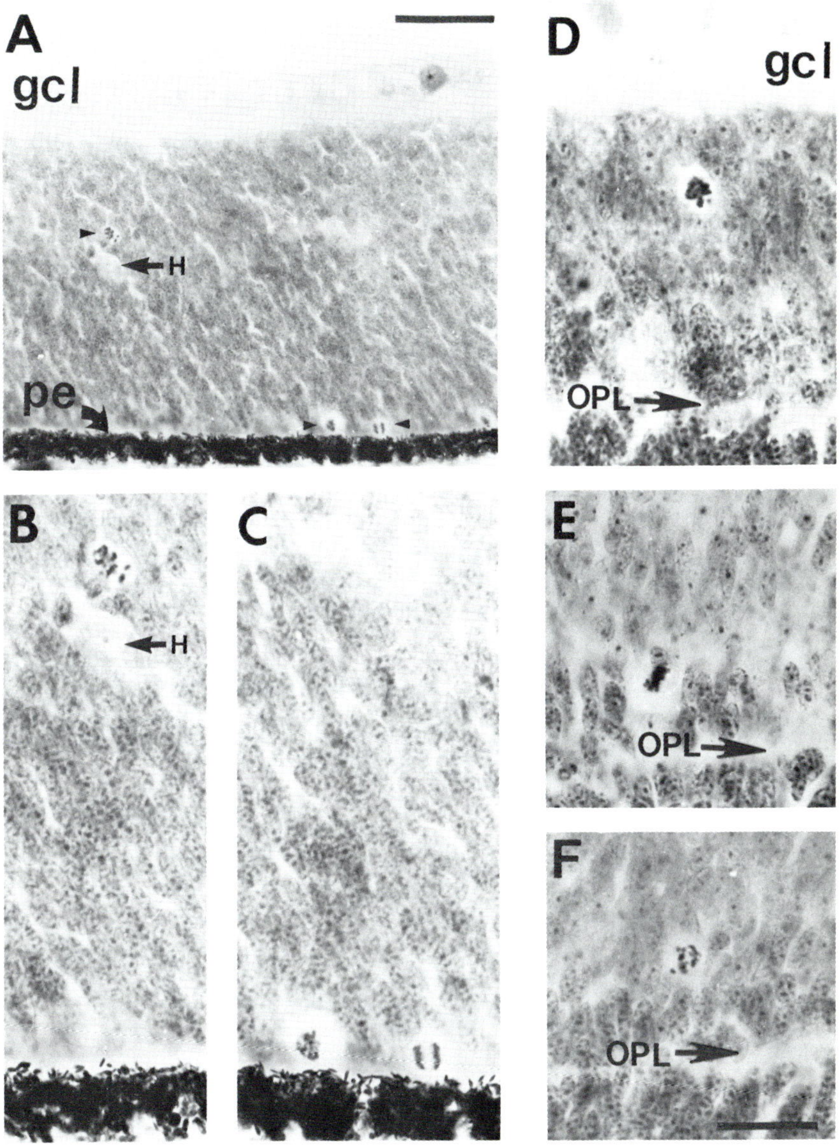

Figure 7A. Photomicrograph of the retina of a P1 kitten stained with nuclear fast red/ crystal violet. Three dividing cells are arrowed. Two (one in prophase, one in ana-/ telophase) are at the ventricular surface, and the third is located internal to a horizontal cell (H) and hence

(Legend continued at foot of page 36)

in regions of retina over which the OPL has not formed. This pattern of cytogenesis may therefore both precede and persist after the formation of the OPL. Cells undergoing division away from the ventricular surface of the retina have been seen by Johns (1982) in the goldfish, but in this species the cells apparently migrate back to the outermost retinal layer and become photoreceptors. In the cat, cells mitosing in the INL appear to stay there (Figure 6). The nature and fate of these cells remain unclear, but they may, by analogy with the late-dividing cell population in the subventricular zone of the developing telencephalon, be neuroglial (Schmechel, Rakic 1979), presumably Müller cells.

SUMMARY

The present observations on cell birth and movement in the developing retina of the cat provide support for three general conclusions:

First, towards the end of retinal cytogenesis, the interkinetic migration of the nuclei slows and then stops. This deceleration is synchronous with the slowing and termination of mitotic activity at the ventricular surface. These observations are consistent with the lengthening of the mitotic cycle during development, described in other regions of the central nervous system.

---

in the presumptive INL. Scale bar= 100µm.

Figures 7B,C. Higher power photomicrographs of the dividing cells in A. B shows the cell dividing in the INL, and C shows the two cells dividing at the ventricular surface. Scale bar= 20 µm.

Figures 7D-F. Higher power photomicrographs of other dividing cells in the INL of a P1 kitten. The cells are in prophase (D), metaphase (E) and ana-/ telophase (F). Scale bar= 20µm.

Abbreviations: gcl- ganglion cell layer; pe- pigment epithelium

Second, the formation of the outer plexiform layer is synchronous with the deceleration of cytogenesis at the ventricular surface. It is possible that the formation of the OPL is a factor in terminating cytogenesis, by blocking the associated movement of nuclei.

Third, some cells in the inner part of the neuroblast layer divide in situ, without their nuclei migrating to the ventricular surface. Further, some cells in the INL continue to divide for up to 10 days after the formation of the OPL, also without their nuclei migrating back to the ventricular surface. This previously unknown region of mitotic activity may correspond to the subventricular zone (Angevine et al. 1970) of other areas of the neural tube.

ACKNOWLEDGEMENTS

We wish to thank Mss. J. Duncan and M. Wegner for technical assistance, and Y. Cheska for comments on the manuscript. Mr. Paul Halasz developed computer software and hardware for the retinal reconstructions. Mr. David Solomon constructed the motorized microscope essential for this work. This research was supported by the NH&MRC of Australia, and by NIH post-doctoral fellowship EY05433 to D.H.R.

REFERENCES

Angevine JB, Bodian, D, Coulombre AJ, Eddes MV, Hamburger V, Jacobson M, Lyser KM, Prestige MC, Sidman RL, Varon S, Weiss P (The Boulder Committee) (1970). Embryonic vertebrate central nervous system: revised terminology. Anat Rec 166:257.

Atlas M, Bond VP (1965). The cell generation cycle of the eleven-day mouse embryo. J Cell Biol 26:19-24.

Edwards SB, Hendrickson A (1981). The autoradiographic tracing of axonal connections in the central nervous system. In Heimer L, Robards MJ (eds): "Neuroanatomical Tract-Tracing Methods", New York: Plenum Press, p171.

Fraser FJ (1982). A selective stain for mitotic figures, particularly in the developing brain. Stain Tech 57:219.

Greiner JV, Weidman TA (1980). Histogenesis of the cat retina. Exp Eyc Res 30:439.

Halasz P, Stone J (1983). Development of retinal vasculature in the cat. Proc Anat Soc Aust N Zeal 35.

Hinds JW, Hinds PL (1974). Early ganglion cell differentiation in the mouse retina: an electron microscopic analysis utilizing serial sections. Dev Biol 37:381.

Jacobson M (1978) "Developmental Neurobiology." New York: Plenum Press.

Johns PR (1982). Formation of photoreceptors in larval and adult goldfish. J Neurosci 2:178.

Johns PR, Rusoff AC, Dubin MW (1979). Postnatal neurogenesis in the kitten retina. J Comp Neurol 187:545.

Maslim J, Stone J (1983). Synaptogenesis in the cat's retina. Proc Anat Soc Aust N Zeal 19.

Nowakowski RS, Rakic, P (1974). Clearance rate of exogenous ^{3}H-thymidine from the plasma of pregnant rhesus monkeys. Cell Tissue Kinet 7:189.

Rapaport DH, Stone J (1982). The site of commencement of maturation in mammalian retina: observations in the cat. Dev Brain Res 5:273.

Rapaport DH, Stone J (1983a). The topography of cytogenesis in the developing retina of the cat. J Neurosci 3:1824.

Rapaport DH, Stone J (1983b). Time course of morphological differentiation of cat retinal ganglion cells: influences on soma size. J Comp Neurol (in press).

Sauer FC (1935). Mitosis in the neural tube. J Comp Neurol 62:377.

Schmechel DE, Rakic P (1979). Arrested proliferation of radial glial cells during midgestation in rhesus monkey. Nature 277:303.

Sidman RL (1961). Histogenesis of mouse retina studied with thymidine-H^3. In Smelser GK (ed): "The Structure of the Eye", New York: Academic Press, p 487.

Sidman RL (1970). Autoradiographic methods and principles for study of the nervous system with thymidine-H^3. In Nauta WJH and Ebbesson SOE (eds): "Contemporary Research Methods in Neuroanatomy", New York: Springer-Verlag, p252.

Stone J, Maslim J, Rapaport DH (1984) The development of the topographical organization of the cat's retina. In Stone J, Dreher B, Rapaport DH (eds.): "Development of Visual Pathways in Mammals", New York: Alan R. Liss (this volume)

Stone J, Rapaport DH, Williams RW, Chalupa L (1982). Uniformity of cell distribution in the ganglion cell layer of prenatal cat retina: implications for mechanisms of retinal development. Dev Brain Res 2:231.

Development of Visual Pathways in Mammals, pages 39–57

THE DEVELOPMENT OF HETEROGENEITIES IN DISTRIBUTION AND SOMA SIZES OF RAT RETINAL GANGLION CELLS

Bogdan Dreher[1], Rebecca A. Potts[1], Solomon Y.K. Ni[1] and M.R. Bennett[2]
Department of Anatomy[1] and Neurobiology Research Centre[2], University of Sydney, Sydney, N.S.W., Australia. 2006

INTRODUCTION

Centro-peripheral Gradients in the Adult Retina

In adult rats, as in almost all adult mammals studied so far (for reviews see Hughes 1977, Stone 1983a,b), the distribution and soma sizes of cells located in the ganglion cell layer are not uniform across the retina (Lashley 1932, Hughes 1977). Thus, in retinal wholemounts of adult rats, both putative ganglion cells (that is cells within the ganglion cell layer with perikarya larger that 6.0um and conspicuous Nissl substance in their cytoplasm, Fukuda 1977) and cells unequivocally identified as ganglion cells by the retrograde labelling after massive injections of enzyme horseradish peroxidase (HRP) into the retino-recipient nuclei (Keens 1981 and the present study) concentrate in the small region located in the superior temporal retina about 1.5mm from the optic disc. Their density declines steeply towards the upper temporal periphery and more gradually toward the lower periphery (Fig. 1A). In the present study in those animals in which virtually all ganglion cells were labelled by HRP (the total number - about 113000 - of labelled retinal cells in those animals was very similar to the number of axons in the optic nerve - Lam, Sefton, Bennett 1982; Potts, Dreher, Bennett 1982; Perry, Henderson, Linden 1983) the ganglion cell density in the highest density region (presumed "area centralis") was between 5000 and 6000 cels/mm^2 while that at the far periphery did not exceed 1000 cels/mm^2 (Figure 1A). The density distribution is virtually the same in albino (Sprague-Dawley) and hooded (Manchester) rats (see however Keens 1981) and the centro-

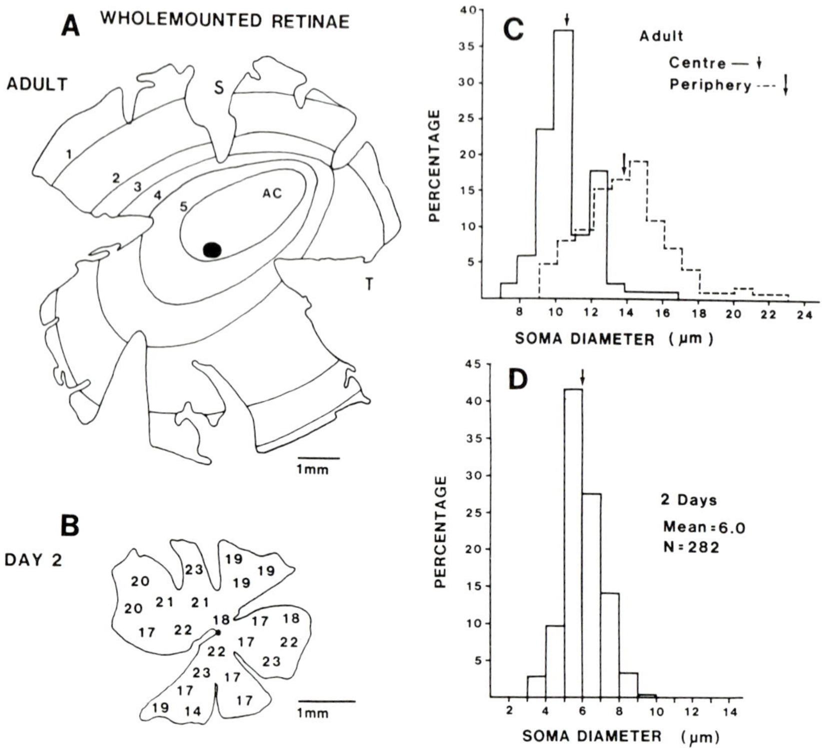

Fig. 1A & B. Maps of ganglion cell (labelled by massive HRP injections into the retino-recipient nuclei) distribution in the retinae of the adult and the newborn (day 2) rat. The elipsoid outlines on the wholemount of the adult retina represent the isodensity lines. The numbers associated with each line are the criterion value of the line in units of 1000 cells/mm^2. The black spot represents the optic disc while AC indicates the location of the "area centralis". Symbols S and T indicate superior and temporal edges of the retina. On the wholemount of the retina of the newborn the numbers represent the cell density in units of 1000 cells/mm^2. Note the uniformity of ganglion cell density. Fig. 1C & D. Frequency histograms of soma diameters of retinal ganglion cells in adult and newborn rats. Sample sizes in adults - 300 cells. Note the substantial centro-peripheral differences in mean soma diameters (indicated by arrows). In the newborn rat the range of soma diameters is small and there is no centro-peripheral gradient in soma sizes.

peripheral density gradient, like that in closely related hamsters (Tiao, Blakemore 1976) and mice (Dräger, Olsen 1981) is rather shallow.

The retina of the adult rat exhibits also a centro-peripheral gradient in soma sizes (Fukuda 1977). This gradient is apparent despite the fact that the ganglion cells of all morphological classes distinguished in the rat (Perry 1979; Ni, Dreher 1981) are present at any retinal location (Figs. 2D, & 5C,D) and the ranges of the perikaryal diameters of cells located in the regions of different ganglion cell density partially overlap. Thus, although the range of soma diameters (6-16.5μm) of HRP labelled cells located at the area centralis partially overlaps with the range of soma diameters (9.5-22.5μm) of labelled cells located in the lower nasal periphery (Figs. 1C & 2C,D), the mean soma diameter (10.7μm) of centrally located cells is significantly smaller (Mann-Whitney U test, $p < 0.05$ two tailed criterion) than that (13.8μm) of cells located in the lower nasal periphery of the retina. A very similar centro-peripheral soma size gradient has been observed among the ganglion cells of adult mice (Dräger, Olsen 1981).

Uniformity of Ganglion Cell Density and Soma Sizes in the Retina of the Newborn Rat

It is unlikely that the centro-peripheral gradients in density and soma sizes result from a localized high rate of proliferation of ganglion cells at the area centralis. Thus, in the newborn rats, that is shortly after the cessation of ganglion cell proliferation (which occurs between the 12th and 17th day of a 21-22 day gestation period - Lund 1978), neither density (Fig. 1B) nor soma size centro-peripheral gradients are apparent. The retinal cells labelled retrogradely by massive (1ul) HRP injections into both cerebral hemispheres, are tightly packed into two or three strata within the ganglion cell layer (Fig. 2A cf Braekevelt, Hollenberg 1970; Perry et al. 1983) and their density varies irregularly between 17000 and 23000 cells/mm^2 (Figs. 1B & 3A). The ganglion cells of newborn rats are much more uniform in size than those of adults (soma size range being only 3.5-9.5μm), the mean soma diameter is small (6.0μm) and virtually the same across the retina (Figs. 1D & 2B).

In an attempt to elucidate the mechanism(s) underlying the ontogeny of the centro-peripheral density and soma size gradients among rat ganglion cells, we followed

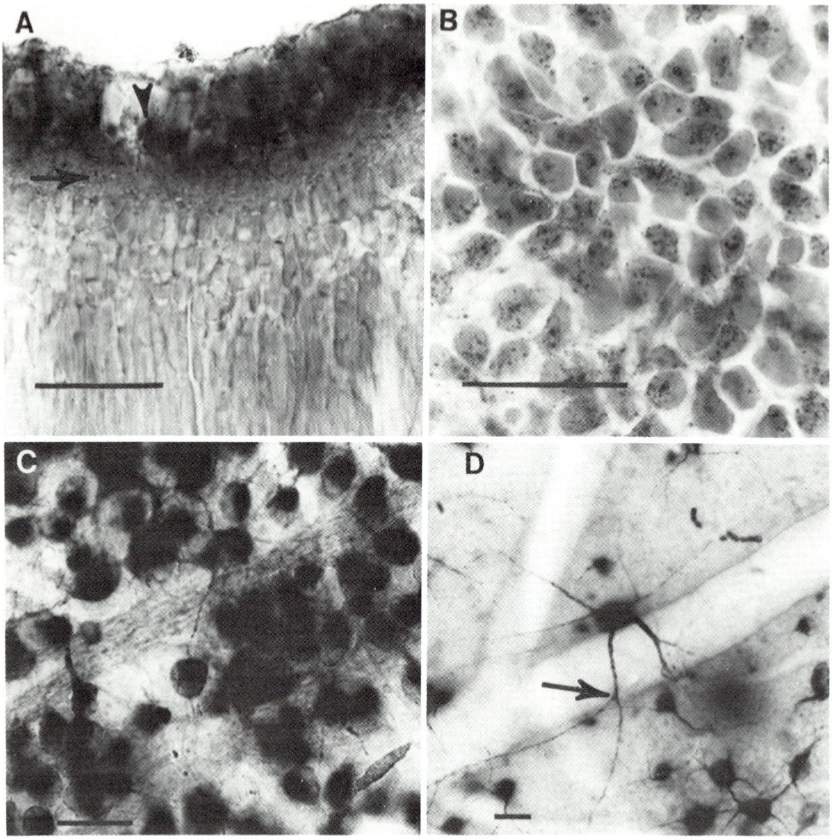

Fig. 2. Photomicrographs of HRP-labelled cells in the ganglion cell layer of the rat. A - cross-section of a 2 day old rat retina. Note that although the HRP injections were massive (1ul into each hemisphere) labelled cells (vertical arrow) are restricted to the ganglion cell layer which is separated from the lower layers by a thin plexiform sheet (horizontal arrow). B - fragment of the wholemount of a 2 day old rat retina. Note homogeneity of soma sizes of labelled cells and presence of unlabelled cells in ganglion cell layer. C - HRP-labelled cells in the area centralis of the retinal wholemount of an adult rat. Note high density of labelled cells. D - HRP-labelled cells in the lower nasal periphery of adult retina. Note wide variety of soma and dendritic tree sizes. In the centre Type IIIb cell of Ni and Dreher (1981). Note medium size soma, large dendritic tree and medium gauge primary dendrites (arrow) with few branches The scale bars - 25μm.

the time course of development of those gradients and related it to (1) the time course and extent of the naturally occurring death of rat retinal ganglion cells and (2) the time course and extent of the areal growth of the rat retina.

METHODS

In a series of albino (Sprague-Dawley) and hooded (Manchester) rats, ranging in age from newborn to adult, retinal ganglion cells were retrogradely labelled by massive (1-3μl) injections of 30% (in Tris-HCl buffer of pH 8.6 with 2-5% diamethylsulphoxide) solutions of HRP (Grade I, lyophilized, Boehringer Mannheim) into the retino-recipient nuclei of each hemisphere. About 15-20h after the HRP injections, animals were reanaesthetized and perfused transcardially. The details of the anaesthesia, injection, perfusion and retinal whole-mounting have been described elsewhere (Potts et al. 1982; Dreher, Potts, Bennett, 1983, Stone 1981). A detailed scale drawing of each retina was made using a camera lucida attachment and its area measured using a computer-linked graphics tablet. The density distribution maps were constructed from counts of 5000 to 10000 labelled cells distributed in 200-500 sampling areas. Counts of labelled cells and measurements of their soma diameters (expressed as the means of the long and short axes) were made while viewing the retinae through a microscope with 100 x (or a 40 x) oil immersion objective.

DEVELOPMENT OF THE CENTRO-PERIPHERAL GRADIENT IN GANGLION CELL DENSITY: RELATION TO GANGLION CELL DEATH, GROWTH OF THE RETINA AND TRANSRETINAL MIGRATION.

Figure 3A shows the variation in density of the labelled ganglion cells as a function of retinal location in representative retinae of albino rats ranging in age from postnatal day 2 (P2) to adulthood. The distributions of ganglion cell density in hooded rats of the same ages are virtually identical with those of albino rats. The density values plotted in Fig. 3A represent the values at 0.5mm (occasionally 0.25mm) intervals along the horizontal line running from the nasal to the temporal edge of the retina across the region of the highest ganglion cell density or (in the animals younger than P13 days) through the region where the highest density area would be expected to develop. In all animals younger than P13 days the density values vary rather irregularly

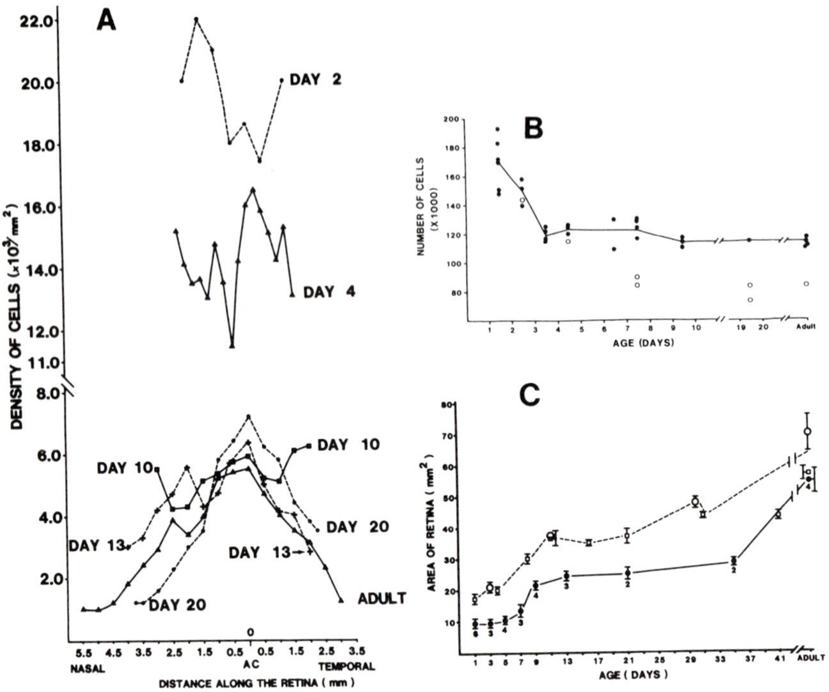

Fig. 3A - Density profiles of HRP-labelled retinal cells along the nasotemporal axes through the retinae of rats of different ages (as indicated). In the younger animals, the area centralis is not apparent and the ganglion cell density varies irregularly. 3B - Estimated number of HRP-labelled cells in the retinae of rats of different ages. The black dots indicate the estimated numbers of labelled cells in the animals in which massive HRP injections into the retino-recipient nuclei were made 15-20h before the perfusion (Potts et al. 1982). The open circles indicate the estimated numbers of labelled cells in the retinae contralateral to the neonatally ablated superior colliculi. HRP was also injected 15-20h before the perfusion (Dreher et al. 1983). 3C - Graph of the retinal areas in animals of different ages. Vertical bars indicate the standard deviation. Solid line - the areas after dehydration. Broken line - the areas before dehydration (cf Perry, Walker, 1980). Note that although dehydration produces a substantial shrinkage, the amount of shrinkage at different ages is virtually the same.

and no localized peak in the ganglion cell density can be discerned. A similar picture emerges in the animals younger than P13 when the density values along any other horizontal line running across their superior retinae are plotted - clearly the area centrales have not yet developed. On the other hand, during the first 10 postnatal days, there is about a fourfold, fairly uniform, reduction in overall density of ganglion cells. Thus, in the retinae illustrated in Fig. 3A, the ganglion cell density on P2 varied irregularly around the value of 20000 cells/mm^2 (17400 to 22000 cells/mm^2). By P4 the density varied between 11400 and 16450 cells/mm^2 while by P10 the ganglion density values ranged between 4250 and 6200 cells/mm^2.

Only by P13 is there some indication (and then only in about 50% of the animals studied) of a peak of ganglion cell density emerging in the region corresponding to the location of the area centralis of the adult animal. However, if present, this peak is very slight - the ganglion cell density (6400 cells/mm^2) being only about twice that at the far periphery (2800-3000 cells/mm^2); frequently an additional peak of almost the same density (5500 cells/mm^2) is present in the nasal retina. It is worthwhile to note that the area centralis starts to emerge at the time when the first of the synaptic complexes are formed in the inner plexiform layer and the first weak responses to light flashes are detectable in the electroretinogram (Weidman, Kuwabara 1968). Furthermore, the rats open their eyes for the first time at about P14.

It is, however, not until postnatal day 20 that a clear-cut, sharply localized concentration of ganglion cells in the upper temporal retina is apparent in all animals. At that age, the centro-peripheral density gradient between the high density region in the upper temporal retina and far nasal periphery is already very similar to that in adult animals (7200 cells/mm^2 at the peak vs. 1200 cells/mm^2 at the nasal edge of the retina). On the other hand, even by P20 the ganglion cell density at the temporal edge of the retina (3500 cells/mm^2) is substantially greater than that at the nasal edge; it is also greater than the density in the peripheral regions of the adult retinae (about 1000 cells/mm^2).

What are the mechanisms underlying the change from uniform to non-uniform distribution of retinal ganglion cells? Three different (not necessarily exclusive) mechanisms have been postulated.

Firstly, it has been hypothesized that in the foetal cat the rapid change in the ganglion cell distribution from uniformity to an 18:1 centro-peripheral density gradient is largely due to "a process of differential maturation in which there is a marked variation between the area centralis and peripheral retina in the proportion of precursor cells which survive and mature as ganglion cells" (Stone, Rapaport, Williams, Chalupa 1982). In support of this hypothesis, one can cite recent observations concerning the rate of naturally occurring cell death in the ganglion cell layer of newborn hamsters and rats. Thus, Sengelaub and Finlay (1982) noted that in hamsters, during the first 10 postnatal days, the rate of cell death differs across the retina, being higher in the peripheral that in the central retina. Similarly, Cunningham, Mohler and Giordano (1982) reported that in 6 and 7 day old rats there are significantly more degenerating cells in the ganglion cell layer of the nasal retinae than in the ganglion cell layer of the temporal retinae (where the future area centralis is located). Indeed, Sengelaub and Finlay (1982) suggested that "some local specializations of retina might be partially sculpted by differential cell death".

Secondly, it has been postulated that the increase in the centro-peripheral gradient in ganglion cell density observed in the retinae of the postnatal cats can be accounted by a "differential growth" of the retina - that is the growth being greatest at the retinal edge and least at the area centralis (Mastronarde, Thibeault, Dubin 1980; Stone et al. 1982).

Thirdly, the uniform distribution of ganglion cells can also be changed to non-uniform by "transretinal migration" of ganglion cells. Thus, Morest (1970) provided evidence indicating that in the neonatal rats, many of the ganglion cells migrate to the ganglion cell layer, only after they send their axons into the optic nerve. More importantly, the transretinal migration of ganglion cells away from the fovea seems to underlie the formation of the foveal pit in the retinae of humans (Mann 1969) and macaque monkeys (Hendrickson, Kupfer 1976). Finally, Beazley and Dunlop (1983) concluded that in the Western Australian marsupial - quokka (Setonix brachyurus) the transretinal migration of ganglion cells is the most likely mechanism underlying the evolution of the area centralis.

Which of these mechanisms is responsible for the development of the centro-peripheral ganglion cell density gradient in the rat?

"Differential Cell Death" - Both our own data (Fig. 3B - Potts et al. 1982) and those of Perry et al. (1983) indicate that in the rat during the first 10 postnatal days, there is a substantial (at least 35-45%) loss of retinal ganglion cells. This loss, is due to ganglion cell death (Dreher et al. 1983; Perry et al. 1983), rather than transformation of ganglion cells into "displaced amacrine cells" (cf Hinds, Hinds, 1978; 1983). The bulk of this loss (27-29% of all ganglion cells) occurs during the first 5-6 postnatal days and by P10, the number of ganglion cells and axons in the optic nerve either stabilizes at (Lam et al. 1982; Potts et al., 1982) or is only very slightly above the adult level (Perry et al. 1983). It appears thus that the loss of ganglion cell contributes strongly to the uniform reduction of the ganglion cell density observed by us during the first 10 postnatal days. However, since the centro-peripheral gradient of ganglion cell density develops only between P10 and P20, that is after the period of massive ganglion cell loss, it appears to us that "differential cell death" (or "maturation") probably does not underlie the development of the centro-peripheral density gradient in the rat retina. Similarly, the constancy in numbers of the ganglion cells during the development of the area centralis (and visual streak) in quokka argues against the idea that the ganglion cell density gradient in this marsupial is brought about by ganglion cell death (Beazley, Dunlop 1983).

"Differential Growth of the Retina" - By P10, the retinal area is 2.1-2.2 times greater than that of the newborn rat (Fig. 3C; cf Perry, Walker 1980). Clearly then the growth of the retina plays an important role in the overall reduction in the ganglion cell density which takes place during the first 10 postnatal days (Fig. 3A). However, between P10 and P20 the retinal area increases only by 15-20% (Fig. 3C); even if it is assumed that the entire growth after P10 occurs at the retinal edges and there is no growth at the area centralis, the differential growth of the retina could not account for the 6:1

centro-peripheral ratio of ganglion cell density observed at P20 and adult animals.

"Transretinal Migration" - Thus, by exclusion, we are left with the hypothesis of transretinal migration. On the one hand the fact that the ganglion cell density at the area centralis of the adult rat is very similar to the approximately uniform density in the retinae of the 10 day old rats argues against this hypothesis. On the other hand however, consistent with the hypothesis of transretinal migration, at P20 that is shortly after the emergence of the area centralis, the ganglion cell density in the area centralis (7000-7500 cells/mm^2) is always about 20-30% higher than that in the area centralis of the adult. The ensuing reduction of ganglion cell density is probably related to the ongoing areal growth of the retina (Fig. 3C).

Although the transretinal migration might underlie (at least partially) the formation of shallow density gradients characteristic of nocturnal rodents (rats, hamsters and mice) as well as some marsupials (quokkas - Beazley, Dunlop 1983; South and North American opossums - Hokoc, Oswaldo-Cruz 1978; Rapaport, Wilson, Rowe 1981) it probably does not underlie the formation of the area centralis in the cat (with its very steep centro-peripheral density gradient of about 100:1 - Stone et al. 1982). It is thus worthwhile to note in this context the evidence suggesting that the area centralis of the rodents is not a homologue of the area centralis of the carnivores or the fovea of the primates (cf Rapaport, Stone 1984). Thus, the area centralis in the nocturnal rodents is located well outside the crescent-shaped region in the temporal retina in which the great majority (Fig. 4A) of ipsilaterally projecting ganglion cells is located (cf for rats Cowey, Perry 1979; Keens 1981, Linden, Perry 1982; Jeffery, Cowey, Kuypers, 1981; for hamsters Tiao, Blakemore 1976; Rhoades, Kuo, Polcer 1982; for mice Dräger, Olsen 1980; Godement 1984). Clearly, the "area centrales" of the rodents (unlike those of the cat and fovea of the primates) do not coincide with the centres of binocular correspondences in the two retinae (Tiao, Blakemore 1976; Dräger, Olsen 1980; Reese, Jeffery 1983).

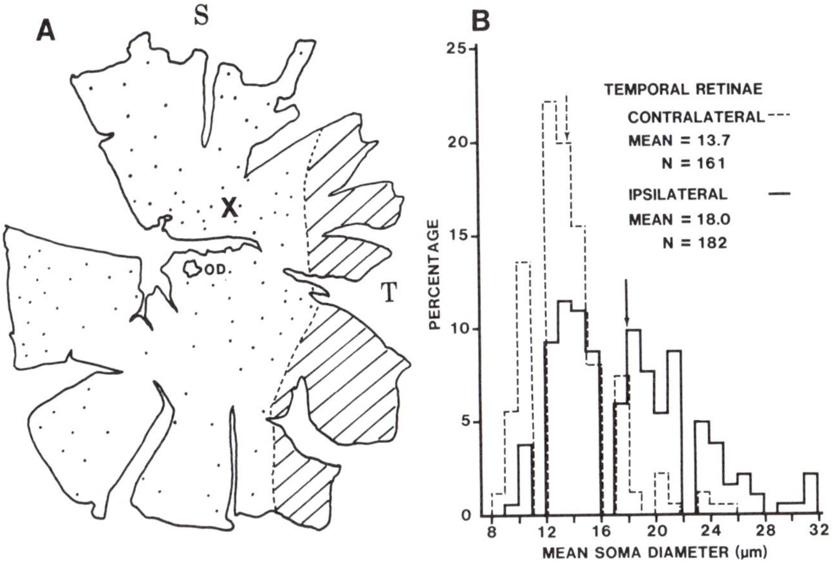

Fig. 4A - Diagram of the retinal wholemount (not to scale) of the albino rat. The hatched area indicates the region (temporal crescent) in which the great majority (93%) of 1500 ganglion cells labelled by massive HRP injection into the ipsilateral LGNd is located. The unusually superior location of temporal crescent is probably related to the difficulty in orienting correctly the eye when making the major cut. S and T denote respectively superior and temporal edges of the retina. OD indicates the position of the optic disc. X indicates approximate position of the area centralis - it is clearly well outside the temporal crescent. 4B - Frequency histograms of soma diameters of HRP-labelled ganglion cells located in the temporal crescents of the retinae ipsi and contralateral to HRP injected LGNd of the hooded rat. Note the substantial difference in the mean soma sizes due mainly to a high proportion of cells with large perikarya projecting to the ipsilateral LGNd. Note also that mean soma size (13.7μm) of cells projecting to the contralateral LGNd is virtually the same as that (13.8μm) of cells in the nasal retina labelled by massive HRP injections (Fig. 1C).

DEVELOPMENT OF HETEROGENEITIES IN GANGLION CELL SOMA SIZES: RELATION TO THE CENTRO-PERIPHERAL GANGLION CELL DENSITY GRADIENT.

As mentioned in the Introduction, in the newborn rat, ganglion cells throughout the retina are fairly homogenous in size (3.6-9.5μm - Fig. 1D) and there is no discernable centro-peripheral soma size gradient. By P10, when a substantial areal growth of the retina has occurred, the range of ganglion cell soma sizes broadens significantly (5-14μm) and the mean soma size increases to 8.7μm. At the same time the principal morphological classes of ganglion cells described in the retina of the adult (Figs. 2D & 5D, cf Perry 1979; Ni, Dreher 1981) start to be distinguishable (Perry, Walker 1980). By P13 there is a further increase in both soma size range (5-16μm) and the mean soma size (9.3μm). However, it is only by P20 that the centro-peripheral soma size gradient becomes apparent (Fig. 5B) and all the morphological classes of ganglion cells are clearly distinguishable, although ganglion cells perikarya have not yet reached the adult size (compare Figs. 5C and D).

It appears that the development of heterogeneity in ganglion cell soma sizes (reflecting the emergence of distinct morphological classes of ganglion cells) is not related to the development of the centro-peripheral density gradient. On the other hand, the development of the centro-peripheral gradient in perikaryal sizes of ganglion cells seems to be closely linked with the development of the ganglion cell density gradient.

In an additional series of experiments, we were able to obtain evidence indicating that indeed the link between the two centro-peripheral gradients is very strong. Thus, in 20 day old and adult rats in which one of the superior colliculi (SC) has been removed within 12h of animal birth (for details see Dreher et al. 1983) the overall density of ganglion cells (labelled by massive injections of HRP into the remaining retino-recipient nuclei) in the retinae contralateral to ablated SC is substantially lower than that in the retinae ipsilateral to the ablation. This "decrowding" results from the additional ganglion cell death induced by the neonatal removal of the SC (Perry, Cowey 1979; Linden, Perry 1982; Dreher et al. 1983). Although the ganglion cell density in the retinae contralateral to the SC is not entirely uniform, there is no clear-cut high density area and nowhere does the density exceeds 2800-3200 cells/mm^2 (Fig.

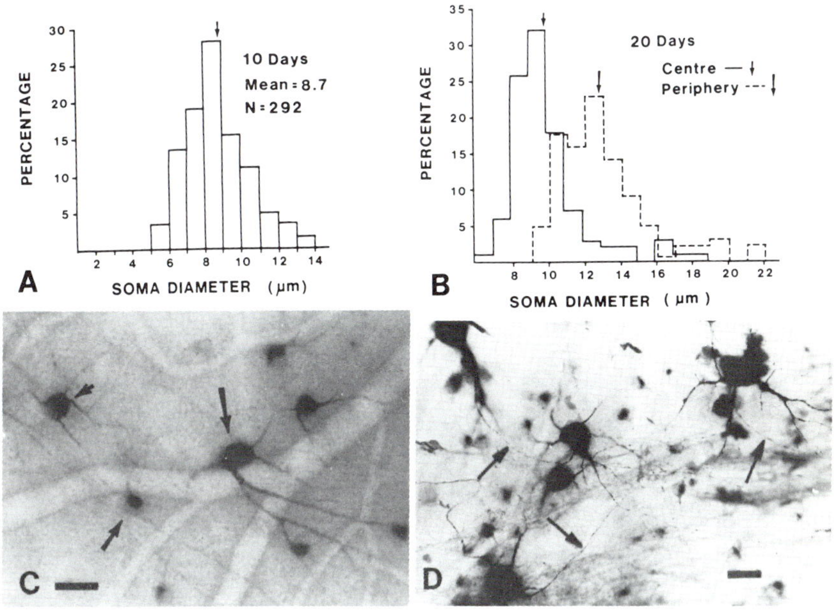

Fig. 5A and B - Fequency histograms of ganglion cell soma sizes of (respectively) 10 and 20 day old rats. Note that by P20 there is a clear-cut centro-peripheral soma size gradient. About 300 cells were sampled in the area centralis and the same number in the nasal periphery. 5C - photomicrographs of HRP-labelled ganglion cells in the nasal retina of a 20 day old rat. Long arrow points to Type I cell (cf Perry 1979; Ni, Dreher 1981). Note relatively large soma and large gauge primary dendrites. Medium length arrow points to Type IIa cell (Ni, Dreher 1981 cf Type II of Perry 1979). Note small perikaryon. Short arrow points to Type IIIb cell (Ni, Dreher 1981). Note medium size soma and medium gauge primary dendrites. 5D - photomicrograph of HRP-labelled ganglion cells in the temporal retina of the adult hooded rat. Note several large Type I cells. Arrows point to their axons (large gauge 0.8-1.2um in diameter). Scale bars in 5C and 5D: 25μm.

6A). The lack of a clear-cut area centralis is accompanied by a substantial increase in mean perikaryal sizes as well as a substantial increase in soma size ranges of cells located in the central region. Thus, while in normal adult animals (or in the retinae ipsilateral to

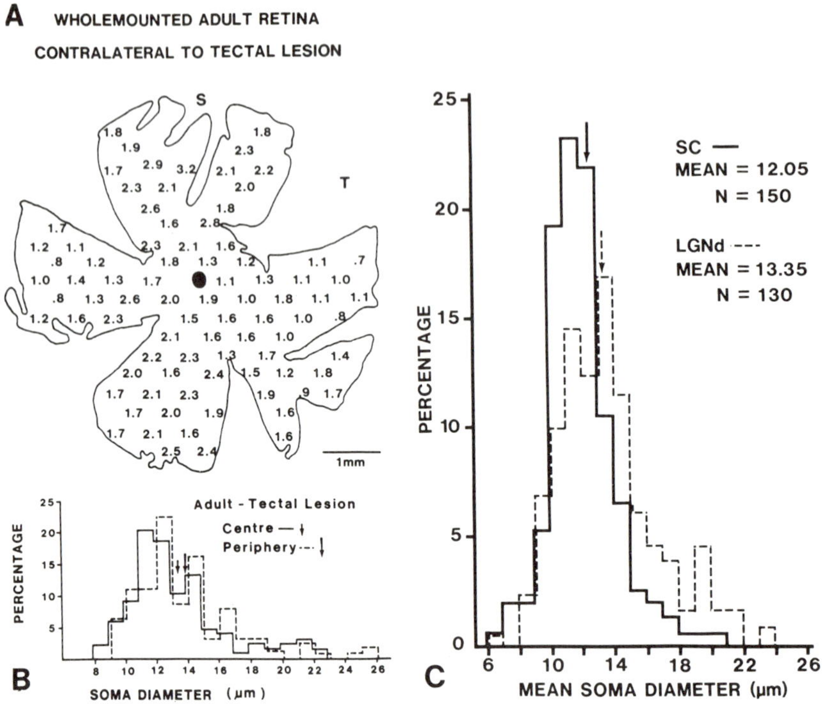

Fig. 6A - Wholemount of adult retina contralateral to neonatal tectal ablation. The numbers indicate density of ganglion cells labelled by massive HRP injections into the retino-recipient nuclei in units of 1000 cells/mm^2. The black spot in the middle - optic disc. Note that the ganglion cell density nowhere exceeds 2800-3200 cells/mm^2. 6B - Frequency histograms of soma sizes of HRP-labelled cells located in the central and peripheral (nasal) parts of the retina contralateral to neonatally ablated SC. Note that both the soma size ranges and the mean soma sizes (indicated by arrows) are very similar. Sample sizes in 6B are 300 cells. 6C - Frequency histograms of ganglion cells in the nasal retinae (3mm away from the optic disc) labelled by HRP injections into SC or LGNd. Note that the cells projecting to SC are significantly (Mann-Whitney U test - $p < 0.05$ two tailed criterion) smaller than those projecting to LGNd.

ablated SC) the mean soma size of ganglion cells located at the area centralis is 10.7μm (range 6-16.5μm - Fig. 1C) in the retinae contralateral to ablated SC it is 13.4μm (range 8-23μm - Fig. 6B). The difference is statistically significant (Mann-Whitney U test $p < 0.05$ two tailed criterion). Furthermore, no centro-peripheral soma size gradient is apparent in the retinae contralateral to the ablated SC; the mean soma sizes (13.9μm) and the soma size ranges (9-26μm) of cells located in the nasal peripheral retina are virtually the same as those of cells located centrally.

Perry and Linden (Perry, Linden, 1982; Linden, Perry, 1982) provided evidence indicating that the sizes of the dendritic trees of rat retinal ganglion cells are partially determined by interactions with neighbouring cells. It appears from our data that the principal factor underlying the centro-peripheral gradient in the ganglion cells soma sizes is the centro-peripheral difference in the ganglion cell density. Similarly in the cat, reducing the ganglion cell density at the area centralis by sectioning the optic tract at P11, results in a significant increase (in comparison with normal animals) in the perikaryal sizes of cells located in the "decrowded" region (Rapaport, Stone 1983). On the other hand, there appears to be some level of density below which the "crowding" effect does not seem to operate. For example, in North American opossum (Didelphis virginiana) in which the ganglion cell density at the area centralis does not exceed 2700 cells/mm^2, there is no clear-cut centro-peripheral soma size gradient, despite the fact that the centro-peripheral ganglion cell density ratio (about 5:1) is very similar to that in the rats (Rapaport, Wilson, Rowe, 1981). In other words, below certain level of ganglion cell density, the perikaryal sizes of retinal ganglion cells are apparently not affected by interactions with their neighbours.

The differences in the mean soma sizes are not exclusively related to the differences in the ganglion cell density. Apart from differences in size related presumably to "pre-programmed" differences between the morphological classes (Perry 1979; Ni, Dreher 1981), there are additional differences which seem to be related to the pattern of projection of ganglion cells. Thus, although in a given region of the adult retina the ranges of some sizes of ganglion cells projecting to the superior colliculus (SC) overlap almost completely with those of cells projecting to the dorsal lateral geniculate nucleus (LGNd) the mean soma sizes of cells projecting to SC are significantly smaller than those projecting to the LGNd (Fig. 6C cf Lund, Land, Boles 1980). Although ganglion cells of all functional classes seem to project to SC and LGNd (Sefton 1968) within a particular morphological class, the cells with larger perikarya project to both SC and LGNd while those with smaller somas project only to SC (Ni, Dreher 1981). Similarly, of the cells located in the temporal crescent, those projecting to the ipsilateral LGNd have substantially larger perikarya than those of cells projecting to the contralateral LGNd (Fig. 4B). The difference is mainly due to the very high proportion of large Type I cells projecting to the ipsilateral LGNd. At the moment however, we do not know the time course or the mechanisms underlying the development of the heterogeneities in ganglion cell soma sizes related to the patterns of their projections to the target nuclei.

CONCLUSIONS

It appears that in the rat "differential cell death" (or "differential maturation") of ganglion cells does not underlie the development of ganglion cell density gradient. On the other hand both "transretinal migration" of ganglion cells and "differential growth" of the retina might play a significant role in the development of this gradient.

The development of the centro-peripheral soma size gradient is closely linked with the development of the centro-peripheral density gradient. Thus the centro-peripheral soma size gradient is largely determined by the intraretinal environment.

ACKNOWLEDGEMENTS

We would like to thank Ann J. Sefton and Richard Bandler for reading the manuscript and providing helpful criticisms. We would also like to thank Murray J. McCall for help in the preparation of figures and Michelle Figon for typing the manuscript. This project has been supported by a grant from N.H.&M.R.C. of Australia.

REFERENCES

Beazley LD, Dunlop SA (1983). The evolution of an area centralis and visual streak in the marsupial Setonix brachyurus. J Comp Neurol 216:211.

Braekevelt CR, Hollenberg MJ (1970). The development of the retina of the albino rat. Am J Anat 127:281.

Cowey A, Perry VH (1979). The projection of the temporal retina in rats, studied by retrograde transport of horseradish peroxidase. Exp Brain Res 35:457.

Cunningham TJ, Mohler IM, Giordano DL (1982). Naturally occurring neuron death in ganglion cell layer of the neonatal rat: morphology and evidence for regional correspondence with neuron death in superior colliculus. Dev Brain Res 2:203.

Dräger UC, Olsen JF (1980). Origins of crossed and uncrossed retinal projections in pigmented and albino mice. J Comp Neurol 191:383.

Dräger UC, Olsen JF (1981). Ganglion cell distribution in the retina of the mouse. Invest Ophthal Visual Sci 20:285.

Dreher B, Potts RA, Bennett MR (1983). Evidence that the early postnatal reduction in the number of rat retinal ganglion cells is due to a wave of ganglion cell death. Neurosci Lett 36:255.

Fukuda Y (1977). A three-group classification of rat retinal ganglion cells: histological and physiological studies. Brain Res 119:327.

Godement P (1984). Development of retinal projections in the mouse. In: Stone J, Dreher B, Rapaport DH (eds). "Development of Visual Pathways in Mammals". New York: Alan R Liss, (this volume).

Hendrickson A, Kupfer C (1976). The histogenesis of the fovea in the macaque monkey. Invest Ophthal 15:746.

Hinds JW, Hinds PL (1978). Early development of amacrine cells in the mouse retina: an electron microscopic serial section analysis. J Comp Neurol 179:277.

Hinds JW, Hinds PL (1983). Development of retinal amacrine cells in the mouse embryo: evidence for two modes of formation. J Comp Neurol 213:1.

Hokoc JN, Oswaldo-Cruz E (1978). Retina and optic nerve of the opossum: quantitative analysis of the ganglion cells and optic fibres. In: Rocha-Miranda CE, Lent R (eds). "Opossum Neurobiology". Rio de Janeiro. Academia Brazileira de Ciencias, p 33.

Hughes A (1977). The topography of vision of mammals of contrasting life style: comparative optics and retinal organization. In: Crescitelli F (ed) "Handbook of Sensory Physiology" VII/5. Berlin Heidelberg New York: Springer Verlag, p 613.

Jeffery G, Cowey A, Kuypers HGJM (1981). Bifurcating retinal ganglion cells axons in the rat, demonstrated by retrograde double labelling. Exp Brain Res 44:34.

Keens JS (1981). Aspects of the retinal projections in rats. Proc Aust Physiol Pharmacol Soc 12 (2):163P.

Lam K, Sefton AJ, Bennett MR (1982). Loss of axons from the optic nerve of the rat during early postnatal development. Dev Brain Res 3:487.

Lashley KS (1932). The mechanism of vision. V. The structure and image forming power of the rat's eye. J Comp Psychol 13:173.

Linden R, Perry VH (1982). Ganglion cell death within the developing retina: A regulatory role for retinal dendrites? Neurosci 7:2813.

Lund RD (1978). "Development and Plasticity of the Brain". New York: Oxford University Press.

Lund RD, Land PW, Boles J (1980). Normal and abnormal uncrossed retinotectal pathways in rats: an HRP study in adults. J Comp Neurol 189:711.

Mann I (1969). "Development of the Human Eye". New York: Grune and Stratton.

Mastronarde DN, Thibeault MA, Dubin MW (1980). How ganglion cells redistribute during postnatal growth of the cat retina. Invest Ophthal Suppl 70.

Morest DK (1970). The pattern of neurogenesis in the retina of the rat. Z Anat Entwgesch 131:45.

Ni SY, Dreher B (1981). Morphology of rat retinal ganglion cells projecting to the thalamus and midbrain. Proc Aust Physiol Pharmacol Soc 12(2): 97P.

Perry VH (1979). The ganglion cell layer of the retina of the rat: a Golgi study. Proc R Soc Lond B 204: 363.

Perry VH, Cowey A (1979). The effects of unilateral cortical and tectal lesions on retinal ganglion cells in rats. Exp Brain Res 35:85.

Perry VH, Henderson Z, Linden R (1983). Postnatal changes in retinal ganglion cell and optic axon populations in the pigmented rat. J Comp Neurol 219:356.

Perry VH, Linden R (1982). Evidence for dendritic competition in the developing retina. Nature 297:683.

Perry VH, Walker M (1980). Morphology of cells in the ganglion cell layer during development of the rat retina. Proc R Soc Lond B 208:433

Potts RA, Dreher B, Bennett MR (1982). The loss of ganglion cells in the developing retina of the rat. Dev Brain Res 3:481.

Rapaport DH, Stone J (1983). Time course of morphological differentiation of cat retinal ganglion cells: influences on soma size. J Comp Neurol 221:42.

Rapaport DH, Stone J (1984). The area centralis of the retina in the cat and other mammals: focal point for function and development of the visual system. Neurosci (in press).

Rapaport DH, Wilson PD, Rowe MH (1981). The distribution of ganglion cells in the retina of the North American opossum (Didelphis virginiana). J Comp Neurol 199:465.

Reese BE, Jeffery G (1983). Crossed and uncrossed visual topography in dorsal lateral geniculate nucleus of the pigmented rat. J Neurophysiol 49:877.

Rhoades RW, Kuo DC, Polcer JD (1982). Effects of neonatal cortical lesions upon retinocollicular projections in the hamster. Neurosci 7:2441.

Sefton AJ (1968). The innervation of the lateral geniculate nucleus and anterior colliculus in the rat. Vision Res 8:867.

Sengelaub DR, Finlay BL (1982). Cell death in mammalian visual system during normal devlopment: I. Retinal ganglion cells. J Comp Neurol 204:311.

Stone J (1981). "The Wholemount Handbook". Sydney: Maitland Publications.

Stone J (1983a). Topographical organization of the retina in a monotreme: Australian spiny anteater Tachyglossus aculeatus. Brain, Behav & Evol 22:175.

Stone J (1983b). "Parallel Processing in the Visual System". New York: Plenum Publishing Corp.

Stone J, Rapaport DH, Williams RW, Chalupa L (1982). Uniformity of cell distribution in the ganglion cell layer of prenatal cat retina: implications for mechanisms of retinal development. Dev Brain Res 2:231.

Tiao YC, Blakemore C (1976). Regional specialization in the golden hamster's retina. J Comp Neurol 168:439.

Weidman TA, Kuwabara T (1968). Postnatal development of the rat retina. Arch Ophthal 79:470.

Development of Visual Pathways in Mammals, pages 59–73

THE DEVELOPMENT OF GANGLION CELL MOSAICS

V. H. Perry

Department of Experimental Psychology
University of Oxford
South Parks Rd., Oxford, U.K.

Introduction

It is now generally recognized that the ganglion cells of the mammalian retina are not a homogeneous class of cells. In the first instance they were subdivided into on- and off-centre cells (Kuffler 1953) but more extensive physiological studies have revealed further subdivisions of cell types in the cat retina. These cells in the cat are usually referred to as X-, Y- and W-cells, the W-cells being an umbrella term for a heterogeneous population of cells but with some properties in common (see Rodieck 1979; Lennie 1980, for reviews). If there were a class of cells which performed an important function in the visual process it would not be surprising to find such cells distributed over the whole retina so that no part of the visual field was, for example, "X-blind". Exactly how the cells of a particular type are distributed in the retina is not a simple matter to determine physiologically, but if it were possible to show which morphological class of ganglion cell was the anatomical substrate of a given physiological type then by staining only this population of cells it would then be possible to show how cells covered the surface of the retina and thus how they sampled the visual field.

There is now good reason to believe that the morphologically identified alpha, beta and gamma cells of the cat retina are the anatomical counterparts of Y-, X- and W-cells (Boycott, Wässle 1974; Peichl, Wässle 1981; Wässle, Boycott, Illing 1981). The evidence that the alpha cells are the morphological substrates of Y-cells is based on somewhat firmer ground than the relationship between beta and X-cells and gamma and W-cells. A recent study has suggested that there may be as many as twenty-three anatomically different types of ganglion cells and mapping each of these cells to a physiologically identified category will be difficult using the sort of indirect approaches of previous studies (Kolb, Nelson, Mariani 1981).

In the retina of the cat it is possible to identify the alpha cells in both Nissl stained and reduced silver stained whole-mounts and it can be shown that not only are the cell bodies regularly spaced within a small patch of retina (Wässle, Reimann 1978) but the dendrites of the alpha cells, visible in reduced silver stains, cover every part of the retina (Wässle, Peichl, Boycott 1981). Furthermore, the alpha cells can be subdivided into two populations of cells, those whose dendrites branch in the inner plexiform layer close to the inner nuclear layer and those which branch in the middle of the inner plexiform layer; these cells correspond to the off- and on-centre Y-cells respectively (Peichl, Wässle 1981). The regular spacing of the alpha cell population is found to be all the more precise when subdivided into on- and off-centre cells and the dendrites of each population of cells covers the retina so that there are no "Y-blind" parts of the visual field (Wässle, Peichl, Boycott 1981). The problem my colleagues and I have been interested in is how these regularly spaced patterns are generated in development.

The regular spacing between different populations of cells in the retina is by no means unique to ganglion cells. It was recognized some time ago that the cones in the retina of the fish and birds are regularly spaced (see Scholes 1976). More recent studies on the cones of the primate and cat and the horizontal cells of the cat (Wässle, Reimann 1978) and amacrine cells in the rabbit (Vaney, Peichl, Boycott 1981) all show a regular spacing between their cell bodies. The regular spacing of cells in the retina is not confined to the neurones and glia of the retina, the Müller cells (Dräger 1983), the microglia (Dräger 1983; Hume, Perry, Gordon 1983) and the astroglia (unpublished observations) all show a regular spacing. The presence of populations of cells which have a constant nearest neighbour distance is less easy to observe in other parts of the nervous system where the experimenter has to contend with the third dimension, but a good example of a

Table 1. Pyknotic nuclei and F4/80-labelled macrophages in the retinal layers at different ages.

Age (days)	Ganglion cell layer	Inner nuclear layer	Outer nuclear layer
16 day embryo	4 (3)	7 (5)	-
0	54 (34)	10 (1)	-
5	31 (24)	68 (16)	-
10	0	10 (2)	2
20	0	0	0

The number of pyknotic nuclei were counted in three sections through the centre of two eyes from two CBAT6/T6 mice of each age. The numbers in brackets indicate the number of pyknotic nuclei with F4/80 positively stained processes surrounding them.(from Hume,Perry, Gordon 1983).

The cell death in the ganglion cell layer and inner nuclear layer is accompanied by an influx of macrophages but no macrophages were identified in the outer nuclear layer and very few pyknotic nuclei have been observed in this layer. In mice homozygous for the rd mutation, which results in massive degeneration of the photoreceptors in the late stages of photoreceptor development, macrophages are readily observed phagocytosing the degenerating cells (Perry, Hume, Gordon unpublished observation). Thus, although cell death is a major feature of the development of the ganglion cell and inner nuclear layers this does not seem to be the case for the outer nuclear layer. If the mosaics of cells in the inner retinal layers develop from an excess of cells we need to know what rules govern the selective loss of some cells and how this might result in the regular spacing of retinal neurones. The following discussion will be largely limited to the ganglion cells.

Terminal Competition Can Regulate Cell Death

The removal of one eye from a newborn rat results in an abnormally expanded projection from the remaining eye to the ipsilateral subcortical visual centres when examined in the mature animal (Lund, Cunningham, Lund 1973). Recent evidence shows that the effect of removing one eye in a developing rodent is to preserve an already extensive ipsilateral projection found in the newborn (Land, Lund 1979) and prevent the loss of a large excess of cells which project ipsilaterally in the newborn rat (Sengelaub, Finlay 1981; Jeffery, Perry 1982). These excess cells which project ipsilaterally arise not only from the lower temporal region of the retina but also from the nasal retina (Lund, Land, Boles 1980; Jeffery, Perry 1982). It has been argued that the reduction in cell death results from an increase in the available size of the target field since the competing axons from the other eye have been removed. While it is clear that competition between the axons from the two eyes may be important for the development of a precise nasotemporal division, it is not clear how such a mechanism could produce a precise spacing of cells within the retina. The sizes of ganglion cell axon terminals are very large and unlikely to produce the resolution required for the cell body spacing observed in the retina (Bowling, Michael 1980). Furthermore, if terminal competition was the only factor regulating cell death then we might find at the nasotemporal border a discontinuity in the retinal mosaic and no such discontinuity has been observed. The most compelling evidence that there must be an intraretinal mechanism is that the on- and off-mosaics of ganglion cells in the retina appear to be independently distributed (Wässle, Peichl, Boycott 1981).

Dendritic Competition Can Regulate Cell Death

If terminal competition is insufficient to produce a retinal mosaic then we must look within the retina for possible mechanisms which regulate the spacing between the cell bodies. Two sorts of experiments provide evidence that dendrites of neighbouring ganglion cells compete for space within the retina and that this competition may regulate which cells degenerate (Linden, Perry 1982; Perry, Linden 1982).

Transection of one optic tract in the newborn rat results in the rapid degeneration of cells in the nasal retina of the contralateral eye. When these animals were mature the intact optic tract was injected with HRP and the ipsilateral eye examined. In contrast to the normal animal a large number of cells which project ipsilaterally are found in the nasal retina. The retina looks rather like that of an animal in which one eye had been enucleated. It was of course possible that the excess cells in the nasal retina had survived because the optic tract lesions had transected sub-cortical commissural fibres and created some excess target space in the intact hemisphere. However, anterograde tracing revealed that the increased ipsilateral projection from the eye to the intact tract was found in regions which did not receive commissural fibres. We suggest that the cells in the nasal retina survived because we had removed their neighbouring ganglion cells and thus reduced the competition between the dendrites. We do not know exactly what the dendrites are competing for but it would be reasonable to assume it is for synaptic contact with their afferents, the bipolar and amacrine cells. In the mouse the onset of synaptogenesis begins at one day after birth (Olney 1968), but is not known in the rat.

Further evidence that ganglion cell dendrites compete within the retina was found at the region of the nasotemporal border. The cells at the border had their dendrites directed towards the nasal retina the region with a reduced ganglion cell density. A more extreme example of the same phenomenon was observed when small retinal lesions were made in the retinae of newborn rats. The lesion transected the axons of ganglion cells from a sector of the retina. At maturity the ganglion cells were retrogradely filled with HRP and those cells along the border of the depleted region were found to have their dendrites directed into this region. Thus the dendrites of ganglion cells appear to be able to modify their orientation to find the region of minimum competition (Fig. 3).

The cells at the border of a region depleted of ganglion cells now offer a good opportunity to study how the neighbouring cells of the same type may or may not constrain the development of the dendritic shape of a cell. In addition, we have now a system where it is possible to offer a cell an excess afferent supply, in contrast to the majority of studies which have the effect of removing the afferents from developing neurons (e.g. Bradley, Berry 1976). We have made two further observations firstly, the modified ganglion cells are clearly recognizable as belonging to a particular class and secondly, the dendrites have not expanded their dendritic fields to occupy all the available space in either the vertical or horizontal plane. Indeed our preliminary quantitative observations reveal only modest changes in cell morphology.

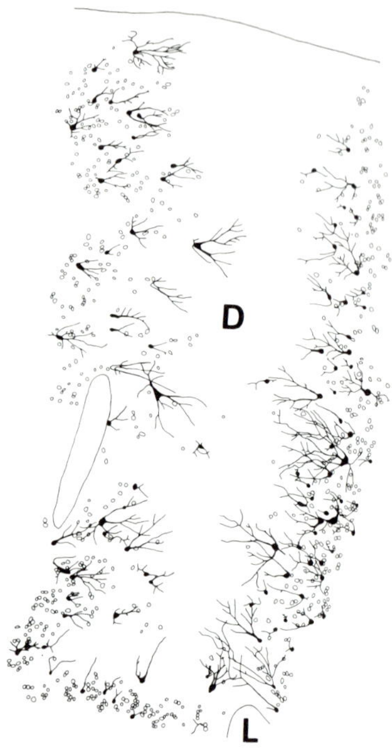

Fig. 3. Ganglion cells retrogradely labelled with HRP around a region devoid of ganglion cells (D). A small lesion (L) was made in the retina on the day of birth. (from Perry,Linden 1982).

Generation of Cell Mosaics

The experiments outlined above suggest that cell death may play a role in the fine tuning of the ganglion cell mosaic. The orientation of ganglion cell dendrites is modifiable so that the dendritic field may be adjusted locally to cover the surface of the retina uniformly but we have little idea of what the pattern of ganglion cells looks like prior to the cell death period. Are the cells generated in a dense but highly ordered mosaic or are they generated randomly? In the absence of a specific marker to label developing Type I ganglion cells in the rat, it is possible to take advantage of the fact that cell death can be partly prevented and a frozen developmental stage can be examined. Removal of one eye at birth in the rat preserves twice the normal number of ipsilaterally projecting cells from the temporal crescent (Jeffery, Perry 1982). In reduced silver stains the Type I cells can be recognized and their distribution in the normal and monocular rat examined at maturity (Jeffery 1982; Jeffery, Perry in preparation). The Type I cells in the normal and monocular rat form significantly non-random distributions but the pattern in the monocular rat is less regularly distributed than in the normal rat. These results suggest that the Type I cells are not generated in a scaled down version of the adult distribution or in a random pattern, and indeed it seems unlikely that a totally random pattern could give rise to a precise mosaic even with selective cell death.

The regulation of the number of cells of a particular type which are produced by a germinal cell is not known although there is suggestive evidence for some sort of feedback mechanism. Intraretinal injections of 6-hydroxydopamine into the eye of the goldfish destroys all the dopaminergic neurons and new cells are added in concentric rings at the margin of the retina. Negishi, Teranishi, Kato (1982) have shown that the dopaminergic cells are produced at a greater rate in treated than in untreated retinae and that the processes of the first dopaminergic cells to be produced are oriented towards the region devoid of dopaminergic cells. These results suggest that the presence of the dopaminergic cells may in some way regulate the number which are produced from the germinal cells. We do not know whether such a mechanism operates during the normal development of the mammalian retina but is an avenue worth exploring. Furthermore these results show that the orientation of the dendrites of a retinal interneurone are also modifiable in much the same way as we have shown for the ganglion cells (Perry, Linden 1982).

Summary

The experiments described above show that during normal development cell death plays a significant role in the development of the inner nuclear and ganglion cell layers. The number of ganglion cells which will die depends upon the availability of target tissue and the afferent supply. It is clear that these two mechanisms interact to produce both a precise nasotemporal division and also a regular matrix of ganglion cells across the surface of the retina. Indeed the obvious explanation for the findings that increasing the target field by monocular enucleation does not completely prevent cell death (Sengelaub, Finlay 1981; Jeffery 1982) is that there are insufficient afferents for all the ganglion cells originally generated. These developing ganglion cells axons and dendrites compete for their connections and it is the latter mechanism which may be partly responsible for the development of a regular ganglion cell mosaic. The present techniques do not allow us to say how precise is the original pattern, it is a less orderly pattern than that found in the adult but unlikely to be a completely random distribution.

In the inner nuclear layer there is a phase of cell death which follows the phase in the ganglion cell layer and it seems plausible that the cells in the inner nuclear layer match their numbers to those of the ganglion cells. In the outer nuclear layer there is virtually no cell death and this may reflect different mechanisms operating in this layer. It has been suggested that the regular cone matrices found in birds and fish retina are produced by the temporal sequence in which the different photoreceptors are generated (Scholes 1976) although at present there is little clear cut evidence to support this suggestion.

Recent studies have shown that the glial cells also lie in regular matrices across the retina. It remains an intriguing question as to whether there is any interaction between the neuronal and glial populations that may be responsible for the development of these patterns.

Acknowledgements: V.H. Perry is a Locke Fellow of the Royal Society. Table 1 was reproduced from the J Cell Biol (1983) 97:253-257 by copyright permission of Rockefeller University Press. Figure 3 was reproduced by permission from Nature (1983) 297:683-685, Macmillan Journals Ltd.

References

Austyn JM, Gordon S (1981). F4/80: A monoclonal antibody directed specifically against the mouse macrophage. Eur J Immunol 10:805.

Bowling DM, Michael CR (1980). Projection patterns of single physiologically characterised optic tract fibers in cat. Nature 286:899.

Boycott BB, Wässle H (1974). The morphologic types of ganglion cells of the domestic cat's retina. J Physiol 240:397.

Bradley P, Berry M (1976). The effects of reduced climbing and parallel fibre input on Purkinje cell dendritic growth. Brain Res 109:133.

Dräger UC (1983). Coexistence of neurofilaments and vimentin in a neurone of adult mouse retina. Nature 303: 169.

Dreher B, Potts RA, Bennett MR (1983). Evidence that the early postnatal reduction in the number of rat retinal ganglion cells is due to a wave of ganglion cell death. Neurosci Letters 36:255.

Hinds JW, Hinds PL (1978). Early development of amacrine cells in the mouse retina: An electron microscopic, serial section analysis. J Comp Neurol 179:277.

Hinds JW, Hinds PL (1983). Development of retinal amacrine cells in the mouse embryo: evidence for two modes of formation. J Comp Neurol 213:1.

Hume DA, Perry VH, Gordon S (1983). Immunohistochemical localization of a macrophage specific antigens in developing mouse retina. Phagocytosis of dying neurons and differentiation of microglial cells to form a regular array in the plexiform layers. J Cell Biol 97:253.

Hume DA, Robinson AP, MacPherson GG, Gordon S (1983). The mononuclear phagocyte system of the mouse defined by immunohistochemical localization of antigen F4/80. The relationship between Langerhans cells, reticular cells and dendritic cells in lymphoid and hematopoietic cells. J Exp Med (in press).

Innocenti GM (1981). Growth and reshaping of axons in the establishment of visual callosal connections. Science 212:824.

Jeffery G (1982). The Development and Plasticity of Primary Visual Pathways. D Phil thesis Oxford.

Jeffery G, Perry VH (1982). Evidence for ganglion cell death during development of the ipsilateral retinal projection in the rat. Dev Brain Res 2:176.

Kolb H, Nelson R, Mariani A (1981). Amacrine cells, bipolar cells and ganglion cells of the cat retina: a Golgi study. Vision Res 21:1081.

Kuffler S (1953). Discharge patterns and functional organization of mammalian retina. J Neurophysiol 16:37.

Lam K, Sefton AJ, Bennett MR (1982). Loss of axons from the optic nerve of the rat during early postnatal development. Dev Brain Res 3:487.

Land PW, Lund RD (1979). Development of the rat's uncrossed retinotectal pathway and its relation to plasticity studies. Science 205:698.

Lennie P (1980). Parallel visual pathways: a review. Vision Res 20:561.

Linden R, Perry VH (1982). Ganglion cell death within the developing retina: A regulatory role for retinal dendrites? Neuroscience 7:2813.

Lund RD, Cunningham TJ, Lund JS (1973). Modified optic projections after unilateral eye removal in young rats. Brain Behav & Evol 8:51.

Lund RD, Land PW, Boles J (1980). Normal and abnormal uncrossed retinotectal pathways in rats: an HRP study in adults. J Comp Neurol 189:711.

Negishi K, Eranishi T, Kato S (1982). New dopaminergic and indoleamine-containing cells in the growth zone of goldfish retinas after neurotoxic destruction. Science 216:747.

Olney JW (1968). An electron microscopic study of synapse, formation, receptor outer segment development and other aspects of developing mouse retina. Invest Ophthal 7: 250.

Oppenheim RW (1981). Neuronal cell death and some related regressive phenomena during neurogenesis: A selective historical review and progress report. In Cowan WM (ed): "Studies in Developmental Neurobiology: Essays in Honor of Victor Hamburger", Oxford University Press, p 74.

Peichl L, Wässle H (1981). Morphological identification of on- and off-centre brisk transient (Y) cells in the cat retina. Proc R Soc Lond B 212:139.

Perry VH (1979). The ganglion cell layer of the retina in the rat: a Golgi study. Proc R Soc Lond B 204:363.

Perry VH (1981). Evidence for an amacrine cell system in the ganglion cell layer of the rat retina. Neuroscience 6:931.

Perry VH, Linden R (1982). Evidence for dendritic competition in the developing retina. Nature 297:683.

Perry VH, Henderson Z, Linden R (1983). Postnatal changes in retinal ganglion cell and optic axon populations in the pigmented rat. J Comp Neurol 219:356.

Potten CS, Allen TD (1974). The fine ultrastructural identification and organization of the epidermal proliferative unit. J Cell Sci 15:291.

Potts RA, Dreher B, Bennett MR (1982). The loss of ganglion cells in the developing retina of the rat. Dev Brain Res 3:481.

Rodieck RW (1979). Visual pathways. Ann Rev Neurosci 2: 193.

Scholes JG (1976). Neuronal connections and cellular arrangement in the fish retina. In Zettler F, Weller R (eds): "Neural Principles in Vision", Springer-Verlag, Berlin, pp 63.

Sengelaub DR, Finlay BL (1981). Early removal of one eye reduces normally occurring cell death in the remaining eye. Science 213:573.

Vaney DI, Peichl L, Boycott BB (1981). Matching populations of amacrine cells in the inner nuclear and ganglion cell layers of the rabbit retina. J Comp Neurol 199:373.

Wässle H, Reimann HJ (1978). The mosaic of nerve cells in the mammalian retina. Proc R Soc Lond B 200:441.

Wässle H, Peichl L, Boycott BB (1981). Morphology and topography of on- and off-alpha cells in the cat retina. Proc Roy Soc B 212:157.

Wässle H, Boycott BB, Illing RB (1981). Morphology and mosaic of on- and off-beta cells in cat retina and some functional considerations. Proc R Soc Lond B 212:177.

Development of Visual Pathways in Mammals, pages 75–88

DEVELOPMENT OF THE AREA CENTRALIS AND VISUAL STREAK IN MAMMALS

S A Dunlop and L D Beazley

Neurobiology Laboratory
Department of Psychology
The University of Western Australia
Nedlands, Western Australia 6009

Cells in the ganglion layer are organized as an area centralis and visual streak in many adult vertebrates such as frog, possum, quokka and cat (Coleman *et al.* 1983; Freeman, Tancred 1978; Beazley, Dunlop 1982, 1983; Stone 1965, 1978; Hughes 1975). In 1981 we reported the first wholemount study of the developmental sequence giving rise to the adult cell distribution using a frog *Heleioporus eyrei* (Dunlop, Beazley 1981). Here we review our recent studies on the formation of this specialization in three mammals. We chose two marsupials, a small ground dwelling wallaby, *Setonix brachyurus* (quokka) and the western grey kangaroo, *Macropus fuliginosus fuliginosus*, as well as placental domestic cat.

In these species we shall describe first the adult topography and then representative stages towards its evolution. When considering the distribution of the total cell population, similar changes were found in all three species but these took place prenatally in cat and post-natally in the marsupials. In prenatal cat, Stone *et al.* (1982) recently described the formation of the area centralis from an even distribution of neuronal precursors identified by depth of staining. We chose to analyse the distribution of the total cell population since although in adults ganglion cells could readily be distinguished morphologically from other cell types such as displaced amacrines and glia, we felt this classification was not possible in immature retinae where all cells appeared somewhat similar. To over-come this, we used the neuronal marker horseradish peroxidase (HRP) not only to distinguish ganglion cells in young quokka joeys but also to confirm our previous identification of cells in mature retinae (Beazley, Dunlop 1983). The immaturity of joeys at birth compared to placental mammals suggested indeed that a marsupial was particularly amenable for an HRP study. We chose quokka in preference to kangaroo since it is a conveniently sized animal which breeds well in captivity. Also, due to delayed implantation of the blastocyst, this species gives birth 28 days after a joey is removed from the pouch, allowing accurate staging of pouch young using external morphological characteristics (Shield, Woolley 1968).

RESULTS

Adults

Cell morphology. In the adults of all three species several cell types were apparent in the ganglion layer (Figs 1, 2, 3). Retinal ganglion cells were distinguishable by their usually larger size, clumped deeply stained Nissl substance and a paler nucleus sometimes with a dark staining nucleolus. Retrograde transport of HRP in quokka confirmed our identification of ganglion cells in this species. Ganglion cells comprised 55%, 50% and 20% of the total cell population respectively, for quokka, kangaroo and cat. Many of the remaining cells, which tended to be smaller, smoother in outline and with a more darkly staining nucleus, contained faint strands of basophilic material in their sparse cytoplasm. These cells were considered likely to be neurons such as displaced amacrines (Hughes, Wieniawa-Narkiewicz 1980; Perry 1981). They were frequently distinguishable and far in excess of a presumed glial population which were round and even more densely stained. Throughout these studies our total cell counts comprised ganglion cells, other neuronal cell types and glia. Where present, the vascular network was readily distinguishable and excluded from our analysis.

Cell distribution and number. In the adults of each species there was a temporally situated area centralis surrounded by a visual streak which was particularly pronounced in kangaroo. Peak densities in the area centralis were 60, 80 and 100 cells/(100 μm)^2 and fell to below 10, 5 and 10 cells/(100 μm)^2 in the periphery of quokka, kangaroo and cat respectively. Our HRP study in quokka indicated that ganglion cells were organized as an area centralis and visual streak but the remaining cells were more evenly distributed. In quokka, kangaroo and cat total cell counts were approximately 350,000, 1.5 million and 900,000 respectively.

Developmental Stages

The distribution of the total cell population was initially uniform in all three species and gradually evolved to the adult patterns. From our more extended series we have here selected two stages for each species to illustrate this transition. These stages were pouch ages 50 and 80 days for quokka, 39 and 106 days for kangaroo; embryonic days E42 and E58 were chosen for cat.

Cell morphology. At the earlier of these stages for each species, when cell distribution was approximately uniform, cells were smaller than in the adult with a gradation of staining (Figs 1, 2, 3). The largest tended to be paler, often with a darkly stained nucleolus. In quokka and cat some cells contained Nissl substance, suggesting these retinae were at a relatively more mature stage than the 39 day kangaroo. In the quokka HRP study, 70% of the total cell population contained HRP granules. Some of these tended to be slightly larger and paler but could not have been recognized as ganglion cells except for their filling with HRP.

By the later of our chosen developmental stages, when topography was no longer even, cells in each species had enlarged and were more differentiated. In quokka, 60% of the total cell population contained HRP, these being cells with a strong Nissl reaction.

Cell distribution and number. At younger stages when retinae were smaller than in the adult, densities of the total cell population were high, the majority being in the ranges of 80-180, 200-300 and 150-200 cells/(100 μm)^2 for quokka, kangaroo and cat respectively. In none of the three species could a pattern be discerned in the distribution of the total cell population. The density of HRP filled cells in quokka varied between 50 and 150/(100 μm)^2 and their topography was approximately even although values dropped somewhat nasally. Total cell counts of 450,000, 1.6 million and 1.4 million in

a
b
a
b
c
ONH
c
> 30
> 30
130
100
ONH
80
60
d
ONH
d
60%
70%
80%
ONH
50%
50%
60%
50%
e
ONH
V
1mm
Quokka - 50 days pouch life, left eye
e
ONH
Quokka - 80 days pouch life, right eye
V
1mm

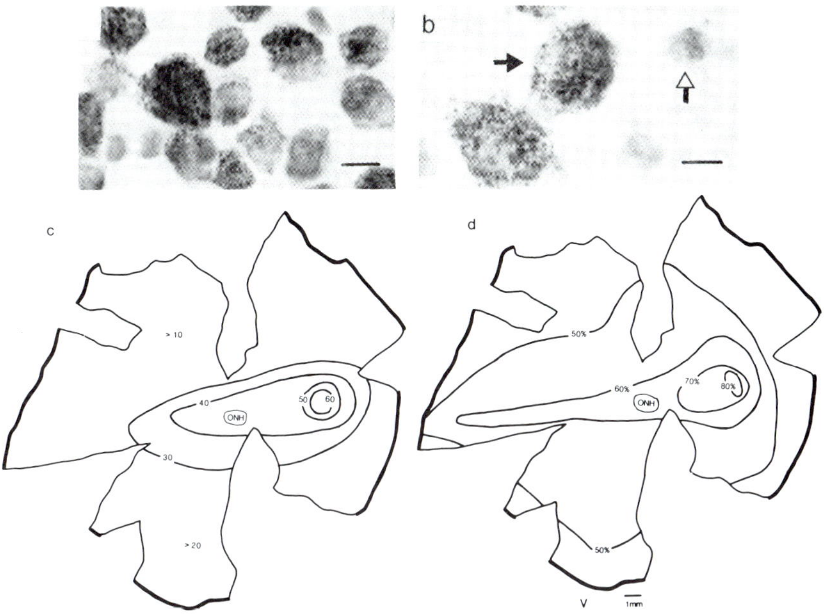

Fig. 1. Quokka - Histology (a & b): HRP-filled ganglion cells (solid arrows) and unfilled displaced amacrines (open arrows) are shown for mid-temporal (a) and mid-dorsal (b) retina at 50 days, and in the area centralis (a) and mid-dorsal (b) retina at 80 days and in the adult. Pycnotic cells (circled) containing HRP granules are shown at 50 and 80 days. Bars are 10 µm. - Cell distribution maps (c, d & e): total cell population (c), numbers referring to cells/(100 µm)^2, and % HRP-filled cells (d) are presented. % pycnotic cells for pouch young (e) are also shown, symbols being · 0.1-0.9%, · 1-3% and • 4-7% of the total cell population. V, ventral; ONH, optic nerve head. Joeys received bilateral pressure injections of HRP under ether anaesthesia to fill the optic tracts and visual centres. Adults were anaesthetised with saffan and halothane, multiple optic tract injections being given stereotaxically (Meyer, unpublished). After 1-2 days transport retinae oriented with a dorsal cut were briefly fixed in glutaraldehyde, processed with cobalt enhanced diaminobenzidine, wholemounted and stained with cresyl violet. Retinae were examined systematically at 1000x magnification sampling between 4-8% of retinal area. In each sample area ((100 µm)^2) total cell number, the number of HRP-filled cells and pycnotic cells were counted.

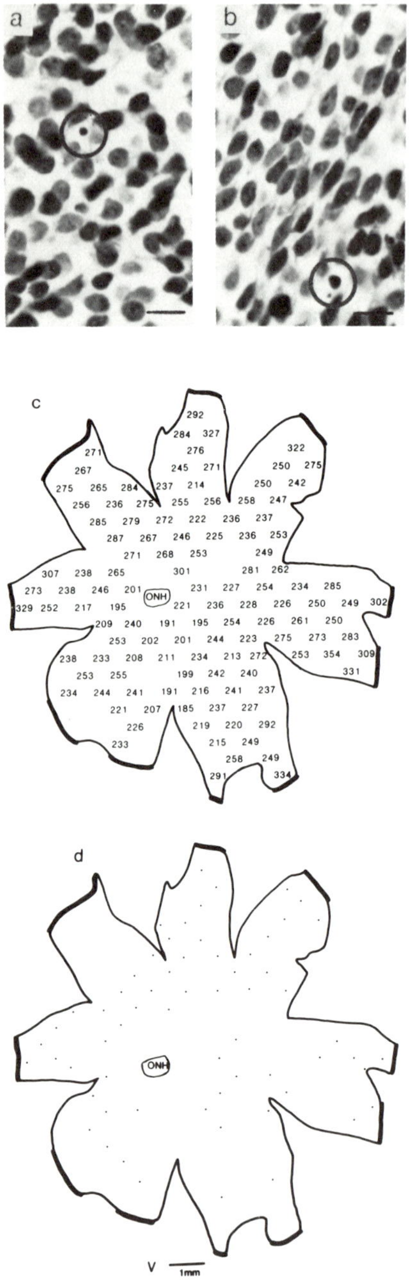

Kangaroo - 39 days pouch life, right eye

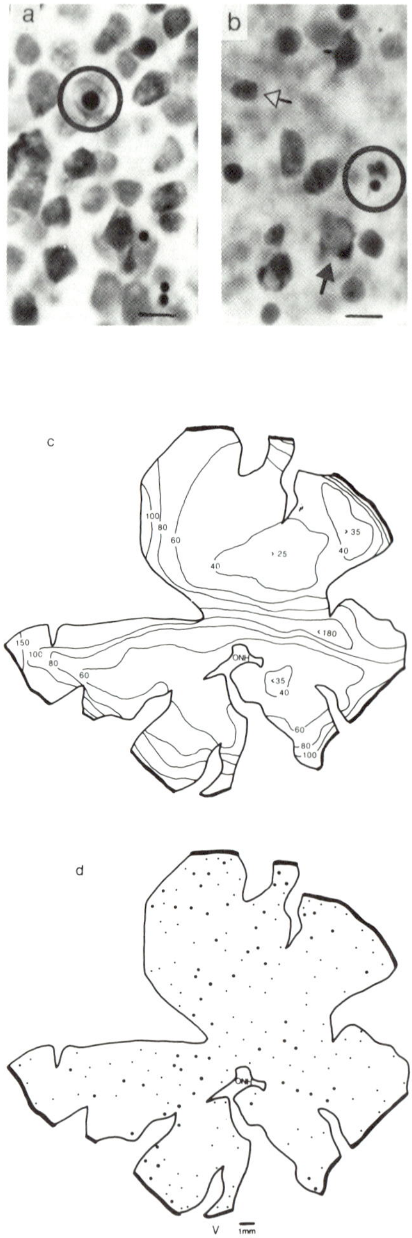

Kangaroo - 106 days pouch life, left eye

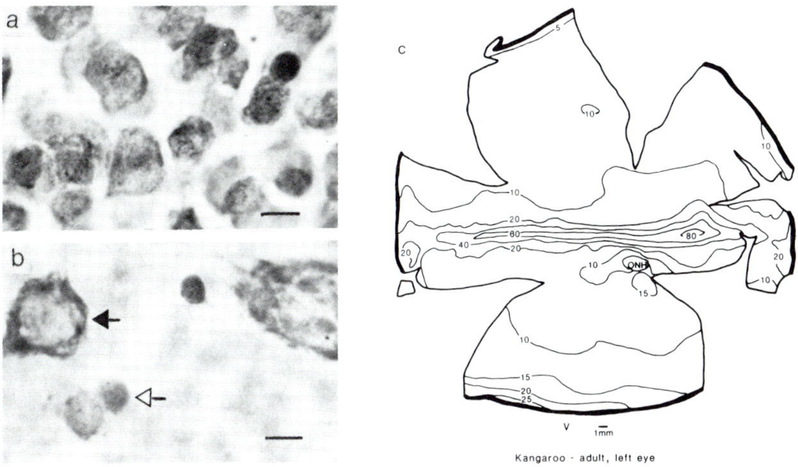

Fig. 2. Kangaroo - Histology (a & b): at 39 days cells appeared similar and at equivalent densities in mid-temporal (a) and mid-dorsal (b) retina. Cells in the area centralis (a) and mid-dorsal (b) retina are shown for 106 days and in the adult, ganglion cells (solid arrows) being distinguished from displaced amacrines (open arrows) particularly in the periphery (b). Pycnotic cells (circled) were seen in the pouch young, one being engulfed by a macrophage (106 days, b). Bars are 10 μm. - Cell distribution maps (c & d): total cell population (c), numbers referring to cells/ $(100\ \mu m)^2$, and % pycnotic cells for pouch young (d) are shown, symbols being · 0.1-0.9%, · 1-3%, • 4-7% of the total cell population. V, ventral; ONH, optic nerve head. Animals were obtained as part of licenced culls. All eyes were oriented with a dorsal cut, adult tissue being removed within 2 minutes of death and fixed in 10% buffered formalin. Joeys were anaesthetised with saffan and the eyes fixed in 5% buffered formalin for a maximum of 4 h. Cresyl-stained wholemounts were examined as described for Fig. 1.

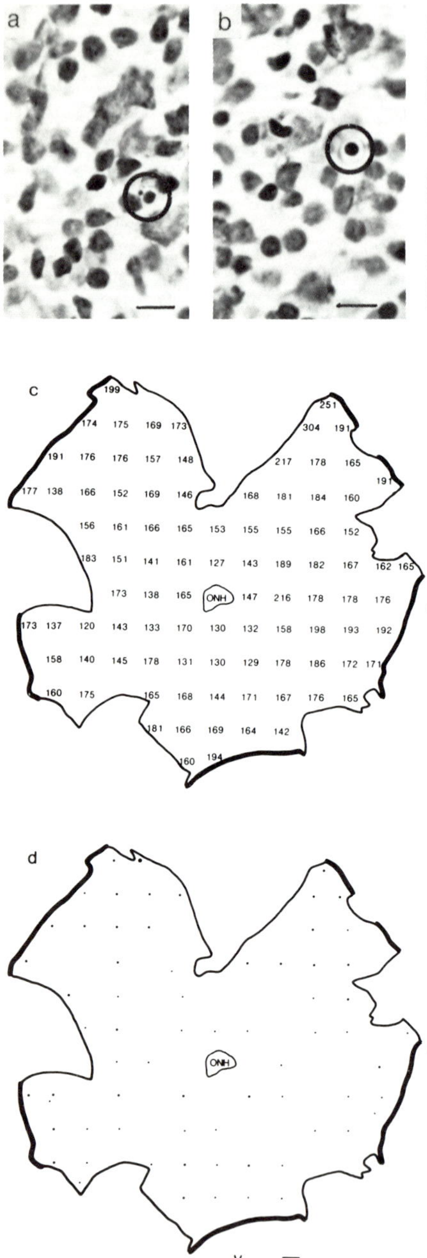

Cat -E42, left eye

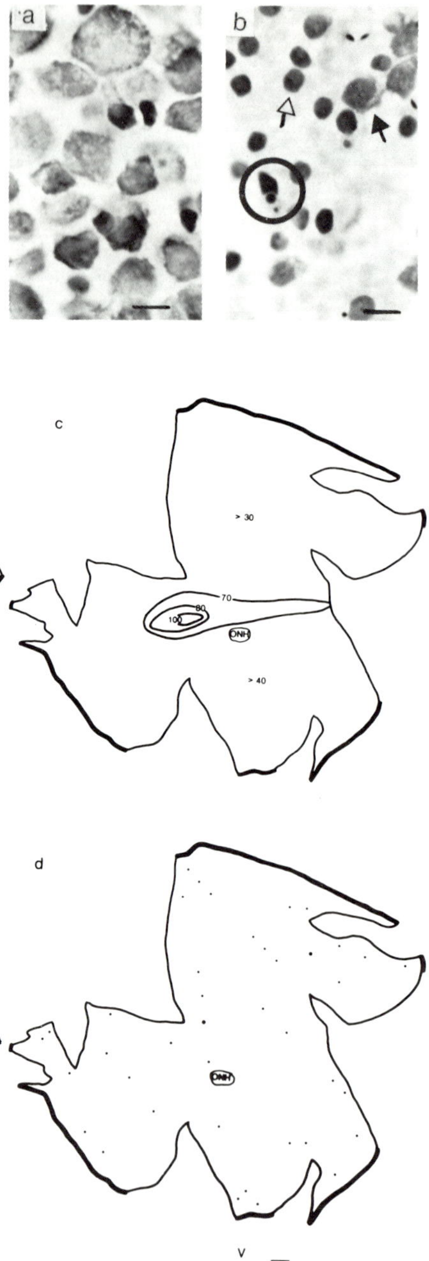

Cat - E58, right eye

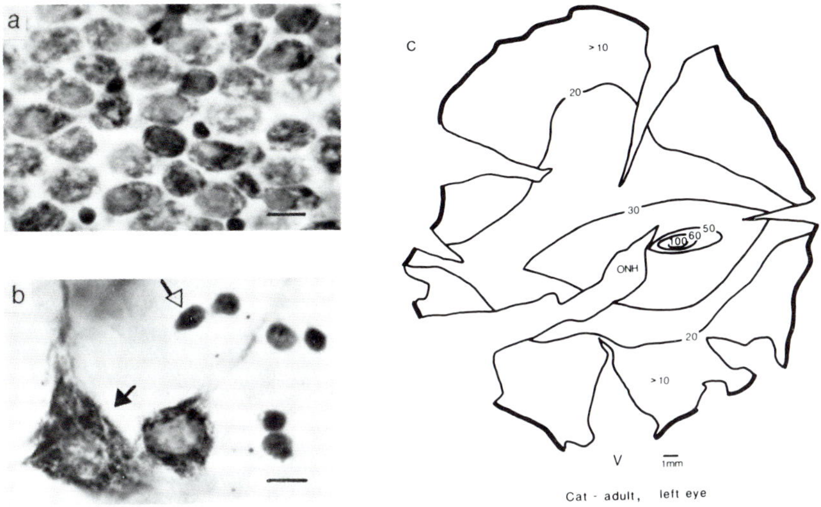

Fig. 3. Cat - Histology (a & b): at E42 cells appeared somewhat similar, although Nissl was observed in a proportion, and at equivalent densities in mid-temporal (a) and mid-dorsal (b) retina. Cells in the area centralis (a) and mid-dorsal (b) retina are shown for E58 and the adult, ganglion cells (solid arrows) being distinguished from displaced amacrines (open arrows) particularly in the periphery (b). Pycnotic cells (circled) were seen in the foetal stages, one being engulfed by a macrophage (E58, b). Bars are 10 µm. - Cell distribution maps (c & d): total cell population (c), numbers referring to cells/(100 µm)^2, and % pycnotic cells for pouch young (d), symbols being 0.1-0.9%, · 1-3%, of the total cell population. V, ventral; ONH, optic nerve head. Staged foetuses were obtained after timed matings by caesarian section under saffan anaesthesia. Foetuses were further anaesthetised with saffan. Eyes were oriented, and fixed as for Fig. 2. Cresyl-stained wholemounts were described as in Fig. 1.

quokka, kangaroo and cat respectively were slightly in excess of adult values, the greatest difference (36%) being found in cat.

By 80 days in quokka, 106 days in kangaroo and E58 for cat, retinae had enlarged and the distribution of the total cell population was essentially adult-like with an area centralis and visual streak. However, densities tended to be elevated particularly in the periphery. In the marsupials peak densities of 130 and 180 cells/(100 μm)^2 respectively were higher than in adults although in cat the value had already stabilized to the adult level of 100 cells/(100 μm)^2. The distribution of HRP filled cells in quokka mirrored that of the total cell population, peak density in the area centralis being 50/(100 μm)^2. There had been a reduction in the total cell number to approximately 400,000, 1.5 million and 1 million in quokka, kangaroo and cat respectively.

Cell death. Unlike adults, the developmental stages in the three species had cells in the ganglion layer with a pycnotic appearance (Sengelaub, Finlay 1982). These cells possessed very darkly stained and condensed nuclei which had in some instances fragmented; frequently a clear halo of cytoplasm was seen. Macrophages appeared to engulf a proportion of pycnotic cells. Such cells were seen across the entire retina with no particular pattern in their distribution. In the quokka joey retinae, HRP granules were seen in approximately half of all pycnotic cells and in some macrophages which appeared to be in the process of ingesting a pycnotic cell. At 50, 39 days and E42, 2.5, 0.4 and 0.7% of the total cell population were pycnotic in quokka, kangaroo and cat respectively. The proportion had fallen to 1.1 and 0.4% in the 80 day quokka and E58 cat but was 2% in the kangaroo at 106 days.

CONCLUSIONS

Summarizing our investigations, first we have demonstrated that in all three species an even cell distribution of the total cell population changed to an area centralis and visual streak. The HRP study in quokka furthermore showed that initially ganglion cells were also evenly distributed. Second, the HRP experiment showed that the proportion of filled cells diminished during development of the ganglion layer. At an even younger stage than reported here,

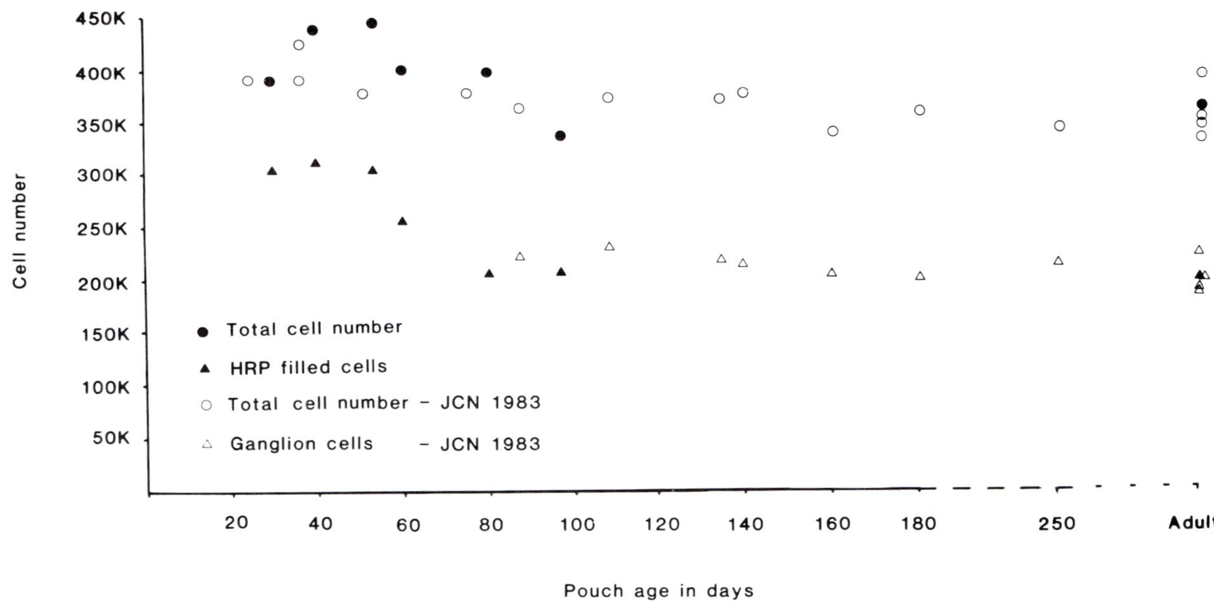

Fig. 4. Quokka - graph of total cell number (closed circles) and number of HRP-filled cells (closed triangles) against age from the HRP experiment (Dunlop, Beazley, in preparation). Total cell number (open circles) and number of ganglion cells (open triangles) from Beazley and Dunlop (1983) are included. Numbers were estimated by proportionality from wholemounts.

filled cells represented 80% of the total cell population (Dunlop, Beazley in preparation), the adult value being 55%. Third, since the slight decrease observed in the total cell population and the number of HRP-filled cells in quokka was comparable, we suggest that the loss occurred within the ganglion cell population. This supposition is supported by the observation that many pycnotic cells contained HRP granules and therefore presumably had sent an axon into the brain.

The total and ganglion cell numbers we have obtained from the HRP study in quokka as well as previously (Beazley, Dunlop 1983) are summarized in Fig. 4. Between 50 and 97 days there is a loss of approximately 100,000 cells. Assuming this is linear, about 100 cells would have to be removed every hour. Using Hughes' (1961) estimate of 3 h for clearance of pycnotic cells, which may be an over-estimate (Perry, personal communication), one would expect to see 300 dying cells at any one time. However, on average we observed approximately 8,000 dying cells suggesting that cell death is by far in excess of that required to reduce cell number. We therefore hypothesize that there is a considerable turnover of cells in the ganglion layer, re-plenishment presumably being from the other nuclear layers. Autoradiographic experiments are now in progress to inves-tigate this. However, whether cell death has a role in changing topography is at present uncertain; it may well relate to other retinal or central events.

We have reached the stage at which changing cell distributions have been described both for the total cell population and cells which fill with HRP. However, the contribution of events such as the generation, death or migration of cells towards the formation of an area centralis and visual streak are only now beginning to be addressed.

ACKNOWLEDGMENTS

LDB is a Senior Research Fellow, National Health and Medical Research Council (Australia). This research was supported by NH&MRC (Australia) grants 77/2087 and 82/0180, and the Muscular Dystrophy Association of Western Australia. We are extremely grateful to R McNeice and D Cuttler for establishing and maintaining our quokka colony in the

Department of Zoology and to Dr Zwicky and Professor Bradshaw for their cooperation and advice. We would like to thank R Prince, D Kitchener and M Box, R McNeice and D Algar for their cooperation in obtaining the kangaroo material and W Longley for collaborating with the analysis. We are indebted to K Duffall for providing the cats and D Nicholls for performing the caesarian sections. We would also like to thank P Meyer and J Darby for technical assistance and H Jurkiewicz for photography. We thank C Pfaff for typing the manuscript and finally would like to convey gratitude to our husbands for their unending patience and support.

REFERENCES

Beazley LD, Dunlop SA (1982). Formation of an area centralis and visual streak in the ganglion layer of the marsupial *Setonix brachyurus*. Proc Phys Soc 326:45.

Beazley LD, Dunlop SA (1983). The evolution of an area centralis and visual streak in the marsupial *Setonix brachyurus*. J Comp Neurol 216:221.

Coleman L-A, Dunlop SA, Beazley LD (1983). Patterns of cell generation in the retina of the frog *Limnodynastes dorsalis*. Neurosci Lett Suppl 11:S36.

Dunlop SA, Beazley LD (1981). Changing retinal ganglion cell distribution in the frog *Heleioporus eyrei*. J Comp Neurol 202:221.

Freeman B, Tancred E (1978). The number and distribution of ganglion cells in the retina of the bush-tailed possum. J Comp Neurol 177:557.

Hughes A (1961). Cell degeneration in the larval central horn of *Xenopus laevis*. J Embryol exp Morphol 9:269.

Hughes A (1975). A quantitative analysis of cat retinal ganglion cell topography. J Comp Neurol 163:107.

Hughes A, Wieniawa-Narkiewicz E (1980). A newly identified population of presumptive microneurons in the cat retinal ganglion cell layer. Nature 284:468.

Perry VH (1981). Evidence for an amacrine cell system in the ganglion layer of the rat retina. Neurosci 6:931.

Sengelaub DR, Finlay BL (1982). Cell death in the mammalian visual system during normal development. I. Retinal ganglion cells. J Comp Neurol 204:311.

Shield RL, Woolley P (1968). Age estimation by measurement of pouch young in the quokka (*Setonix brachyurus*). Aust J Zool 9:14.

Stone J (1965). A quantitative analysis of the distribution of ganglion cells in the cat's retina. J Comp Neurol 124: 337.

Stone J (1978). The number and distribution of ganglion cells in the cat's retina. J Comp Neurol 180:753.

Stone J, Rapaport DH, Williams RW, Chalupa L (1982). Uniformity of cell distribution in the ganglion layer of prenatal cat retina: Implications for retinal development. Dev Brain Res 2:231.

Development of Visual Pathways in Mammals, pages 89–102

PRENATAL DEVELOPMENT AND REORGANIZATION IN THE VISUAL SYSTEM OF THE CAT

L. M. Chalupa and R. W. Williams

Department of Psychology,
University of California,
Davis, California 95616, U.S.A.

Four years ago we initiated a research program which sought to answer two questions: (1) what are the events in fetal life which give rise to mature patterns of connection in the mammalian visual system? and (2) to what degree is the maturation of the visual system perturbed by interrupting binocular interactions at an early stage of development?

Before summarizing the progress we have made in answering these questions, it is worth considering why we chose to study cats. First, since the introduction of the monocular deprivation paradigm by Wiesel and Hubel (1963), much has been learned about normal and abnormal postnatal development of vision in cats, and this information provides a foundation for investigating the early development of this species' visual system. Second, the cat is multiparous, and it is therefore possible to acquire a sufficient number of fetuses of known gestational age. Third, the 63-65 day gestation of the cat is long enough to allow adequate temporal resolution of various events that occur during prenatal development. Fourth, the size of the fetal cat's brain is comparatively large, and thus accessible to various types of manipulations.

We began our studies by tracing the development of retinal projections to the main visual centers, using the anterograde transport of horseradish peroxidase and tritiated leucine. One of the tracers was injected into the right eye, and the other tracer into the left eye, of fetuses of known gestational age. Pregnancy was timed by exposing an estrous female to a potent male for 24 hours. Several matings were observed during this period, and the end of the 24 hour exposure was considered to mark the first day of gestation (embryonic day 1, or E-1).

Our earliest successful injections were made on E-38 (Fig. 1). At this age virtually all ganglion cells have been generated (Walsh et al. 1983), ganglion cell growth cones have extended through the optic nerve (Williams et al. 1983a), and at least some of the axons have reached as far as the posterior margin of the superior colliculus (Williams, Chalupa 1982a).

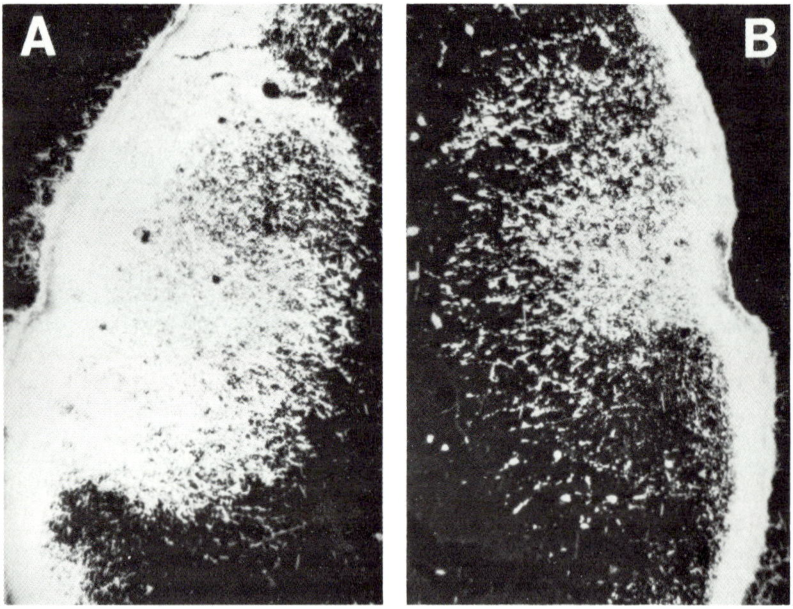

Fig. 1. Retinogeniculate projections in an E-38 cat fetus demonstrated by the anterograde transport of HRP. **A.** The contralateral projection; **B.** the ipsilateral projection. Darkfield illumination; X 100

As shown in figure 1, as early as E-38 there is substantial overlap of projections from right and left eyes. The contralateral fiber influx, however, is much heavier (Fig. 1A) than the ipsilateral (1B). Furthermore, it appears that only a fraction of the retinal ingrowth has yet entered the terminal fields by this age.

Between E-38 and E-56 there is a gradual elaboration and restructuring of retinal projections. The density of terminal label becomes much greater, and by E-46 virtually the entire lateral geniculate nucleus (Fig. 2), pretectum (Williams, Chalupa 1983a), and superior colliculus (Williams, Chalupa 1982a) receive heavy projections from both eyes. Overlap of projections from right and left eyes is 100% in the geniculate nucleus (Fig. 2). Virtually complete overlap has also been described in the lateral geniculate of another carnivore, the ferret (Linden et al. 1981), and in the monkey (Rakic 1977). (Our findings in the cat differ, however, from those of Shatz (1983), who reported that at all fetal ages, including E-46, most of the cat's geniculate nucleus is innervated only by fibers from the contralateral eye.) It is crucial to note that at E-46 the pattern of labeling is not uniform. In the geniculate nucleus the densities of the crossed and uncrossed retinal projections appear to vary inversely. This pattern foreshadows the development of discrete layers which contain terminals of either the right or the left eye. As shown in figure 3, by E-56 segregation is essentially complete.

There are significant differences in the gestational age at which segregation is achieved in different nuclei. Segregation starts and ends more than a week earlier in the geniculate nucleus than in either the superior colliculus or the pretectum (see Williams and Chalupa, 1983a). Furthermore, the separation of retinal fibers takes place later in the C laminae than in the A laminae of the geniculate, as was originally suggested by Shatz (1983). Since only the A laminae receive a dominant input from the beta class of ganglion cells, these differences may be related

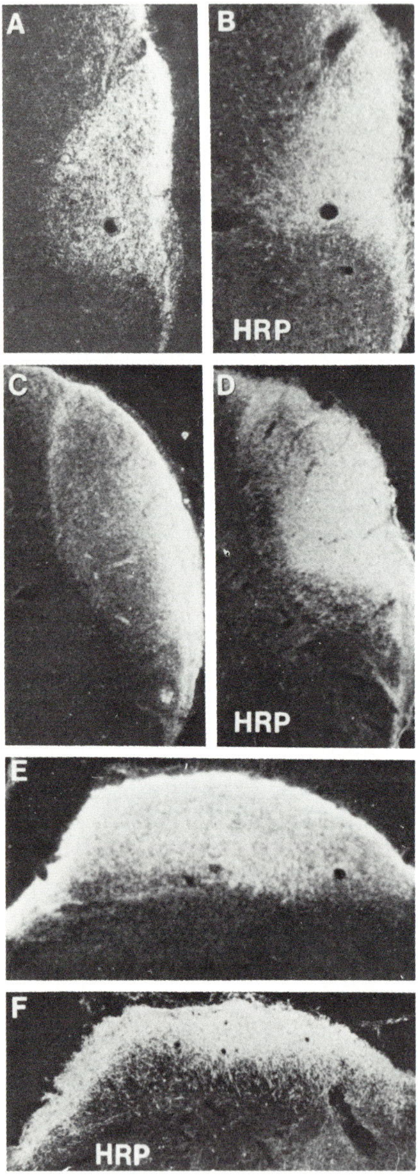

Fig. 2. Overlap of projections from right and left eyes to the lateral geniculate nucleus at E-46. The contralateral eye received an injection of ^{3}H-leucine; the ipsilateral eye received an injection of HRP. Therefore, panels **A**, **C**, and **E** show the crossed input and the other 3 (**B**, **D**, and **F**) show the uncrossed input. Pairs of adjacent sections are shown, under darkfield at 3 representative rostro-caudal levels: **A** and **B** are at the caudal pole; **C** and **D** are central; **E** and **F** are rostral. Shrinkage during processing was greater in HRP sections than of autoradiographic sections. Therefore, the magnification of each pair of micrographs was adjusted so as to obtain the most accurate match between tissue landmarks. Although overlap is virtually 100%, note that the density of labeling is not uniform. Segregation is incipient at this age. **A B**, **E** and **F** at X 40-60; **C** and **D** at X 16-20.

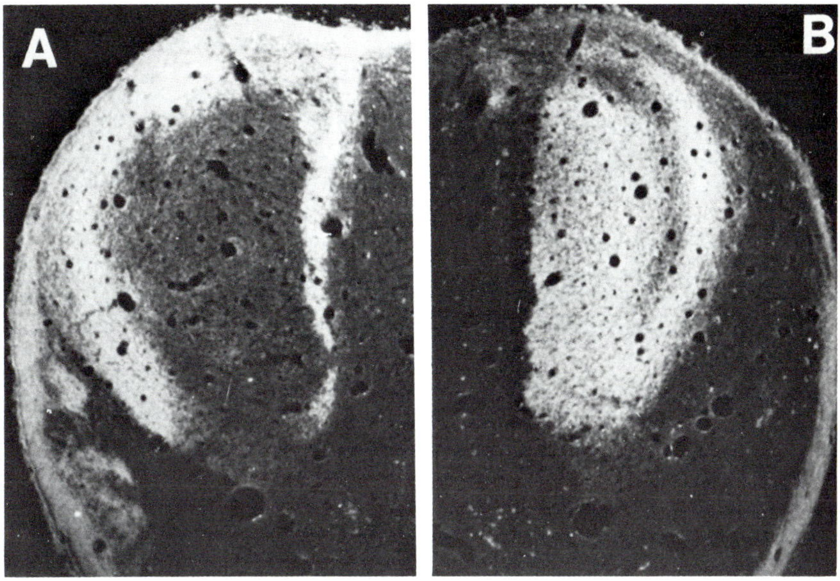

Fig. 3. Segregation of the crossed and uncrossed retinogeniculate influx at E-56. Darkfield micrographs of peroxidase reaction product. **A.** Crossed projection; **B.** Uncrossed projection.

to the differential maturation of the major ganglion cell types. Medium-sized retinal ganglion cells--presumably the beta class, which project heavily to the dorsal layers of the adult geniculate--are mostly generated before the small cells of the gamma class, which project to the superior colliculus, pretectum and C layers of the geniculate (Polley et al. 1981; Kliot, Shatz 1982; Walsh et al. 1983).

During gestation there is a marked overproduction of ganglion cell axons (Ng, Stone 1982; Williams, 1983; Williams et al., 1983 a,b; R.W. Williams, M.J. Bastiani, and L.M. Chalupa, in progress). Using a quantitative electron microscopic method, we have found that the first 100 ganglion cell axons venture into the optic stalk on E-19. Four days later, about 1,000 fibers are present. By E-28 the number of axons has increased to approximately 40,000. At E-28 and E-33 the counts include a substantial number of growth cones (Fig. 4). For instance, at E-33, 3% (n = 8,000) of the neurites are growth cones. These are characterized by extensive veil-like processes, some of which extend radially 4 microns from the core (Fig. 4). Growth cones tend to grow around the perimeter of the nerve, in apposition to glial processes. However, some extending axons are also found in the central region, without apparent connection to glial precursors.

The peak axon count of 660,000 is attained at E-39. At this age only a few hundred growth cones were noted. A population of more than 500,000 is maintained until E-44. Between E-44 and E-48 there is a precipitous loss of fibers. By E-48 the axon complement has eroded to 330,000. Thereafter the loss of axons is more gradual; the adult population of about 160,000 is not reached until at least several weeks after birth.

The excess axons produced early in development could potentially contribute to the wide distribution of retinal projections observed during fetal life. However, the initial massive loss of axons, which begins even before E-44, precedes by several days the onset of segregation in the geniculate nucleus. Moreover, fiber loss is still occurring during the first postnatal month, long after the segregation of all retinal efferents is complete. Thus, while fiber loss may underlie the segregation of early retinal projections, we suspect that there are other features of the developing visual system which are shaped by this phenomenon.

Early Termination of Binocular Competition

Our second objective was to determine the contributions made by prenatal binocular competition to the development of the cat's visual pathways. For this purpose we removed one eye from fetuses at known gestational ages (between E-

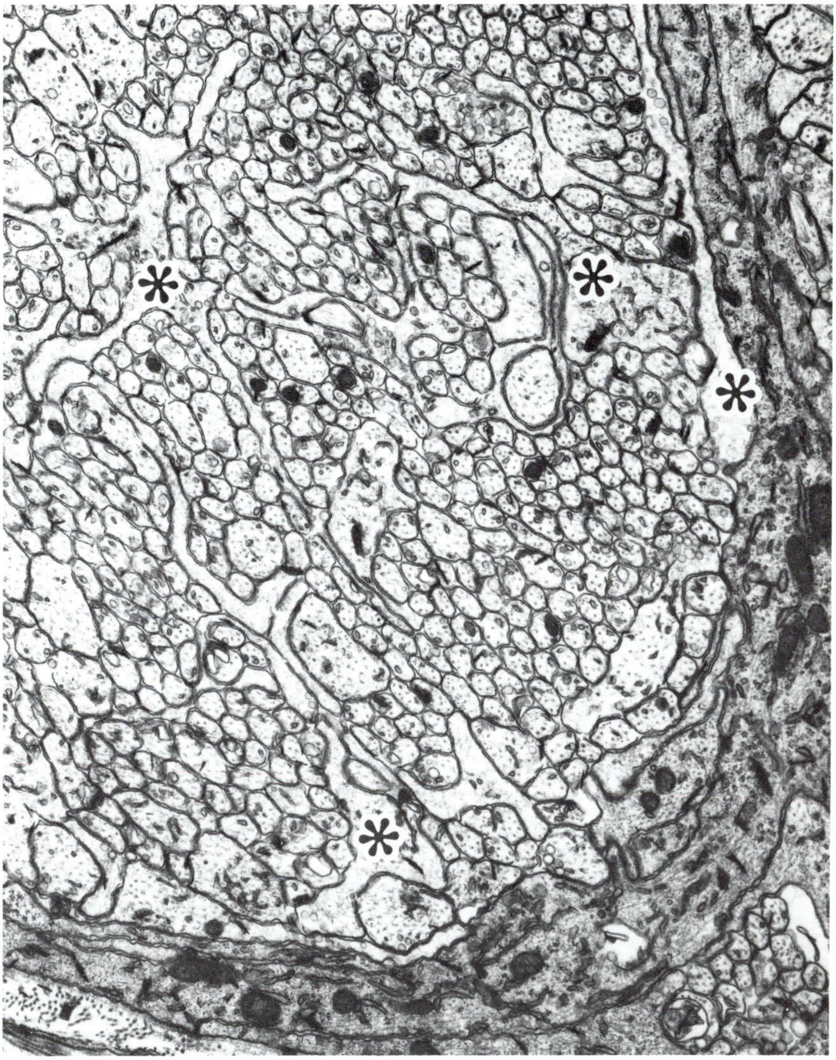

Fig. 4. Growth cones and axons in a peripheral fasciculus of the optic nerve at E-33. To the right, dark astroblastic processes separate the fascicles and form a limiting membrane around the nerve, the edge of which crosses the lower left corner. Some of the very large growth cones are marked with asterisks. X 16,000

40 and E-56), and when these animals reached maturity we studied the organization of their visual systems.

In these enucleated animals, the remaining retina innervates the entire ipsilateral and contralateral geniculate nuclei (Fig. 5) (Williams, Chalupa 1982b & 1983b; Williams 1983). However, as shown in figure 5B, the geniculate ipsilateral to the remaining eye is substantially smaller than that on the contralateral side. The morphology of the lateral geniculate nuclei of these one-eyed cats is characterized by two laminae: a dorsal magnocellular layer and a ventral layer. A similar result has been described by Rakic (1981) in prenatally enucleated monkeys. Presumably, the magnocellular layer corresponds to what would normally have been layers A and A1, whereas the ventral layer corresponds to what would have been subdivided into layers C, C1, C2, and C3. Thus, the 6 laminae of the normal cat's geniculate are supplanted by two composite layers.

We have also shown that there are significantly more optic nerve fibers in prenatally enucleated cats than is normal (Williams et al. 1983b). Furthermore, the number of fibers within the optic nerve of these animals matches the number of ganglion cells in the remaining retina (Henderson et al. 1983; Chalupa et al., in review). The results of these experiments, summarized in table 1, show that the prenatally enucleated cats have about 20% or 30,000 more ganglion cells than normal animals. Obviously this excess could contribute to the widespread retinal projections found in one-eyed cats.

An unexpected result of this study was that the number of ganglion cells saved does not depend critically upon the gestational age at which enucleations are performed. Removal of an eye at E-42, when there are more than 500,000 fibers in the optic nerve, is no more effective in saving ganglion cells from imminent death than is enucleation at E-51, when there are 300,000 axons.

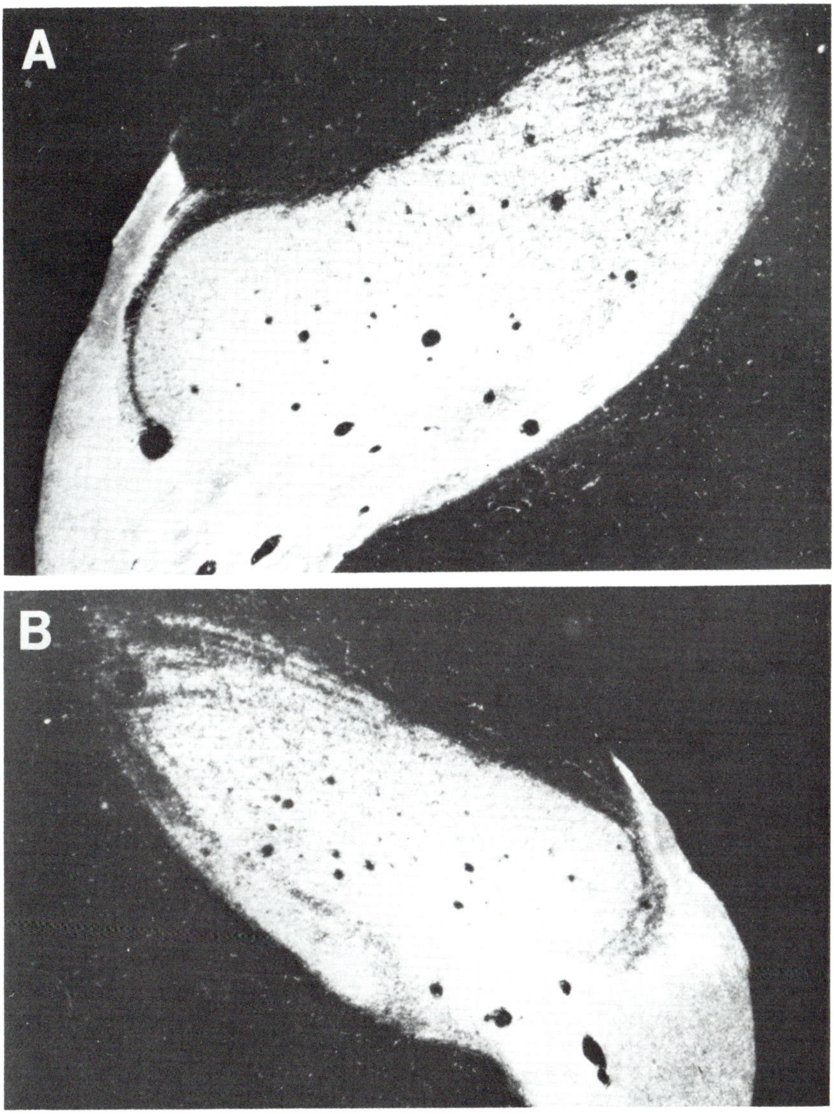

Fig. 5. Retinogeniculate projections in an adult cat from which one eye was removed on E-49. The crossed projection is shown in **A**; the uncrossed in **B**. All parts of both nuclei are labeled with the peroxidase chromogen, however, the ipilateral nucleus and the ipsilateral retinal projection are considerablly smaller.

TABLE 1

Effects of Prenatal Unilateral Enucleation upon Ganglion Cell and Optic Axon Number

Animal*	Ganglion Cell Number	Axon Number
Control 1	151,000 (R)**	---
	152,000 (L)	---
Control 2	---	158,000
Control 3	---	159,000
Control 4	---	164,000
E-42 Enucleate	180,000	178,000
E-51 Enucleate	182,000	190,000
E-45 Enucleate	---	198,000
E-46 Enucleate	---	200,000

*All animals were mature
**(R) right retina; (L) left retina

Early eye removal has been shown to maintain widespread retinal projections in a number of mammalian species (eg., Chow et al. 1973; Sanderson 1978; Land, Lund 1979; Frost, Schneider 1979; Rakic 1981); however, relatively little is known about the functional organization of the altered retinal connections (but see Rhoades, Chalupa 1980). Accordingly, we sought to determine the functional properties of the retinal influx to the geniculate of prenatally enucleated cats. For this purpose, extracellular single cell recordings were made within the lateral geniculate, contralateral and ipsilateral to the remaining eye of adult animals from which an eye had been removed more than 2 weeks before birth. The results are clear-cut: all regions of the geniculate nuclei of these cats are functionally innervated by the retinal input, and the topographic organization is similar to that of the normal cat (Williams, Chalupa 1982b & 1983b; Williams 1983). Furthermore, injections of peroxidase into the medial region of the geniculate, which contains the area centralis representation, revealed a normal decussation pattern in the remaining retina (Williams, Chalupa 1983b). This result rules out the possibility that an aberrant retinal input from the inappropriate hemiretina was either functionally suppressed or ineffective in driving thalamic units. Parenthetically, it should be noted, that recently we have found a clear decussation pattern, following unilateral injection of HRP, as early as E-44 (Lia et al., 1983).

Even though the functional organization of the lateral geniculate nucleus appeared normal in the prenatally enucleated cats, it seemed important to also examine the properties of the visual cortex in these animals. Studies of the visual cortex were carried out in collaboration with Drs. Brenda Shook and Lamberto Maffei (B.L. Shook, L. Maffei, and L.M. Chalupa, 1983). Small iontophoretic deposits of horseradish peroxidase conjugated to wheat germ agglutinin were made into the A lamina of normal cats and into the most dorsal portion of the geniculate nuclei of enucleated animals. As expected, discrete patches of label were found in layer IV of normal cats, whereas continuous label was found in the enucleated cats. Thus, in agreement with the previous work of Rakic (1981), ocular dominance columns fail to develop following early removal of one eye.

In the same cats long, tangential microelectrode penetrations were made through area 17. Most of these penetrations extended 2.5-3.0 mm at an oblique angle down the medial bank of the marginal gyrus, and thus the electrode traversed a region which in the normal cat is occupied by a number of discrete ocular dominance columns. In the prenatally enucleated cats, all neurons could be activated by the remaining eye, and the activity did not show any signs of waxing and waning. In contrast, in animals enucleated as adults, regions of reduced visual activity were recorded when similar penetrations were made. It is particularly noteworthy that cortical neurons of the prenatally enucleated animals exhibited an orderly progression in orientation selectivity. In several penetrations, sequences of cells were encountered which had a 180° cycle of preferred orientation. This is characteristic of hypercolumns in the visual cortex of normal animals (Hubel, Wiesel 1962). Therefore, this finding indicates that orientation columns can develop independently of ocular dominance columns.

There was one significant difference in the visual receptive field properties between prenatally enucleated and normal cats. Within 5° of the area centralis representation--where most penetrations were confined--the dimensions of receptive fields were significantly smaller in prenatally enucleated animals than in normal cats. One possible explanation for this intriguing finding is that dendritic fields of ganglion cell may be smaller than normal. We know that the remaining retina of these animals has about 30,000 more ganglion cells than normal, yet the retinal area is not appreciably larger. One would anticipate that exacerbated dendro-dendritic competition among ganglion cells in the remaining retina (see Perry, Linden 1982) would result in a reduction in the size of dendritic fields. Such a reduction would permit accommodation of the excess ganglion cell complement without disrupting the retinal mosaic.

REFERENCES

Chalupa LM, Williams RW, Henderson Z. Binocular competition in the fetal cat regulates the size of the ganglion cell population. In review.

Chow KL, Mathers LH, Spear PD (1973) Spreading of uncrossed retinal projection in superior colliculus of neonatally enucleated rabbits. J comp Neurol 151:307.

Frost DO, Schneider GE (1979) Plasticity of retinofugal projections after partial lesions of the retina in newborn Syrian hamsters. J comp Neurol 185:517.

Henderson Z, Williams RW, Chalupa LM (1983) Removal on an eye in the fetal cat increases the number of ganglion cells maintained in the remaining retina. Invest Ophthal Vis Sci 24:290.

Hubel DH, Wiesel TN (1962) Receptive field, binocular interaction, and functional architecture in the cat's visual cortex. J Physiol (Lond) 160:106.

Kliot M, Shatz CJ (1982) Genesis of different retinal ganglion cell types in the cat. Soc Neurosci Abst 8:815.

Land PW, Lund RD (1979) Development of the rat's uncrossed retinotectal pathway and its relation to plasticity studies. Science 205:698.

Lia B, Williams RW, Chalupa LM (1983). Early development of retinal specialization: the distribution and decussation patterns of ganglion cells in the prenatal cat demonstrated by retrograde peroxidase labelling. Soc Neurosci Abstr 9:702.

Linden DC, Guillery RW, Cucchiaro J (1981) The dorsal lateral geniculate nucleus of the normal ferret and its postnatal development. J comp Neur 203:189.

Ng AYK, Stone J (1982) The optic nerve of the cat: Appearance and loss of axons during normal development. Develop Brain Res 5:263.

Perry VH, Linden R (1982) Evidence for dendritic competition in the developing retina. Nature 297:683.

Polley EH, Walsh C, Hickey TL (1981) Neurogenesis in cat retina: A study using ^{3}H-thymidine autoradiography. Soc Neurosci Abstr 7:672.

Rakic P (1977) Prenatal development of the visual system in rhesus monkey. Phil Trans R Soc Lond B 278:245.

Rakic P (1981) Development of visual centers in the primate brain depends on binocular competition before birth. Science 214:928.

Rhoades RW, Chalupa LM (1980) Effects of neonatal enucleation on receptive-field properties of visual neurons in superior colliculus of the golden hamster. J Neurophysiol 43:595.

Sanderson KJ, Pearson LJ, Dixon PG (1978) Altered retinal projections in brushtailed possum, *Trichosurus vulpecula*, following removal of one eye. J comp Neurol 143:101.

Shatz CJ (1983) The prenatal development of the cat's retinogeniculate pathway. J Neurosci 3:482.

Shook BL, Maffei L, Chalupa LM (1983) Functional organization of the cat's visual cortex after prenatal unilateral enucleation. Soc Neurosci Abst 9: in press.

Walsh C, Polley EH, Hickey TL, Guillery RW (1983) Generation of cat retinal ganglion cells in relation to central pathways. Nature 302:611.

Wiesel TN, Hubel DH (1963) Single-cell responses in striate cortex of kittens deprived of vision in one eye. J Neurophysiol 26:1003.

Williams RW, Chalupa LM (1982a) Prenatal development of retinocollicular projections in the cat: An anterograde tracer transport study. J Neurosci 2:604.

Williams RW, Chalupa LM (1982b) The effects of prenatal unilateral enucleation upon the functional organization of the cat's lateral geniculate nucleus. The Physiologist 25:223.

Williams RW (1983) Prenatal Development of the Cat's Visual System. Thesis, University of California, Davis.

Williams RW, Chalupa LM (1983a) Development of the retinal pathway to the pretectum of the cat. Neuroscience, in press.

Williams RW, Chalupa LM (1983b) Expanded retinogeniculate projections in the cat following prenatal unilateral enucleation: Functional and anatomical analyses of an anomalous input. Soc Neurosci Abstr 9: in press.

Williams RW, Bastiani MJ, Chalupa LM (1983a) Addition and attrition of axons within the optic nerve during fetal development: Appearance of growth cones and necrotic axons. Invest Ophthal Vis Sci Suppl 24:8.

Williams RW, Bastiani MJ, Chalupa LM (1983b) Loss of axons in the cat optic nerve following fetal unilateral enucleation: An electron microscopic analysis. J Neurosci 3:133.

Acknowledgement: Supported by grants EY03991 and EY05670 from the National Eye Institute of NIH.

Development of Visual Pathways in Mammals, pages 103–114

MORPHOLOGY AND DEVELOPMENT OF RETINAL DOPAMINERGIC CELLS AS DEMONSTRATED BY IMMUNOCYTOCHEMISTRY OF TYROSINE-HYDROXYLASE

J. Nguyen-Legros[1], C. Versaux-Botteri[1], A. Vigny[2]

[1] INSERM U-154, Hôpital St-Vincent de Paul, 75674 Paris 14, France
[2] Institut de Biologie Physico-Chimique, 75005 Paris, France

Since their discovery by Malmfors (1963), dopaminergic amacrine cells have been demonstrated in all vertebrate retinas so far examined. Until today, they have been visualized on account of their endogenous dopamine content, by means of the Falck-Hillarp histofluorescence reaction. A progress in the visualization of catecholaminergic neurons has been recently introduced by the immunohistochemical detection of their biosynthetic enzymes. This method offers the double advantage of being available for both light and electron microscopic observations and of being applicable to post-mortem samples, thus enabling the visualization of human retinal catecholaminergic neurons.

We have recently shown that an anti-serum directed against tyrosine-hydroxylase (TH), the first enzyme of catecholamine biosynthesis, can demonstrate dopaminergic neurons in the rat retina, with a good specificity (Nguyen-Legros, Berger, Vigny, Alvarez 1981). Among all vertebrate retinas examined, no significant amount of noradrenaline or adrenaline has been detected biochemically (Da Prada 1977), the bovine retina excepted (Nesslhut, Osborne 1982). We have thus used this technique to visualize dopaminergic (DA)-neurons in the retina of adult rodents and primates (Nguyen-Legros, Botteri, Le Hoang, Vigny, Gay in press) as well as to study the appearance and development of DA-neurons in young rats (Nguyen-Legros, Vigny, Gay 1983).

We refer in the present descriptions to the retina of rat, monkey (Cynomolgus), ape (chimpanzee) and human. The

DA-neurons have been demonstrated by means of an anti-TH anti-serum, either with indirect immunofluorescence (Coons 1958) or with unlabelled antibody technique (PAP) (Sternberger 1979) as described elsewhere (Nguyen-Legros, Berger, Vigny, Alvarez 1981). The postnatal development of DA-retinal cells has been studied in rats. From each animal cryostat sections and floating pieces of retina, free from pigment epithelium, were processed simultaneously. The pieces of retina were flat-mounted in glycerine at the end of the reaction and observed in this position. This technique leads to the visualization of Golgi-like images of DA-neurons.

ADULT RETINAS

The retinal TH-immunoreactive cells exhibit common characteristics among the four species.

According to the terminology of Ehinger (1977), four types of amacrine cells can be distinguished in sections, on account of their position in the retinal layers. The most numerous are true amacrines, located in the innermost row of the inner nuclear layer (INL). Their processes extend in the outermost sublayer (sublayer 1 of Cajal) of the inner plexiform layer (IPL), where they form a continuous plexus. The three other types are displaced amacrines whose somas are located, i) more sclerally in the INL (pleomorph cells), ii) in the middle of the IPL (eremite cells), iii) in the ganglion cell layer (alloganglion cells). The processes of these displaced cells join the dopaminergic plexus after a more or less long course through the IPL.

A special type of DA-cells must be added to the four types of amacrines: the interplexiform cells. Although they are lying in the amacrine layer, these cells send processes to both the IPL and the outer plexiform layer (OPL). The observation of such cells in their entirety is rare because the sclerally directed processes generally branch from a dendritic trunk running within the DA-plexus and is not emerging directly from the cell body. How these processes are axons remains unclear. Cajal (1972) described these cells as "ascending axis cylinder" cells in spite of their location among the amacrines. We have obtained some obvious images of interplexiform DA-cells in the rat retina

(Nguyen-Legros, Berger, Vigny, Alvarez 1982). Their existence remains questionable in primates. Although some processes emerging from the DA-plexus in a sclerally direction are visible, there is no evidence that they all reach the OPL.

In flat-mounted pieces of retina, the Golgi-like images obtained with immunohistochemical staining allow for the classification of DA-neurons according to their morphological type.

The great majority of DA-neurons in all four species appear to be stellate cells with 2-4 extremely long and poorly branched dendritic processes. Generally, the principal dendritic trunks do not overlap, but the branches and terminals are intermingled, resulting in a network of tiny varicose processes lying in a plane of the retina and corresponding to the DA plexus observed in sections. Some of these principal cells, in the primate as well as in the rat retina, show an axon-like process (Figs. 2, 4). How these cells correspond to the interplexiform cells will be discussed later.

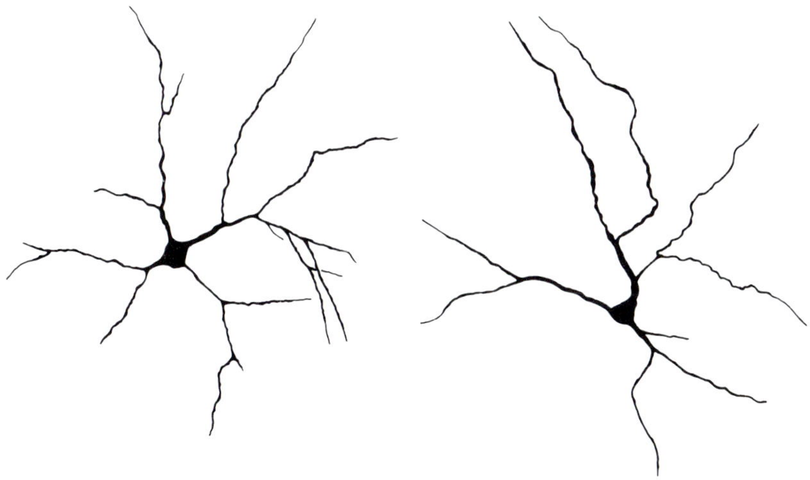

Fig. 1. DA-stellate neurons as seen in flat-mounted pieces of adult retina. Left, human retina; right, chimpanzee retina. Camera lucida drawing.

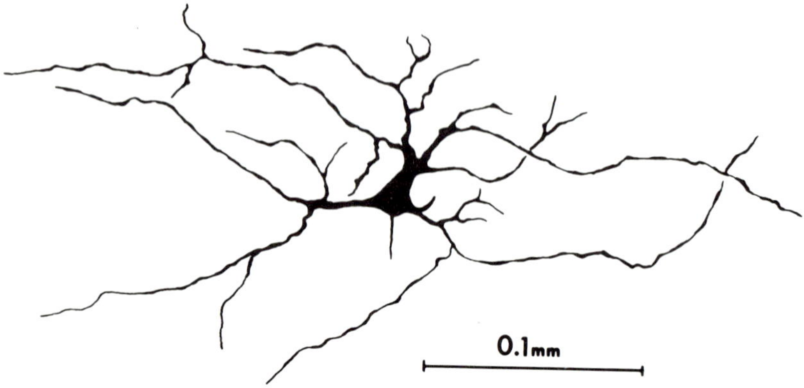

Fig. 2. Camera lucida drawing of a human stellate DA-neuron showing an axon-like process.

Other cells, observed in upper or deeper planes of the retina, correspond to the displaced amacrines. In primate retinas, these neurons have the same stellate morphology, in spite of a weaker staining and shorter processes (Fig. 3). Sometimes these cells and the principal cells appear to be really superposed and their dendrites are intermingled.

In the rat retina, another morphological type of DA-neuron is observed: the "bouquet cells". They are small neurons with a single strong dendritic trunk ramifying in a tuft of tiny beaded processes (Fig. 4). Although these cells are observed in mature retina, it seems to be a characteristic of young animals.

RETINAL DEVELOPMENT

The first DA-neurons are observed in flat-mounts by the third postnatal day. They are very few and exhibit poorly developed processes

By the 5th postnatal day, weakly stained cells can be observed in sections. They are located in the vitreal portion of the INL which is still very thick and of neuroblastic aspect, although it is separated from the outer nuclear layer (ONL) by the newly formed OPL. The young

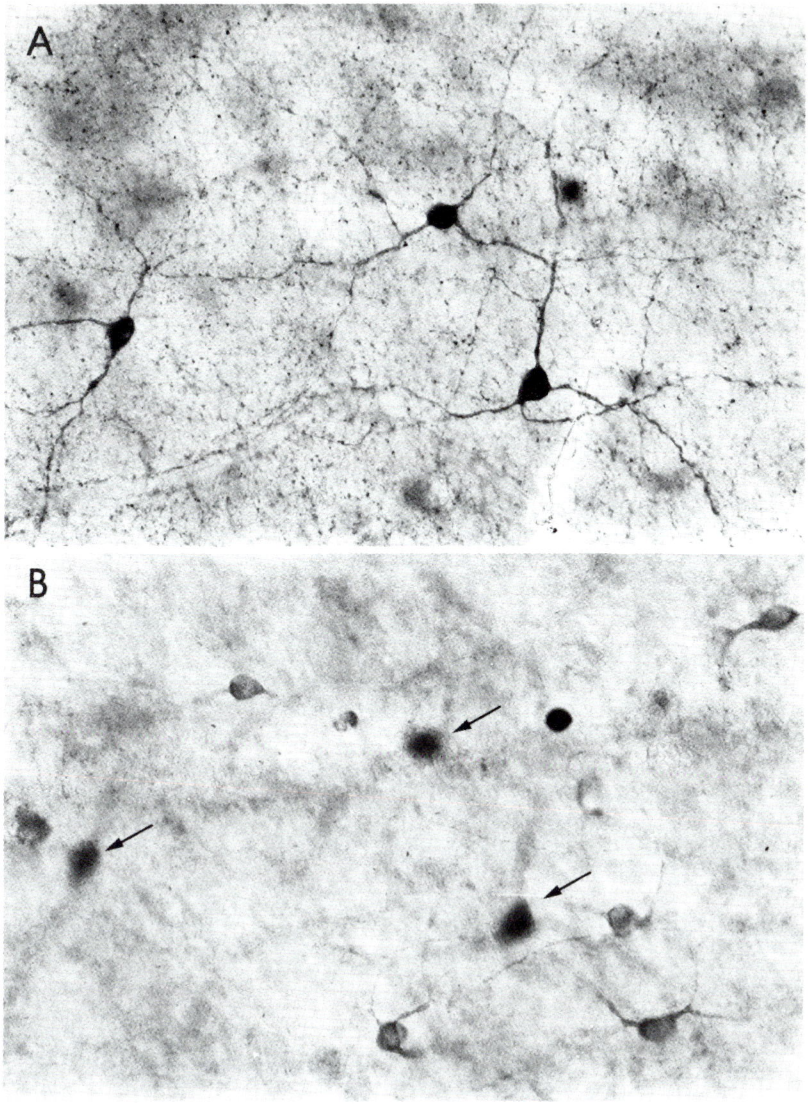

Fig. 3. The same portion of flat-mounted human retina is observed on two different focal planes.
A. focus on principal DA-stellate neurons.
B. focus on displaced cells. The somas of principal neurons are seen as shadows (arrows).

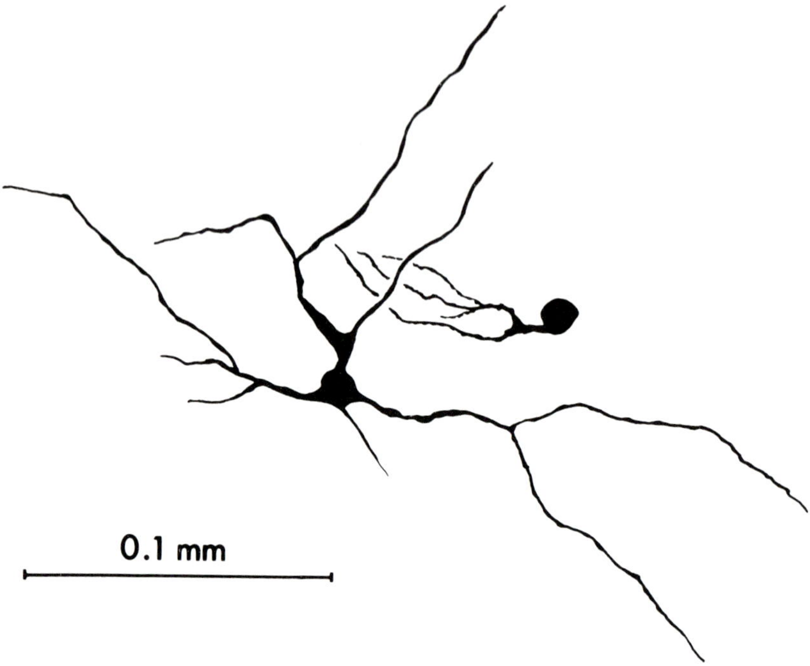

Fig. 4. Camera lucida drawing of the two morphological types of DA-neurons observed in the mature rat retina. The stellate principal neuron shows an axon-like process. The bouquet cell is more deeply located within the retina.

DA-neurons observed at this stage have a pear-shaped soma and a single vitreally directed process ramifying in T-form in the middle sublayer of the IPL.

From the 6th to the 10th postnatal day, an increasing number of DA-cells can be observed. Two types are clearly identified in flat-mounts as well as in sections. The most numerous are the previously described pear-shaped small neurons. The others are larger and round neurons. They are more heavily stained and send processes to the outermost sublayer of the IPL.

Until the opening of the eyes (14-15th postnatal day), the processes of the large neurons develop to form the DA-plexus. The first sclerally directed processes of inter-

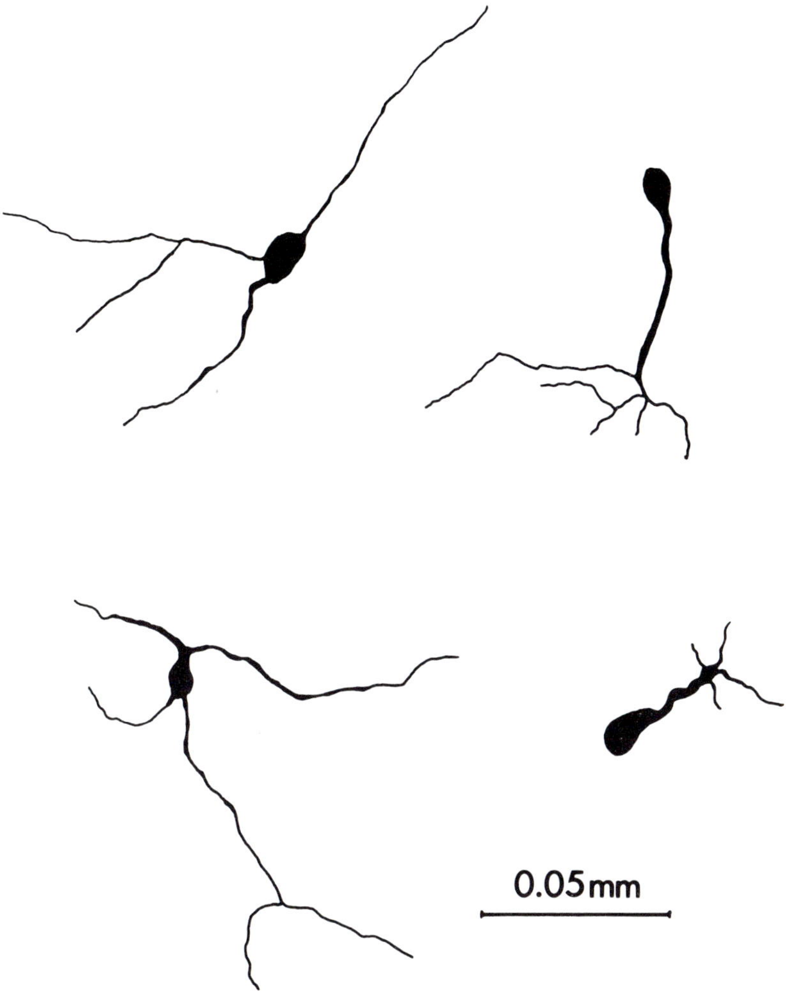

Fig. 5. Examples of young TH-immunoreactive neurons observed in flat-mounts of the rat retina by the 5-6th postnatal day. Camera lucida drawing.

plexiform cells are identified about the 15th postnatal day. During the same period, the small pear-shaped neurons are still observed but their number diminishes.

By three weeks of age, the DA neurons and DA-plexus

are fully developed, while the pear shaped neurons (probably bouquet cells) rarefy. The DA-system of the rat retina seems to be mature at this date.

DISCUSSION

Technical Comments

From a technical point of view, it seems necessary to process immunohistochemical reactions of free-floating pieces of retina rather than on stuck whole-mounted retina. It is likely that the main route of penetration for antibodies and reactives is the sectioned edges of each fragment. The dimensions of these fragments must be adjusted with the thickness of retina, i. e. in different species or according to the peripheral or central sampling of the pieces in primate retinas. The processing of over-large pieces leads to a weaker staining or no staining at all in the central part of the fragment. Consequently, quantitative or localization studies might be hazardous.

Adult Retinas

The major problem of the retinal dopaminergic system is the detection of DA-interplexiform cells, because of the important role devoted to these cells in horizontal cell inhibition (Cohen, Dowling 1983). DA-interplexiform cells were clearly demonstrated in rats (Nguyen-Legros, Berger, Vigny, Alvarez 1981, 1982) but their presence remains questionable in primates. Although there are good presumptions for their existence (Frederick, Rayborn, Laties, Lam, Hollyfield 1982; Hendrickson, Floren, Patterson 1981; Nguyen-Legros, Botteri, Le Hoang, Vigny, Gay in press), they have not been obviously observed in sections and their identification is not possible in flat-mounts.

In the adult primate retina, only one type of DA-cell can be observed in flat-mounts: the true amacrines and the displaced amacrines have the same stellate morphology. So, it is likely that the interplexiform cells are also stellate neurons and differ from the other amacrines only by the direction of their processes. A few cells among stellate neurons exhibit an axon-like process but, in spite of many efforts, we were unable to observe the terminals of this process, which is observed over a short distance from the soma. According to the theory of the dual formation of amacrine cells by both in situ differentiation and dedifferentiation of primitive ganglion cells (Hinds, Hinds 1983), it could be assumed that the axon-like bearing DA-cells may represent some dedifferentiating ganglion cells. This is in agreement with the observation of Hinds and Hinds (1983), which establishes that the majority of displaced amacrines originate from ganglion cells and sometimes exhibit a remnant of axon. Moreover, these cells belong to the class of large dendritic field amacrines, which is in accordance with the very long processes observed in stellate DA-neurons.

Retinal Development

The first TH-immunoreactive neurons observed by the 3rd to 5th postnatal days are already in position of amacrine neurons and exhibit developed processes within the IPL. The appearance of biosynthesis function is preceded by the morphological differentiation which is in agreement with reports on the development of central catecholaminergic neurons (Imai, Iijima 1981). However, in contrast with the brain catecholaminergic nuclei, the catecholaminergic neurons of the retina are identified only after birth. This may be due to the immaturity of the retina in newborn rats whose eyes remain closed until the 14th postnatal day. The same delay in detection of DA is observed in the rabbit retina (Lam, Fung, Kong 1982) but not in the guinea pig whose eyes are open at birth. Fluorescent cells have been detected by their endogenous DA content as early as birth in this animal (Kato, Nakamura, Negishi 1980) whereas in the rat, the first DA-neurons could be identified by the Falck-Hillarp technique by the 10th postnatal day (Kato, Nakamura, Negishi 1980). There is a sequence in the establishment of selective properties of DA-neurons during the development. In the rabbit for example, the DA-uptake properties are established prenatally, while endogenous DA is detected only several days after birth. It could be interesting to investigate the appearance of biosynthetic enzymes in this animal.

However, the striking feature of the postnatal development of TH-immunoreactivity in the rat retina is the abundance of pear-shaped cells with processes in the middle sublayer of the IPL during the period preceding the opening of the eyes, and their decrease in number after this event. In the adult retina, these cells and their processes in the medial sublayer of the IPL are rarely encountered. Whether these cells disappear by neuronal death accompanying the thinning down of the INL, or undergo a modification of their transmitter synthesis capabilities still remains a problem. The recent report, in the rat retina, of noradrenaline-accumulating neurons (Fukuda, Ishimoto, Kuwayama, Shimizu, Shiosaka, Takagi, Tohyama 1982), whose morphology resembles the small pear-shaped TH-immunoreactive neurons of young rats, argues in favour of the second hypothesis. Moreover, the noradrenaline-accumulating neurons are thought to be devoid of noradrenaline synthesizing enzymes. This hypothesis is in agreement with the absence of DBH-like immunoreactivity in the adult retina (unpublished observations).

Finally, our results are in accordance with biochemical and physiological data. It has been established that the response to light of DA-neurons begins at 17-20 days after birth in the rat (Morgan, Kamp 1982), the date we observe the maturity of TH-immunoreactivity. The same date of maturity has been observed for TH activity (Cohen, Neff 1982).

Cajal SR (1972)."Histologie du système nerveux de l'homme et des vertébrés." Madrid: Consejo Superior de Investigaciones Cientificas, p 296.

Cohen JL, Dowling JE (1983). The role of the retinal interplexiform cell: effects of 6-hydroxydopamine on the spatial properties of carp horizontal cells. Brain Res 264:307.

Cohen JL, Neff H (1982). Retinal amacrine cell system tyrosine-hydroxylase: the development of responsiveness to light and neuroleptic drugs. Dev Brain Res 3:160.

Coons AG (1958). Fluorescent antibody methods. In Danielli JF (ed): "General cytochemical methods," New York: Acad Press, p 399.

Da Prada M (1977). Dopamine content and synthesis in retina and nucleus accumbens septi: pharmacological and light-induced modifications. In Costa E, Gessa GL (eds): "Advances in biological psychopharmacology," New York, Raven Press, p 311.

Ehinger B (1976). Biogenic monoamines as transmitters in the retina. In Bonting SL (ed): Oxford, Pergamon, p 145.

Frederick JM, Rayborn ME, Laties AM, Lam DMK, Hollyfield JG (1982). Dopaminergic neurons in the human retina. J Comp Neurol 210:65.

Fukuda M, Ishimoto I, Kuwayama Y, Shimizu Y, Shiosaka S, Takagi H, Tohyama M (1982). Monoamine accumulating neuron system in the rat retina with special reference to noradrenaline accumulating neurons. Exp Eye Res 34:487.

Fung SC, Kong YC, Lam DMK (1982). Prenatal development of GABAergic, glycinergic and dopaminergic neurons in the rabbit retina. J Neurosci 2:1623.

Hendrickson A, Floren I, Patterson R (1981). Neurotransmitter localization in the macaca monkey retina. Invest Ophthalm 20:237.

Hinds JW, Hinds PL (1983). Development of retinal amacrine cells in the mouse embryo: evidence for two modes of formation. J Comp Neurol 213:1.

Imai K, Iijima K (1981). Prenatal differentiation of monoamine-containing neurons in the locus coeruleus of the rabbit: a histofluorescence study. Acta Histochem Cytochem 14:7.

Kato S, Nakamura T, Negishi K (1980). Postnatal development of dopaminergic cells in the rat retina. J Comp Neurol 191:227.

Lam DMK, Fung SC, Kong YC (1981). Postnatal development of dopaminergic neurons in the rabbit retina. J Neurosci 1:1117.

Malmfors T (1963). Evidence of adrenergic neurons with synaptic terminals in the retina of rats demonstrated with fluorescence and electron microscopy. Acta Physiol scand 58:99.

Morgan WW, Kamp CW (1982). Postnatal development of the light response of the dopaminergic neurons in the rat retina. J Neurochem 39:283.

Nesselhut T, Osborne NN (1982). Is noradronaline a major catecholamine in the bovine retina? Neurosc Lett 28:41.

Nguyen-Legros J, Berger B, Vigny A, Alvarez (1981). Tyrosine-hydroxylase-like immunoreactive interplexiform cells in the rat retina. Neurosc Lett 27:255.

Nguyen-Legros J, Berger B, Vigny A, Alvarez C (1982). Presence of interplexiform dopaminergic neurons in the rat retina. Brain Res Bull 9:379.

Nguyen-Legros J, Vigny A, Gay M (1983). Postnatal development of TH-like immunoreactivity in the rat retina. Exp Eye Res 37:23.

Nguyen-Legros J, Botteri C, Le Hoang P, Vigny A, Gay M (1983). Morphology of primate's dopaminergic amacrine

cells as revealed by TH-like immunoreactivity on retinal flat mounts. Brain Res in press.
Sternberger LA (1979). "Immunocytochemistry." New York: John Wiley.

Development of Visual Pathways in Mammals, pages 115–124

DEVELOPMENT OF INHIBITORY TRANSMISSION AT THE RETINAL GANGLION CELLS IN CATS

H. Ikeda, M.A., Ph.D., D.Sc. & J. Robbins, B.Sc.

Vision Research Unit of Sherrington School
The Rayne Institute
St. Thomas' Hospital, London SE1 7EH

In the fully light adapted condition, the receptive fields of retinal ganglion cells of adult cats are made up of a sharply defined small excitatory centre and a strong narrow inhibitory surround. This endows the adult cells with the property of high spatial resolution. In contrast, the receptive fields of cells of kittens are large with a widespread excitatory centre and an ill-defined weak inhibitory surround. These kitten cells have poor spatial resolution (Ikeda 1979). Meanwhile, our microiontophoretic studies on inhibition of retinal ganglion cells of adult cats, suggested that γ-*aminobutyric acid (GABA)* mediates inhibition of *on* cells, while *glycine* mediates inhibition of *off* cells in the area centralis (Ikeda, Sheardown 1983).

We therefore asked the question, whether or not the *GABA* and *glycine* actions are fully developed at *on* and *off* retinal ganglion cells in the kitten. To answer this question, the effect of iontophoretically applied *GABA*, *glycine* and their antagonists were studied on visually evoked response of *sustained* cells within five degrees of the area centralis in 7-9 week old kittens and 18-20 week old cats.

The visual stimuli used were spots or annuli generated on a television screen (background illumination, 100 cd/m^2). *On* cells were stimulated by stimuli brighter than the background (400 cd/m^2), while *off* cells, by those darker than the background (8 cd/m^2). The size of the spot or of an annulus was chosen for each cell so that it produced optimal excitation or inhibition respectively from the cell. They are called 'optimal spot' and 'optimal annulus'. As in our previous study (Ikeda, Sheardown 1983), inhibition which occurred when the optimal spot at the receptive field centre was switched *off* (post excitatory inhibition) and which occurred when the optimal annulus was switched *on* (surround inhibition) were studied.

All cells included in this study were *sustained* cells which produced vigorous firing for at least 20 seconds to a stationary optimal spot under fully light adapted conditions. The *X* or *Y* test was not carried out, but in our experience (Ikeda, Tremain 1979), such *sustained* cells in the area centralis are usually *X* cells.

The intraretinal microiontophoretic technique using multibarrelled pipettes in the optically intact eye of anaesthetized cats has already been described elsewhere (Ikeda, Sheardown 1982). Since our aim was to compare the drug actions on adult cells with those on kitten cells, we had to attempt to quantify the iontophoretic drug effect. To do this, we obtained a threshold current of each drug for each cell.

The threshold current for *GABA* (500 mM, pH 3.0), *glycine* (500 mM, pH 3.0) and *muscimol* (50 mM, pH 3.0) was defined as that level which extinguished the firing of the cell to the optimal spot at the receptive field centre. For *bicuculline methobromide* (2.5 mM, pH 3.0) and *strychnine HCl* (0.2 mM, pH 3.0), the threshold current was defined as that level which completely abolished the cell's visually evoked inhibtions i.e. the post excitatory and surround inhibition.

To obtain the threshold current, we began with an initial current of 5 nA and this was raised in 5 nA steps until the threshold effect was obtained. Each iontophoretic current was applied for exactly one minute, and a further minute was allowed for recovery from the application before the next increased dose was given. For *strychnine* and *muscimol*, 2-3 minutes of recovery period was allowed for each dose. Cells which did not show complete recovery from the drugs were not included in this study.

EFFECT OF BICUCULLINE AND STRYCHNINE ON LIGHT ADAPTED 'SUSTAINED' CELLS IN THE AREA CENTRALIS OF ADULT CATS AND KITTENS

In adult cats, both the post excitatory inhibition to the optimal spot and the surround inhibition to the optimal annulus in *on* cells were blocked by *bicuculline*, whereas those of *off* cells were blocked by *strychnine*. This confirms the findings of Ikeda and Sheardown (1983). Although the same antagonistic effects were generally true in kitten cells, lower doses were required than those needed to block inhibition in adult cells, as shown in Fig. 1.

The upper pen-recorder traces in Fig. 1A and B. were obtained from *on* and *off sustained* cells in the area centralis of a 20 week old cat, whilst the lower traces are from *on* and *off sustained* cells in the area centralis of an 8 week of kitten. Each trace shows the cell activity *before*, *during* and *after* visual stimulation by an optimal spot (indicated by the open circles in each trace for a white spot stimulation to the *on* cells and by the filled circles for a black spot stimulation to the *off* cells).

In the adult *on* cell, *bicuculline* 30 nA, virtually abolished the inhibition of the cell during spot *off*, while the response is totally unaffected by *strychnine* at 40 nA. In the kitten *on* cell, *strychnine* at 20 nA had no effect but *bicuculline* as low as 5 nA abolished the inhibition. Fig. 1B. shows that *strychnine* at 30 nA abolished the inhibition of the adult *off* cell, whereas as little as 5 nA was enough to produce the same effect on the kitten *off* cell. *Bicuculline* had little effect on both the adult and the kitten *off* cells.

Similar results to those illustrated in Fig. 1. were obtained in 36 adult *on* cells and 16 kitten *on* cells (see Fig. 3. left). The mean threshold current for *bicuculline* in the adult cells was 29.6 nA (S.E.±1.8 nA), whereas that for the kitten cells was 9.1 nA (S.E.± 2.0 nA). The mean threshold current required for *strychnine* in 56 adult *off* cells and in 20 kitten *off* cells was 30.5 nA (S.E.±1.5 nA) and 9.1 nA (S.E.±1.3 nA) respectively.

Thus considerably lower current of the antagonists for *GABA* and *glycine* are required to block inhibition of *on* and *off* cells respectively in kittens than required by adult cells. The smaller doses needed to affect the responses of the kitten cells might, however, have been due to the fact that the ganglion cells are smaller and more densely packed in the kitten retina and thus the transmitter receptors may have been nearer to the drug barrels.

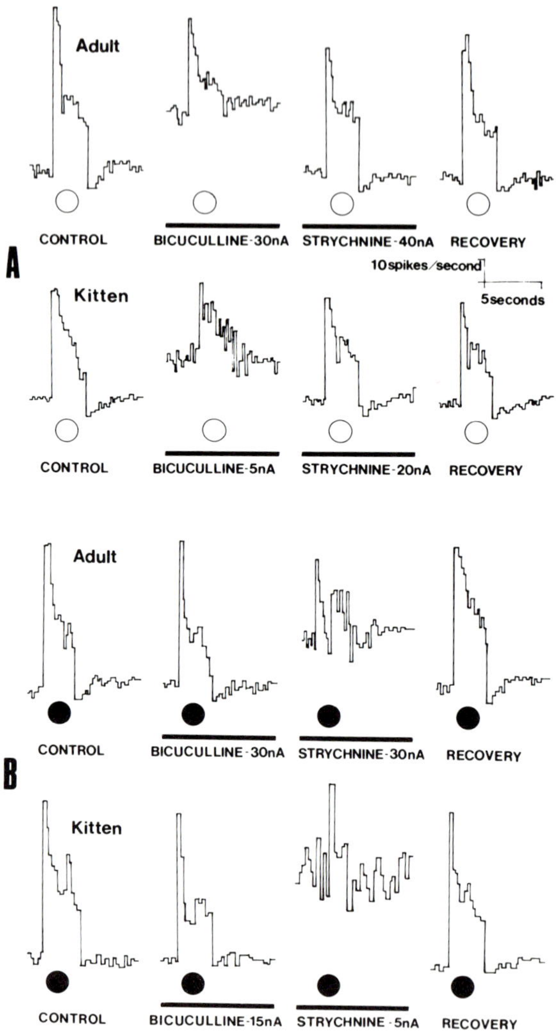

Fig. 1. Pen-recorder traces showing the effects of iontophoretically applied *bicuculline* and *strychnine* on visually evoked responses of *sustained* cells in the area centralis from an adult cat and those from an 8 week old kitten. Firing rate is on the vertical axis. Records of *on* cells (A) and *off* cells (B) to an optimal spot for 3 seconds.

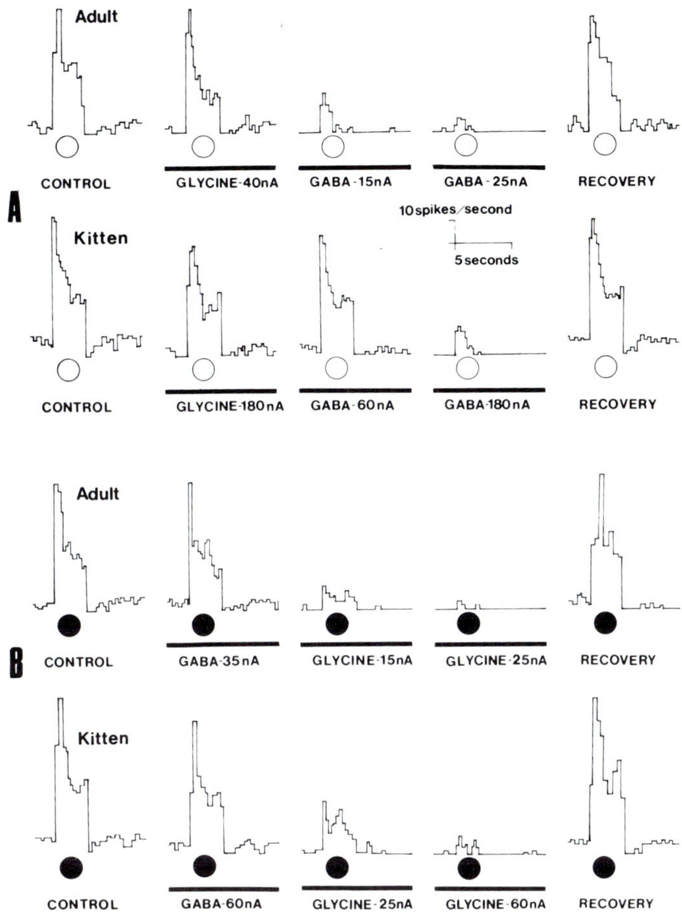

Fig. 2. Pen-recorder traces showing the effects of iontophoretically applied *GABA* and *glycine* on visually evoked responses of *sustained* cells in the area centralis from an adult cat and those from an 8 week old kitten. Firing rate is on the vertical axis. Records of *on* cells (A) and *off* cells (B) responding to an optimal spot for 3 seconds.

EFFECT OF GABA AND GLYCINE ON LIGHT ADAPTED 'SUSTAINED' CELLS IN THE AREA CENTRALIS OF ADULT AND KITTENS

The above hypothesis was tested in the following experiments using *GABA* and *glycine*. If the greater potency of *bicuculline* and *strychnine* in the kitten cell was mainly due to the physical proximity of the inhibitory transmitter receptor in the kitten retina, then also a lower dose of *GABA* and *glycine* would be required by kitten cells than that required by adult cells. On the other hand, a greater dose of *GABA* and *glycine* might be necessary in the kittens to achieve the same inhibitory effect as that on adult cells, if the chemical configuration of *GABA* and *glycine* receptors on the kitten cells were not yet adult like and were more readily accessible to the antagonists but not to the transmitters themselves.

Fig. 2A and B. illustrate the effect of iontophoretically applied *GABA* and *glycine* on the responses of an adult and a kitten cell.

The firing of the *on* cell of the adult cat (Fig. 2A. upper line) is virtually abolished by *GABA* at 25 nA, whereas that of the kitten's *on* cell (Fig. 2A. lower line) needed as much as 180 nA to be inhibited. The relative ineffectiveness of *glycine* on both adult and kitten *on* cells is also illustrated in Fig. 2A.

The firing of the adult *off* cell in Fig. 2B. (upper line) is completely inhibited by *glycine* at 25 nA, whereas that of the kitten cell (lower line) was inhibited at 60 nA. Again, the relative ineffectiveness of *GABA* to *off* cells can be illustrated by the two cells shown in Fig. 2B.

Studies of *GABA* were made on 16 *on* cells in cats aged 18-20 weeks and 11 *on* cells in kittens aged 7-9 weeks (see Fig. 3. right). The mean threshold current for *GABA* for the adult cells was 14.8 nA (S.E.±2.7 nA) and that for the kitten cells was 75.0 nA (S.E.±12.2 nA). The effect of *glycine* were studied on 18 adult *off* cells and 21 kitten *off* cells. The mean threshold current was 23.1 nA (S.E.±2.1 nA) and 82.6 nA (S.E.±13.8 nA) respectively.

The effect of *muscimol* was studied on only 8 adult *on* cells and 6 kitten *on* cells. The results were essentially the same as those obtained from *GABA* except that *muscimol* (50 mM) was more potent than *GABA* (500 mM). Again a less current of *muscimol* was required to inhibit adult *on* cells than kitten *on* cells.

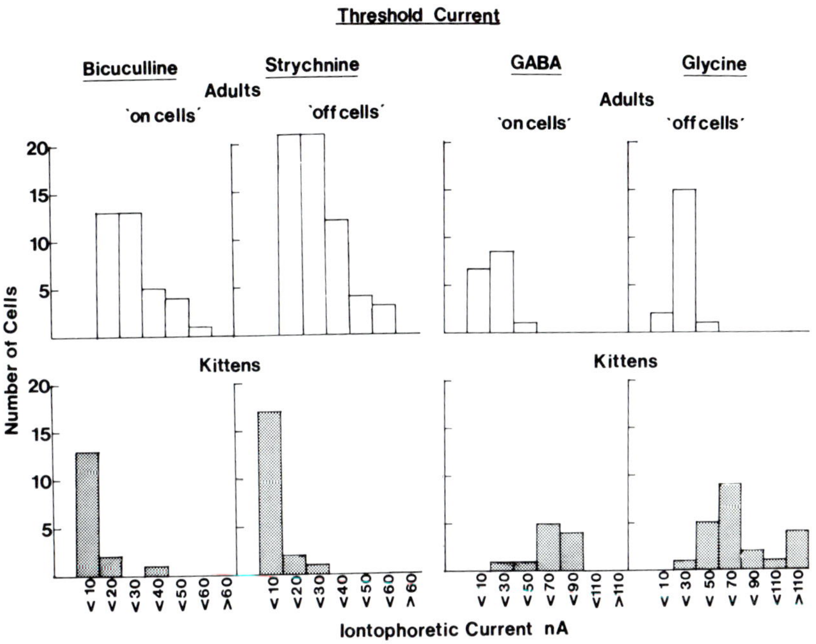

Fig. 3. The strength of iontophoretic current required for *GABA* and *glycine* to inhibit the visual excitation and for *bicuculline* or *strychnine* to block visually induced inhibition completely.

A higher dose of *GABA* and *glycine* was, therefore, necessary to inhibit kitten *sustained* cells than required in adult cells. These experiments, thus, rule out the possibility that the difference in iontophoretic current required for *bicuculline* and *strychnine* between the adult and kitten cells might be due to *GABA* and *glycine* receptors being nearer to the drug barrels in the more densely packed cells of the kitten retina. It is interesting to note that *GABA* and *glycine* doses required to reach threshold are very similar.

This suggests that the characterisitics of our multibarrelled pipettes are not widely different when the drugs are administered at the same concentration (500 mM), similar transport numbers (*GABA* 0.30; *glycine* 0.28) and for the same application time (1 minute) on only those cells with well isolated large spikes.

ARE GABA AND GLYCINE BINDING SITES AT THE RETINAL GANGLION CELLS IN THE AREA CENTRALIS NOT DEVELOPED FULLY IN THE KITTEN RETINA?

As Fig. 3. summarized, a much higher current of *GABA* (and much lower current of *bicuculline*) was needed to affect *on* cells of kittens than required in adult *on* cells. Also, a much higher current of *glycine* (and much lower current of *strychnine*) is needed to affect *off* cells of kittens than required by adult *off* cells.

These results suggest that *GABA* binding sites at *on* cells and *glycine* binding sites at *off* cells in the area centralis of kittens are not as fully developed as in the adult retina.

A criticism may, however, be raised to our conclusion that *GABA* and *glycine* binding sites are not fully developed. One might postulate that the kitten cells may need a higher dose of *GABA* or *glycine* because the uptake system of these drugs in the kitten retina are more efficient than those of adult retina.

We can, however, rule out this criticism for the following two reasons. Firstly, *GABA* uptake in the rabbit retina has been shown to be very poor at birth but *increases* with age and reaches the adult level at about 10 days after birth. Therefore, it is likely that the *GABA* uptake is *less* rather than more efficient in early postnatal life. Secondly, our experiments with *muscimol* suggest that kitten *on* cells are inhibited by much higher doses of *muscimol* than that required by an adult *on* cells. *Muscimol* binds exclusively with *GABA* receptors and there is no endogenous uptake of *muscimol* in the retina or in any other part of the central nervous system (Johnston, Kennedy, Lodge, 1978; De Feudis, 1980).

SIMILARITY OF GABA AND GLYCINE EFFECTS ON KITTEN CELLS IN THE AREA CENTRALIS AND THOSE ON THE ADULT CELLS IN THE PERIPHERAL RETINA

Finally, it may be noteworthy that in the adult retina, *on* cells in the area centralis show no effect to *glycine*, whilst *off* cells show no effect to *GABA* (Ikeda, Sheardown 1983). However, if recordings are made in cells in peripheral zones of adult retina, some *on* cells could be inhibited partially by *glycine*, though with a much higher threshold than that of *GABA*. Similarly some *off* cells in the periphery of adult retina could be inhibited by *GABA* at a much higher current than that required for *glycine* to inhibit the cell.

Although we have not studied many cells systematically in the kitten retina, *on* cells in the area centralis were partially inhibited by *glycine* at a very high current. This may be seen in the response of the *on* cell from a kitten to *glycine* at 180 nA in Fig. 2A. Similarly some *off* cells in the area centralis of kittens were also partially inhibited by a higher dose of *GABA* than that of *glycine*. The minor inhibitory effect of *GABA* can be seen in the *off* cell of the kitten in Fig. 2B. It is possible therefore, that the clear differentiation of *GABA* and *glycine* receptors for *on* and *off sustained* retinal ganglion cells in the area centralis found in the adult cat may not be fully developed in the kitten. A further study will be necessary to clarify this point.

CONCLUSION

The results of our iontophoretic studies suggest that the specificity of *GABA* and *glycine* recognition sites are not fully developed at *on* and *off* retinal ganglion cells respectively in the area centralis of 7-9 week old kittens. Nevertheless, the receptors are already present in a configuration which favour the bindings of its antagonists. Thus the neurophysiological observation that inhibition of retinal ganglion cells in kittens is not fully developed may have a neurochemical basis.

ACKNOWLEDGEMENT

H.I. wishes to dedicate this work to her late husband, John Wolstencroft, one of the pioneers of the micro-iontophoretic technique in the central nervous system. We are greatful to the *Medical Research Council* and the *Special Trustees of St. Thomas' Hospital* for their financial support and Mr. T. Allen for technical assistance.

REFERENCES

De Feudis FV, (1980). Binding studies with muscimol: relation to synaptic γ-amino butyrate receptors. Neurosci 5:675.

Ikeda H, (1979). Visual acuity, its development and amblyopia. J Roy Soc Med 73:546.

Ikeda H, Tremain KE, (1979). Amblyopia occurs in retinal ganglion cells in cats reared with convergent squint without alternating fixation. Exp Brain Res 35:559.

Ikeda H, Sheardown MJ, (1982). Aspartate may be an excitatory transmitter mediating visual excitation of sustained but not transient cells in the cat retina: iontophoretic studies *in vivo*. Neurosci 7:25.

Ikeda H, Sheardown MJ, (1983). Transmitters mediating inhibition of ganglion cells in the cat retina: iontophoretic studies *in vivo*. Neurosci 8:387.

Johnston GAR, Kennedy SME, Lodge D, (1978). Muscimol uptake, release and binding in rat brain slices. J Neurochem 31:1519.

Lam DMK, Fung SC, Kong YC, (1980). Postnatal development of GABA-ergic neurones in the rabbit retina. J Neurol 193:89.

Development of Visual Connections and Neuronal Structure

Development of Visual Pathways in Mammals, pages 127–143

DEVELOPMENT OF RETINAL PROJECTIONS IN THE MOUSE

Pierre Godement

Laboratoire de Neurobiologie du Développement
Bat 440, Université de Paris-Sud
91405 Orsay Cedex (France)

During development, retinal axons must solve a number of problems in order to achieve their precise adult distribution within the subcortical visual centers. The first task - finding the correct target structure- is accomplished with little error, in the sense that structures which are devoid of retinal innervation in the adult are also, in general, not innervated by retinal axons during development (though exceptions exist). Abnormal patterns of retinal innervation in non-visual structures can however result from appropriate surgical manipulations during the early development (Frost 1981). On the other hand, in all mammals studied so far, inputs from each retina are initially mixed within the visual centers and separate only later. The characteristics of those separations depend on the particular structure and species considered. A most striking example of this is that of the development of the monkey lateral geniculate nucleus, in which the overlap between the inputs from each eye is thought to be initially complete (Rakic 1977). In this species there is little difference between the proportions of ipsi and contralateral fibers innervating the geniculate. But on the other hand, in rodents, the contralateral input predominates. It appears that in the dorsal lateral geniculate nucleus (dlGn) of the rodents (i.e. hamster) even during the early stages of the development there is relatively little overlap between the retinal terminals originating from each eye. Specifically, anterograde transport of tritiated amino-acids revealed that the uncrossed fibers even at the early stages terminate in the appropriate sector of the nucleus (So et al. 1978; Frost et al. 1979). More recently through the use of horseradish peroxidase as an anterograde tracer, the immature pattern of these projections in the early stages was revealed even more clearly (So, Schneider 1978; Land, Lund 1979; Bunt et al. 1983). Thus, in the superior colliculi and dlGn of the newborn rat, uncrossed projections have a more widespread distribution than in older animals, both in terms of their topographical as well as laminar patterns of termination (Land, Lund 1979; Maxwell, Land 1981).

The experiments described here were performed on the strongly pigmented C57BL/6 strain of the mouse. The development of the primary visual projections has been investigated in embryonic and postnatal mice, using intraocularly injected horseradish peroxidase (HRP). It is concluded that concerning the retino-geniculate projections the arrival of uncrossed fibers is delayed relative to the arrival of crossed fibers. Our data indicate that in the early stages of the development the distribution of growing uncrossed fibers within the dlGn corresponds approximately to their later distribution. Thus, the retinal fibers distribute in a topographically correct manner at an early stage. Furthermore, the depth-distributions of crossed and uncrossed projections seem to be different throughout development, suggesting that special constraints on the growth of fibers from each eye exist prior to their separation. In a second set of studies, the role of interactions during embryonic life between fibers from each eye on the specification of the laterality of projection of retinal ganglion cells has been investigated. Following unilateral eye destructions performed at embryonic days 12 (E12), 13, or 16, and at birth, in the remaining eye the majority of ipsilaterally-projecting ganglion cells is located in the appropriate region of the retina. Enucleations at E16 or birth lead however to increases in the numbers of ipsilaterally-projecting cells both in the temporal and nasal retina. Enucleations performed at E12 or E13 have more variable effects and in several cases lead to a decrease in the numbers of ipsilaterally-projecting cells in the remaining eye. Preliminary accounts of this work have been published (Godement, Salaün, Saillour, Imbert 1981; 1982).

METHODS

Details of the techniques for intrauterine surgery and tracing connections will be described elsewhere. For tracing retinal projections, HRP (60% in sterile saline) was injected into one eye of embryonic (E14-E18) and neonatal mice. Following appropriate survival times (9-18 hours in embryos, and 18-24 hours in postnatal mice), the animals were perfused through the aorta and serial 40 micron thick frozen sections were processed according to the protocol of Mesulam (1978), using tetramethylbenzidine as the chromogen. In the second study, one eye of embryos age E12, E13, or E16 was destroyed with a microcauther inserted through the uterine wall (see Godement, Salaün, Saillour 1981). When these mice were adult, and in mice with one eye destroyed at birth, HRP (30%) dissolved in polyacrylamide gel (Griffin et al. 1979) was inserted under visual control in the optic tract ipsilateral to the remaining eye, ventral to the ventral geniculate. The optic tract was cut in order to label as many of the optic tract fibers as possible. The same procedures were employed in normal mice of the same strain, C57BL/6. The mice survived for 48 hours after which they were perfused. Their retinas were flat-mounted and reacted for HRP with the Hanker-Yates reagent (Hanker et al. 1977) in pH 5,1 cacodylate buffer with nickel-cobalt intensification (Adams 1981), and kept in glycerine jelly for analysis of cell numbers and sizes. Drawings of retinas ipsilateral to the side of HRP-placement were made with the help of a drawing-tube attachment, and the position of every labelled cell was plotted. The brains were reacted with diaminobenzidine or Hanker-Yates reagent to check the central spread of HRP: in each case it was limited to one side of the brain.

DEVELOPMENT OF RETINOGENICULATE PROJECTIONS.

It has not yet been possible to study with experimental tract-tracing methods the earliest stages of the development of the optic projections in the mouse. Studies at the ultrastructural level (Silver, Sidman 1980) indicate however that ganglion cell axons leave the optic cup at E12 and that the first fibers reach the floor of the diencephalon by E13. In an embryo injected intraocularly at E14, we have observed labelled retinal axons at the level of the chiasma, but none at thalamic or mesencephalic levels, suggesting that few, if any fibers have reached such central levels at this age. Another embryo was injected with HRP at E16: in this case, very strong projections were observed contralaterally, reaching as far as the collicular plate. The optic tract contralateral to the injected eye was very strongly labelled, and many fibers coursed through the anlage of the dlGn and adjoining regions (vlGn, and a region corresponding to the later lateral posterior nucleus). In this embryo, many

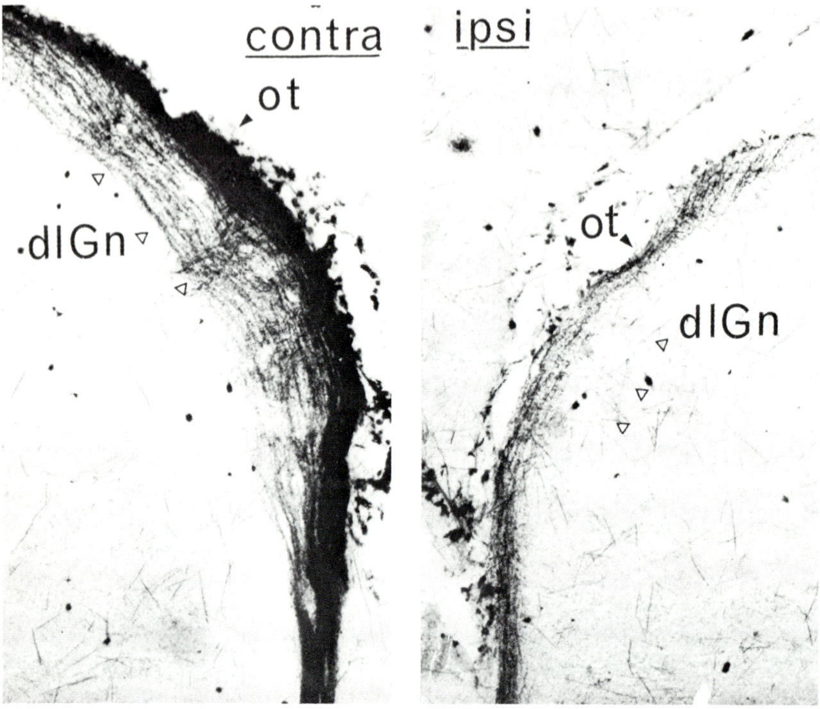

Figure 1. Photographs of two coronal sections through the brain of the embryo injected intraocularly with HRP at E18. On the contralateral side (left) heavily labelled optic fibers course in the optic tract (ot) and within the dlGn (between the optic tract and the empty triangles). Fibers also course through the adjoining ventral geniculate and lateral posterior nuclei (respectively ventral, and dorsal to the dlGn). On the ipsilateral side (right), labelled retinal axons are confined to the optic tract, superficial to the dlGn. (x120).

fibers were observed on the side ipsilateral to the injected eye, between the chiasma and the level of the internal capsule, in the external margin of the optic tract. Only very few ipsilateral fibers could be seen at more dorsal levels, as if the bulk of these fibers had yet to arrive in the vicinity of the visual centers. In an embryo injected with

HRP at E18 (that is one day before birth), the contralateral dlGn was laden with fascicles of labelled transverse fibers, which coursed almost throughout its entire thickness (Fig. 1). Labelled fibers on the ipsilateral side now extended as far caudally as the collicular plate, but remained mostly confined to the optic tract (Fig. 1); only a few entered the external part of the dlGn in some sections.

These observations suggest that in the mouse visual system (as previously suggested for other species: Cavalcante, Rocha-Miranda 1978; Frost et al. 1979) crossed projections develop, on the whole, in advance of the uncrossed ones. A similar conclusion has been recently reached in relation to the time course of the development of retinogeniculate projections in the cat (Shatz 1983), a species where the imbalance between contra- and ipsilateral retinogeniculate projections is much smaller than in the mouse. It appears that in the mouse the delay is not limited to a later ingrowth or timing of elaborating dense projections by ipsilateral fibers, it seems that uncrossed fibers arrive in the vicinity of the visual centers later than crossed afferents.

This delay in the development of uncrossed fibers is in fact still apparent after birth. At this time, dense crossed projections are well established within the dlGn, as evidenced by the addition of a sprinkling of reaction-product granules to the pattern of transverse fiber labelling that can be observed in the embryos. In most brains the density of this labelling is too high to allow a clear observation of the pattern of fiber ingrowth into the dlGn (Fig. 2). However two features can be easily recognized: the transverse fibers tend to occupy with age progressively more external positions, as the nucleus enlarges, and the inner border of the nucleus is, then, relatively devoid of these fascicles, and in many brains of younger animals it contains a slightly higher density of reaction-product granules. Some fibers extend deeper to the dlGn. Until the third postnatal day, uncrossed labelling in the dlGn is rather light, consisting of strands of reaction-product granules directed from the surface to deeper levels of the nucleus, indicative of an ingrowth of these fibers into the dlGn (Fig. 2). In horizontal sections, there was a clear collective orientation of these strands of labelling, from latero-posterior (along the optic tract) to antero-medial. The quantity of these fibers increased during the first three postnatal days, but

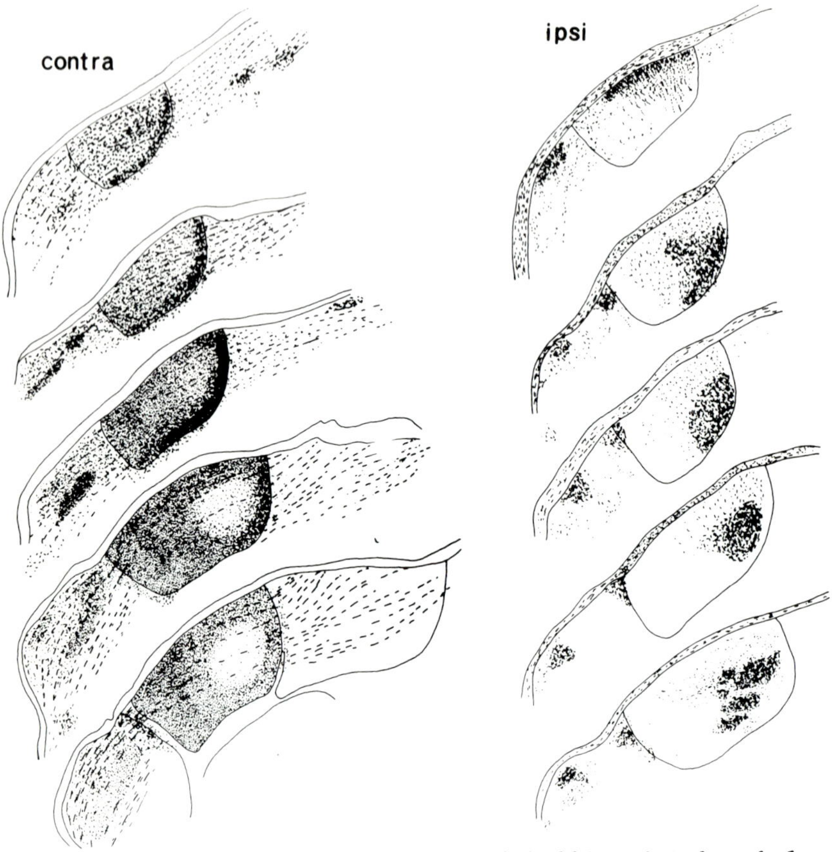

Figure 2. Drawings of anterograde labelling in the thalamus contralateral (left) and ipsilateral (right) to intraocular HRP injections in mice aged, from top to bottom, 1, 3, 4, and 8 days postnatal, and in an adult. The outlines of the optic tract and thalamus are drawn for all ages, and those of the ventral geniculate, lateral posterior, and ventrobasal nuclei are indicated in the drawing of the adult showing contralateral label (lower left). Fiber labelling is indicated by dashes (mainly on the contralateral side). For each drawing, lateral is towards the left, and dorsal is up.

were mostly grouped in a dorso-medial to more ventral sector of the dlGn, corresponding approximately to its binocular sector (Fig. 2). Elsewhere, the labelling was much less dense, except at the surface of the dlGn where there was in the first two postnatal days a narrow band of labelling which later on disappeared: this labelling probably represents at least in part optic fibers that are growing in the outer rim of the dlGn, but not going further into the nucleus, and parts of fibers terminating deeper.

Evidence of a segregation of afferents from each eye was reliably seen in brains of animals injected on the fourth postnatal day (Fig. 2). The uncrossed fibers which shortly after birth spanned the medial extent of the dlGn on the third postnatal day established a (fine-grained) projection along the medial border of the nucleus, away from its surface. The distribution of this projection along the antero-posterior axis of the nucleus was similar to that observed in adults. The labelling was especially dense along the ventro-medial margin of the nucleus (i.e. in its depth). Labelling in the rest of the nucleus was very sparse, except between the surface of the dlGn and the deeper projection (Fig. 2), which was probably due in great part to fibers terminating deeper. At this stage the density of the crossed projections was appreciably greater than that on the ipsilateral side, but in some brains, and more reliably so on the fourth day after birth, there was a slight decrease in the density of projection in the medio-dorsal sector of the dlGn -corresponding, on the other side, to a region of heavy uncrossed projection. In the following days, there occurred a further condensation of the territory of uncrossed termination, concurrent with an increase in the density of these projections. Contralaterally, during the same period, the contralateral fibers disappeared from this region. By the eighth postnatal day, most features of contralateral and ipsilateral projections present in the adult could be recognized, and in animals injected at fifteen days after birth the pattern looked almost adult-like. Globally then, in the mouse dlGn, the separation of fibers from each eye occurs in the same fashion as described for the hamster dlGn (So et al. 1978), that is through a withdrawal of crossed fibers from the region where uncrossed fibers have accumulated. This process in the mouse appears however to be rather slow, and does not only involve a retraction of crossed fibers (or their elimination), but also some retraction of the uncrossed fibers (as visible in the drawings of Figure 2; it is especially the case in the ventro-medial corner of the dlGn).

As in other mammalian species (Rakic 1977; Cavalcante, Rocha-Miranda 1978; Linden et al. 1981; Sanderson et al. 1982; Shatz, DiBerardino 1980; Shatz 1983), each eye's projections are initially mixed in the dlGn, and only later they become separated.

The overlap in the mouse dlGn is however far from total, which is due to two factors. One is that the crossed fibers extend tangentially throughout the nucleus at birth, but the uncrossed fibres are preferentially located in the appropriate sector, along the medial border of the dlGn. This, together with the fact that in rodents at birth the majority of the ganglion cells that project ipsilaterally are densely grouped in the normal temporal region of the retina (cf for rat: Land et al. 1981; Bunt et al. 1983; Martin et al. 1983; and unpublished observations in the mouse), suggests that these projections are at least on a macroscopic level organized in a roughly appropriate manner at a very early stage when they are just growing into the dlGn, and prior to the separation of afferents from each eye. The finding of lower-density projections throughout the dlGn fits well with the finding indicating that at early stages many ipsilaterally-projecting ganglion cells are scattered throughout the nasal retina in the rat (Land et al. 1981; Martin et al. 1983). It remains to be seen, how precise the map of the retinogeniculate projection is at early stages. From our observations, it appears possible that many fibers which are directed towards inappropriate regions of the ipsilateral dlGn do not in fact grow into it to any great extent. The pattern of HRP-labelling suggests in addition that uncrossed fibers in the mouse grow along a particular direction in the nucleus. That this direction corresponds approximately to that of the retinal "projection lines" defined in the adult may not be purely coincidental: it may be that these fibers follow such lines, established by earlier crossed afferents or connections intrinsic to the dlGn. It remains to be seen if crossed fibers, in particular those arriving early, grow into the dlGn in the same manner as later arriving uncrossed (and maybe crossed) fibers.

Another observation is that, just prior to their separation, the crossed fibers seem to be more or less evenly distributed, but the uncrossed fibers project clearly most densely in the depth of the dlGn (e.g. Figure 2). This is in agreement with studies on the hamster retinogeniculate proje ctions (So et al. 1978; Frost et al. 1979) and it appears that a related situation may exist in other species as well (in particular, in the cat; Shatz 1983). These early heterogeneities in the depth-distribution of each eye's projections in the dlGn imply that the process of segregation of terminals from each eye, which involves binocular competition (Rakic 1981), could occur along a direction that is already prespecified in gross form by characteristics of the early pattern of these projections. In mice removing one eye at birth results in a more diffuse than normal projection from the remaining eye to the dlGn ipsilateral to it. However there still is a preferential termination of these fibers in the depth of the dlGn (Godement et al. 1980) -a pattern that in fact resembles that of these projections in normal mice aged three days, described here. Non competitive constraints could thus be involved in this case. One possibility is that optic fibers when growing into the dlGn tend to ramify to a greater extent in its depth, prior to competing for terminal space. This could be especially the case for the fibers that are growing into the dlGn in the neonatal period, among which is the bulk of the uncrossed fibers. In turn it would lead in piling up of the fibers in the depth of the dlGn around the third postnatal day. Following this, competitive interactions with fibers from the other eye could produce a further condensation of the uncrossed terminals and a retraction or elimination of the crossed fibers from the region where uncrossed projections terminate.

EFFECTS OF UNILATERAL EYE DESTRUCTIONS PERFORMED AT OR BEFORE BIRTH ON THE ORIGIN OF IPSILATERAL PROJECTIONS IN THE RETINA

In normal mice injected with HRP in the optic tract, the characteristics of labelling in the retinas ipsilateral to the injections were essentially the same as those described by Dräger and Olsen (1980): labelled ganglion cells were concentrated along the temporo-inferior border of the retina inside a crescent-shaped region amounting to about 20% of the total retinal area (Fig. 3). Some ipsilaterally-projecting ganglion cells were located in the inner nuclear layer. The density of the labelled cells was homogeneous throughout the crescent. Throughout much of its length, the width of the crescent was 1000 microns (in non-dehydrated retinas), and its inner border was sharp (Fig. 3). Numbers of labelled

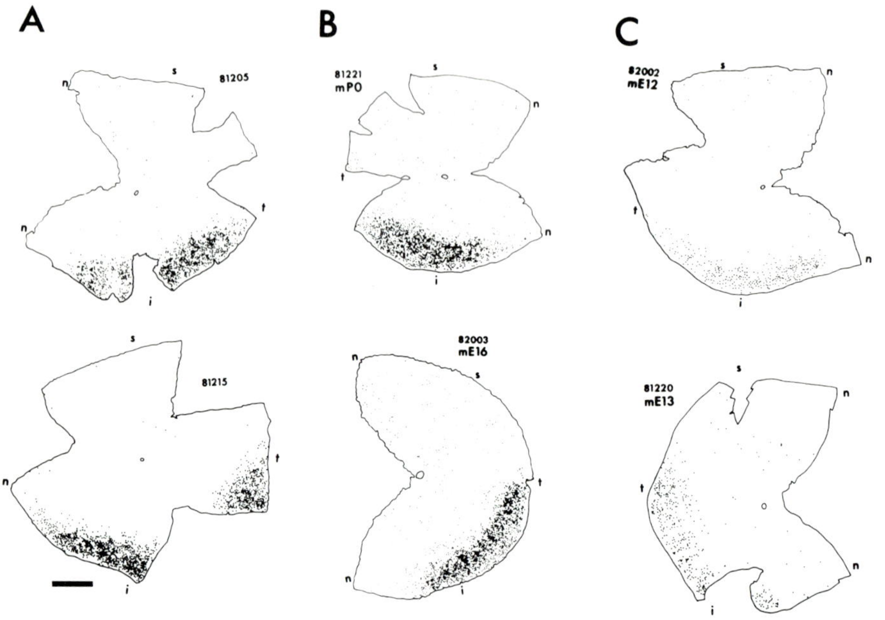

Figure 3. Drawings of flat-mounted retinas ipsilateral to optic tract injections of HRP, with all labelled cells shown (each dot represents a cell). Column A represents retinas of two normal mice, column B retinas of a mouse unilaterally enucleated at birth (top) or at E16 (bottom), and column C retinas of a mouse unilaterally enucleated at embryonic day 12 (top) or 13 (bottom). Figures beside each drawing of a retina refer to the experimental cases listed in Table 1. The optic disk is indicated in each retina. Prior to dissection of the retinas, a large cut was made into them from the nasal corner -after having marked it in the anesthetized animal- towards the optic disk so as to orient them: n=nasal, t=temporal, i=inferior, s=superior. Not all retinas are oriented in the same way in the figure. Scale bar, 1000 microns.

cells in the crescent region were fairly close from mouse to mouse (Table 1); a mean of 932 cells was counted. In the rest of the retina ipsilaterally-projecting ganglion cells were also present, apparently randomly-scattered and few in number:between 10 and 18, depending on the case.

In mice enucleated at birth or at E16, labelling in the retinas ipsilateral to optic tract injections was virtually identical in both cases. Substantial numbers of labelled ganglion cells were found scattered throughout the nasal retina, in higher than normal numbers. These cells were apparently homogeneously distributed throughout the retina, in particular they were not noticeably more numerous near the temporal crescent, or in the region of higher ganglion cell density (slightly temporal, and inferior to the optic disk). A mean of 192 such cells was counted. As in normal mice, the temporo-inferior crescent of the retina contained the highest quantity of ipsilaterally-projecting ganglion cells. Its position was the same as in normal mice, and its internal border was almost as sharp as in normal mice (Fig. 3). The numbers of cells in this part of the retina were increased, relative to normal values, by 37% when comparing the means, and by 58% when comparing the retinas with the highest ganglion cell counts in each case (i.e. 81215 and 82008, see table 1). This increase was achieved mainly through an increase in the cell density, especially of small-diameter cells.

Five mice operated at E13, and three mice operated at E12, were injected with HRP ipsilateral to the remaining eye when they were adult. Two of the mice operated at E12 were in fact microphtalmic on the operated side: a small optic globe was buried in the orbit, and a very thin nerve joined this eye remnant to the chiasma; this was probably the result of an incomplete destruction of the optic vesicle in the embryo. All other mice were monocular. In one E13-operated mouse (case 82005 in table 1) a small branch left the remaining optic nerve before its junction with the floor of the brain, and entered the ipsilateral side of the brain not far from the main optic nerve, which went on to the contralateral side.

In all such mice operated at E12 or E13, as in other mice, most ipsilaterally-projecting ganglion cells were located along the temporal inferior border of the retina (Fig. 3, and Table 1). There was some variability from mouse to mouse in the shape and size of this ipsilaterally-projecting region of the retina, and in the numbers of cells

Table 1: numbers of ipsilaterally-projecting cells

	case:	numbers of cells	
		crescent	outside
normal	81200:	925	12
	81202:	970	
	81204:	908	
	81205:	784	11
	81206:	936	18
	81215:	1073	10
	mean:	932	
ablation at birth	8144:	938	166
	81221:	1318	99
	82008:	1697	216
	82025:	1006	139
	mean:	1239	
ablation at E16	82003:	1215	312
	82005:	1484	218
	mean:	1349	
ablation at E13	81216:	761	8
	81220:	361	26
	82023:	566	49
	82024:	875	275
	82050:	838	233*small ipsi nerve
	mean:	680	
ablation at E12	82002:	349	31
	82049:	899	93§ micro-
	83006:	729	121§ phtalmics

lying within and outside of this region. In some cases (as shown in the retinas of Figure 3), the numbers of cells found within this region were inferior to the values found in normal mice, and few cells were counted also in the rest of the retina. The shape of the crescent was slightly distorted in some of these mice (for example, in the case of 82002, the width of the crescent was reduced: see figure 3), but not in any systematically obvious way from retina to retina. Other mice had higher counts of ipsilaterally-projecting ganglion cells in the temporal crescent. These approached normal values (cases 82024, 82050 in table 1), but were inferior to the values found in later-enucleated mice. The number of cells outside of the crescent was more variable. In cases where relatively high numbers of ganglion cells were counted in and out of the crescent, and in the two E12-operated microphtalmics, the shape of the temporal crescent resembled closely that of normal and later-enucleated mice, and its internal border was sharp. Cells outside of the crescent region, in the nasal retina, were apparently homogeneously scattered.

The results of this study confirm that unilateral enucleation at birth results in higher numbers of ipsilaterally-projecting ganglion cells in the remaining retina both from appropriate and inappropriate retinal sectors (Dräger,Olsen 1979; Jeffery, Perry 1982). Given the considerable incidence of cell death of centrally-projecting ganglion cells during development in various species, in particular after birth in rodents (Potts et al. 1982; Dreher et al. 1983; Perry et al. 1983), it seems likely that the excess cells in the retinas of monocular animals are ganglion cells that should normally have died (Martin et al. 1983). Removing the contralateral input before (E16) or during (birth) ingrowth of the ipsilateral fibers into the visual centers does not appear to result in major differences in numbers and repartition of ipsilaterally-projecting ganglion cells in the retina. In any case it seems that a balance is reached in spite of the fact that through the elimination of contralateral fibers a very large volume of the visual structures is available to the uncrossed fibers. It seems that this balance in the numbers of cells reached falls short of what it should be, if this operation resulted simply in an arrest of loss of ipsilaterally-projecting ganglion cells due to their death: estimates of the incidence of postnatal ganglion cell death place it at more than 50% in rats (Perry et al. 1983), while the values that we find for monocular mice are on the average only 37% above normal, and 58% when comparing the best-filled retinas. Lower estimates of the incidence of postnatal ganglion cell death are however reported by Dreher et al. (1983). Further studies are therefore needed comparing the total number of ipsilaterally-projecting ganglion cells in neonatal and adult mice to settle this problem. Preliminary evidence shows however that the increase in the numbers of cells in the temporal crescent of mice enucleated at E16 or birth is disproportionately large for the small ganglion cells (diameter less than 12 microns). And the ganglion cells that occur outside of the crescent region (the number of which, is much more increased in the monocular mice, relative to the increase in the number of cells in the crescent region) are also mostly small cells in normal and monocular mice. It could be therefore that in the mouse retina the loss of centrally-projecting ganglion cells due to binocular competition is more marked for the smallest ganglion cells.

Since in the E12 and E13 cases the eye was destroyed before arrival of optic fibers at the chiasma, the finding that most ipsilaterally-projecting ganglion cells are in the appropriate region of the retina demonstrates that the specification of which area of the retina gives rise to undecussated axons is largely, maybe totally, independent of interactions at the chiasma or further centrally between fibers from each eye. Possibly, the course of optic fibers through the region of the chiasma is related to their position within the optic nerve, which would not be affected by the enucleations. Those fibers that are the most laterally placed in the optic stalk may have a strong tendancy to continue (laterally) on the same side of the brain. Indeed, if there is some straightforward retinotopy in the embryonic optic nerve, these lateral fibers should belong to temporal retinal regions. A drawing of the optic chiasma of a kitten by Cajal (1909, p.371) illustrates the point clearly that in this species laterally-placed fibers in the optic nerve tend to go ipsilaterally. Similarly in the mouse we have evidence that ipsilaterally-directed fibers appear for the most part to go directly from the lateral edge of the optic nerve to the lateral edge of the optic tract just dorsal to the chiasma (unpublished observations). If such mechanical constraints are responsible for determining the course of optic fibers at the chiasma, then a drastic operation such as eliminating one optic nerve and tract at E12 or E13 should however lead to some distortion of the arrangement of fibers from the remaining eye as they grow through the region of the chiasma: indeed, such a distortion in the shape of the ipsilaterally-projecting crescent is apparent in some cases (Fig. 3). However in other cases in which higher numbers of ganglion cells were counted such a distortion is not apparent.

The presence of crossed fibers from the opposite eye may however be important at early stages for the growth of the uncrossed fibers, as suggested by cases in which prenatal enucleations at E12 or E13 led to abnormally small numbers of ipsilaterally-projecting ganglion cells in the remaining eye relative to normal values, and by the finding that in all cases there were fewer ipsilaterally-projecting ganglion cells in these early enucleates than in the later (E16 or birth) enucleates. Previous investigations (Godement, Salaün, Saillour 1981) showed that the uncrossed central projections were diminished in mice enucleated at E13 and the present study confirms this point on a quantitative level -though more variability is indicated. Although it was argued above that the origin of ipsilateral fibers is determined by mechanisms intrinsic to each retina or optic nerve, it is possible that, especially if it develops earlier, the crossed pathway could provide normally a substrate which helps in the development of the uncrossed pathway. Eliminating the crossed fibers at the earliest stage of optic fiber outgrowth could result in a failure of the normal complement of uncrossed fibers from the other eye to develop, while eliminating it later (i.e. at E16) would have an opposite effect because at this stage a strong (and even excessive) complement of the uncrossed fibers has already bypassed the chiasma.

Acknowledgements. The work described here was done in collaboration with Josselyne Salaün (Institut d'Embryologie du CNRS et du Collège de France). Paulette Saillour provided excellent technical assistance. I thank Michel Imbert for advice and support. This research was supported by grants from the CNRS and the INSERM. Parts of the experiments were conducted at the Laboratoire de Neurophysiologie, Collège de France.

REFERENCES

Adams JC (1981). Heavy metal identification of DAB-based reaction product. J Hystochem Cytochem 29:775.

Bunt SD, Lund RD, Land PW (1983). Prenatal development of the optic projections in albino and hooded rats. Develop Brain Res 6:149.

Cajal SR (1909). Histologie du système nerveux de l'homme et des vertébrés. t.2. Madrid: Instituto Ramon y Cajal 1955.

Cavalcante CA, Rocha-Miranda CE (1978). Postnatal development of retinogeniculate,retinopretectal, and retinotectal projections in the opossum. Brain Res 146:231.

Dräger UC, Olsen JF (1979).Retinofugal projections in pigmented, albino and one-eyed mice. Soc Neurosc Abstr 6:773.

Dräger UC, Olsen JF (1980). Origins of crossed and uncrossed retinal projections in pigmented and albino mice.J Comp Neurol 191:383.

Dreher B, Potts RA, Bennett MR (1983). Evidence that the early postnatal reduction in the number of rat retinal ganglion cells is due to a wave of ganglion cell death. Neurosc Lett 36:255

Frost DO (1981). Orderly anomalous retinal projections to the medial geniculate, ventrobasal and lateral posterior nuclei of the hamster.J Comp Neurol 203:227.

Frost DO, So K-F, Schneider GE (1979). Postnatal development of retinal projections in syrian hamsters: a study using autoradiographic and anterograde degeneration techniques. Neurosc 4:1649

Godement P, Saillour P,Imbert M (1980). The ipsilateral optic pathway to the dorsal lateral geniculate nucleus and superior colliculus in mice with prenatal or postnatal loss of one eye. J Comp Neurol 190:611.

Godement P, Salaün J, Saillour P, Imbert M (1981). Development of the retinofugal pathways in the mouse. Doc Ophtal Proc Ser 30:11.

Godement P, Salaün J, Saillour P (1981). Absence de projections ipsilatérales après destruction très précoce in-utero d'un oeil chez la souris. C R Acad Sci Paris Serie III 293:625.

Godement P, Salaün J, Saillour P,Imbert M (1982). Origine des projections rétiniennes controlatérales et ipsilatérales chez des souris énucléées unilatéralement à la naissance ou in-utero. J Physiol (Paris) 78:4B.

Griffin G, Watkins LR, Mayer DJ (1979). HRP pellets and slow-release gels:two new techniques for greater localization and sensitivity. Brain Res 168:595.

Hanker JS, Yates PE, Metz CB, Rustioni A (1977). A new specific, sensitive and non-carcinogenic reagent for the demonstration of horseradish peroxidase. Histochem J 9:789.

Jeffery G, Perry VH (1982). Evidence for ganglion cell death during development of the ipsilateral projection in the rat. Develop Brain Res 2:176

Land PW, Lund RD (1979). Development of the rat's uncrossed pathway and its relation to plasticity studies. Science 205:698.

Land PW, Hargrove K, Eldridge J, Lund RD (1981). Differential reduction in the number of ipsilaterally projecting ganglion cells during the development of retinofugal projections in albino and pigmented rats. Soc Neurosc Abstr 7:141.

Linden DC, Guillery RW, Cucchiaro J (1981). The dorsal lateral geniculate nucleus of the normal ferret and its postnatal development. J Comp Neurol 203:189.

Martin PR, Sefton A, Dreher B (1983). The retinal location and fate of ganglion cells which project to the ipsilateral superior colliculus in neonatal albino and hooded rats. Neurosc Lett 41:219.

Maxwell PE, Land PW (1981). Development of the retinogeniculate projections in albino and pigmented rats. Anat Rec 199:165A.

Mesulam M-M (1978). Tetramethylbenzidine for horseradish peroxidase neurohistochemistry: a non-carcinogenic blue reaction-product with superior sensitivity for visualizing neural afferents and efferents. J Histochem Cytochem 26:106.

Perry VH, Henderson Z, Linden R (1983). Postnatal changes in retinal ganglion cell and optic axon populations in pigmented rats. J Comp Neurol 219:356.

Potts R, Dreher B, Bennett MR (1982). The loss of ganglion cells in the developing retina of the rat. Develop Brain Res 3:481.

Rakic P (1977). Prenatal development of the visual system in rhesus monkey. Phil Trans R Soc Lond B 278:245.

Rakic P (1981). Development of visual centers in the primate brain depends on binocular competition before birth.Science 214:928.

Sanderson KJ, Dixon PG, Pearson LP (1982). Postnatal development of retinal projections in the brushtailed possum, Trichosorus vulpecula. Develop Brain Res 5:16.

Shatz CJ (1983). The prenatal development of the cat's retinogeniculate pathway. J Neurosc 3:482.

Shatz CJ, DiBerardino AC (1980) Prenatal development of the retinogeniculate pathway in the cat. Soc Neurosc Abstr 6:485.

Silver J, Sidman RL (1980). A mechanism for the guidance and topographic patterning of retinal ganglion cell axons. J Comp Neurol 189:101.

So K-F, Schneider GE, Frost DO (1978). Postnatal development of retinal projections to the lateral geniculate body in syrian hamsters. Brain Res 142:343.

Development of Visual Pathways in Mammals, pages 145–154

DEVELOPMENT OF THE VISUAL SYSTEM IN THE BRUSHTAILED POSSUM

K.J. Sanderson

School of Biological Sciences
The Flinders University of South Australia
Bedford Park, S.A. 5042

The marsupial nervous system is a logical choice for development studies, since the brain is very immature at birth and the postnatal development of the brain goes on over many weeks of pouch life where the young are readily accessible. In addition there is a range of marsupials to choose from with a variety of different lifestyles and considerable variation in the structure of the dorsal lateral geniculate nucleus.

Species in which visual system development is being studied are the American opossum *Didelphis marsupialis aurita* (Cavalcante and Rocha-Miranda 1978 a, b, c; Lent 1981; Lent and Mendez-Otero 1980; Lent and Rocha-Miranda 1978) and from Australia two members of the kangaroo family: the quokka *Setonix brachyurus* (Beazley and Dunlop 1983) and Tammar wallaby *Macropus eugenii* (Wye-Dvorak 1982), a dasyurid marsupial: the Northern native cat *Dasyurus hallucatus* (Robb et al 1982) and the most common member of the possum family: the brushtailed possum *Trichosurus vulpecula* (Warner 1969, 1970; Sanderson et al 1978, 1982). The quokka, Tammar wallaby and brushtailed possum have only one young at a time and the opossum and native cat have at least six young at a time which is an advantage for development studies.

The brushtailed possum is probably the most commonly observed marsupial in Australia, and in New Zealand where it was introduced about 100 years ago; it thrives in the presence of man and finds its way into the roofs of many city houses. It is a cat sized marsupial, with good binocular vision and is at ease both climbing and on the ground. It is mainly a herbivore but will take a wide variety of foods, including meat when offered and it breeds readily in captivity. In Tasmania the brushtailed possum has a peak breeding period in May (autumn or fall in Australia) and breeds very little during the rest of the year. In Adelaide breeding occurs throughout the whole year but is less during winter than the rest of the year. The gestation period is about 17½ days (Lyne 1967) and the single young is about 13 mm long at birth. The young remains in the pouch, tightly attached to a nipple for about 3½ months and is first seen out of the pouch at about 4 months of age. Eye opening occurs at 3-4 months of age (Lyne and Verhagen 1957) also.

At birth the cerebral hemispheres are not present; they gradually form over a period of about 2 months following birth, by which time most of the neurons in the cerebral cortex have formed. Considerable postnatal cell birth occurs also in the thalamus and mesencephalic tectum, as well as in the cerebellum. The postnatal cell birth in these regions can be recognised both from an inspection of the normal histology of the brain in early postnatal young (Warner 1969, 1970; Sanderson et al 1982) and from ^{3}H thymidine autoradiography.

BIRTHDAYS OF NEURONS IN BRUSHTAILED POSSUM

The lateral geniculate nucleus (LGd) can be recognised at 19 days postnatal, but not at 9 days and it achieves an adult pattern of lamination by about 70 days postnatal. Our ^{3}H thymidine autoradiographic studies reveal some labelled neurons in the LGd for an injection made at 9 days postnatal with virtually no labelled cells in the LGd for an injection made at 12 days postnatal. These results suggest that formation of neurons in the LGd is complete soon after the 9th postnatal day.

The superior colliculus (SC) or optic tectum is little more than a thin walled tube at nine days postnatal, with a considerable zone of cell proliferation around the ventricle. By about 19 days postnatal the superior colliculus has well defined cell layers and by about 40 days the pattern of layers is rather like that of the adult (Sanderson et al 1982). The ^{3}H thymidine autoradiography is in accordance with these observations. Many labelled cells can be seen in the superior colliculus when the ^{3}H thymidine is injected at 9 days (Fig. 1), fewer at 12 days postnatal and almost none for an injection at 39 days postnatal.

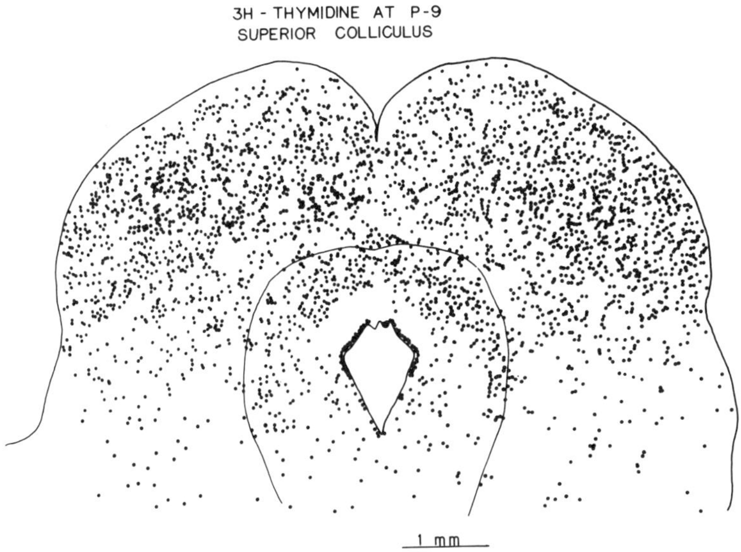

Fig. 1. Distribution of labelled neurons in a frontal section through the superior colliculus of a 4 month old possum which had received a ^3H thymidine injection at postnatal day 9.

Visual cortical neurons are formed after most of the cell formation in the LGd and SC has ceased. The ^3H thymidine injections carried out at 12, 39 and 60 days postnatal revealed cell birth restricted to layer 2 of visual cortex at 60 days postnatal, cell birth in layers 3-4 at 39 days postnatal (Fig. 2) and no cell birth in the visual cortex at 12 days postnatal. Further experimental material will fill in the gaps in this study. However based on these results and inspection of other areas of cerebral cortex it is probable that cell formation in the visual cortex extends from day 15-20 (layer 6) to about day 60-65 postnal (layer 2).

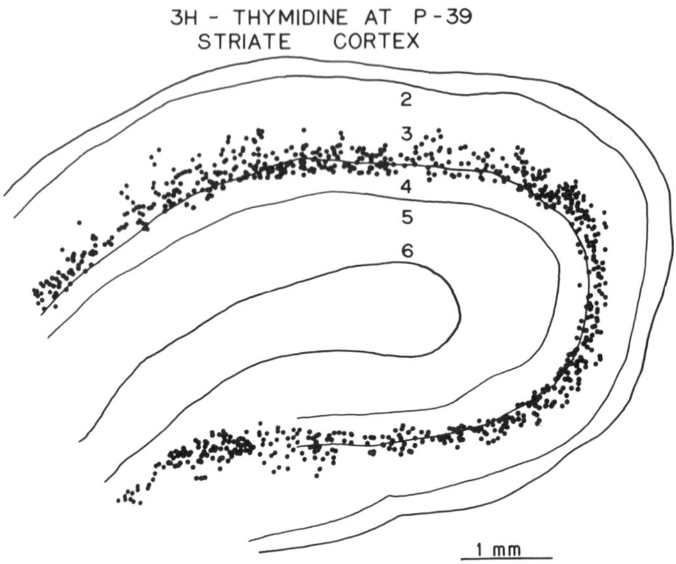

Fig. 2. Labelled neurons in a frontal section through striate cortex of a 5 month old possum which had received a ^{3}H thymidine injection at postnatal day 39.

DEVELOPMENT OF THE RETINAL PROJECTIONS IN BRUSHTAILED POSSUM

Because of technical difficulties we could not determine when the optic nerve axons first reach their targets in the brain. At the earliest stage we could observe (postnatal day 24) the optic nerve axons have reached all of their targets but the pattern of terminations is immature and there is considerable overlap of retinal terminals in areas of the LGd and SC which in the adult receive input only from one eye. Figure 3 (adapted from Sanderson et al 1982) shows the gradual progression in retinogeniculate projections from the juvenile condition (postnatal day 24) to the adult pattern, which is achieved about postnatal day 70, a month or so before eye opening occurs. The adult pattern of retinotectal connections is formed by about postnatal day 50 (Sanderson et al 1982).

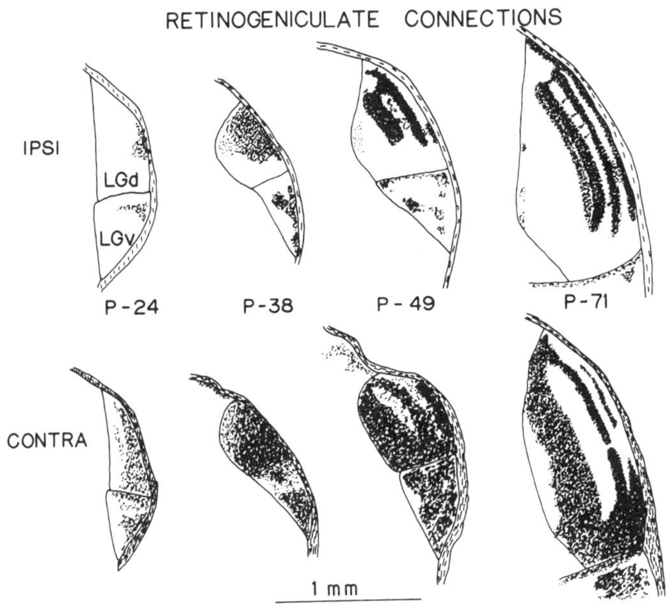

Fig. 3. Frontal sections showing the development of the retinogeniculate connections in the possum from postnatal day 24 to 71 as displayed by intraocular injections of ^{3}H proline.

EFFECT OF EYE REMOVAL IN BRUSHTAILED POSSUM

Because there is a long postnatal period of brain development before eye opening occurs, about 3-4 months, it is possible to study the effects of eye removal carried out at a very early stage in the possum. We found altered retinal projections to LGd, SC and the medial terminal nucleus (MTN) of the accessory optic system with the most striking effects for eye removals carried out before 50 days of age (Sanderson et al 1978). In particular we found that following eye removal in infancy, there was an apparent increase in the projection field of the remaining eye, to take over the territories of the eye which had been removed. Figure 4 (adapted from Sanderson et al 1978) compares the retinal projections to the LGd and SC in a normal adult possum and an adult which had one eye removed at about 25 days postnatal.

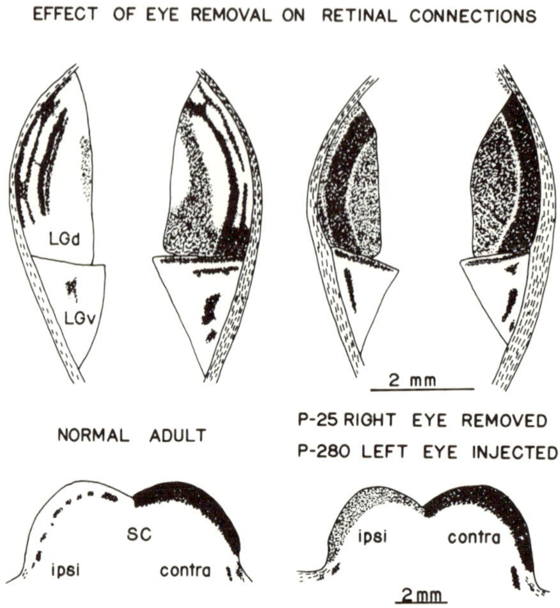

Fig. 4. Frontal sections of the LGd and SC in possums showing the retinogeniculate and retinotectal connections in a normal adult possum (left) and a 10 month old possum which had one eye removed at 25 days postnatal (right). Note that in the normal possum each eye projects to separate layers in the LGd whereas in the one eyed possum the remaining eye projects to all parts of both LGds. Similarly in the one eyed possum there is a considerable retinal projection to the ipsilateral SC in addition to the normal projection which fills the contralateral SC.

Several interpretations for the expanded projections are possible (i) that the remaining eye has expanded its field of projection to take over the regions which would be deafferented by eye removal (ii) that some of the deafferented regions die and there is an increase in size of the remaining regions and (iii) that some of the increased projection results from preservation of the juvenile condition where there is considerable overlap of the projections of the two eyes (Sanderson et al 1982). It seems likely that both (i) and (iii) are contributing factors to the altered projections in one eyed possums since at least some of the expanded projections of the remaining eye seem to represent a genuine expansion which cannot be explained any other way (expanded projections to SC and MTN) and some of the expanded projections are similar to the condition found in very young possums (LGd and possibly SC). Finally it can be noted that early postnatal eye removal produces large alterations in the projections of the remaining eye in the possum and opossum (Lent and Rocha-Miranda 1978) and has less profound effects in mammals such as the cat (Hickey 1975) in which the prenatal period of development is much longer.

COMPARISON WITH OTHER MAMMALS

Our results show that the lateral geniculate neurons have all formed soon after postnatal day 9. The lateral geniculate neurons then migrate to their lateral position in the thalamus where they can be recognised by postnatal day 19, as a uniform collection of cells. Shortly after this, at postnatal day 24 it is possible to observe the retinal fibres entering the lateral part of the LGd and by about 40 days postnatal the retinal fibres fill the contralateral LGd and a good part of the ipsilateral LGd. Between postnatal days 40 and 70 the cellular laminae of the LGd gradually form and the fibres from the two eyes segregate into separate ipsilateral and contralateral terminal bands. All of this maturation process occurs before eye opening and is similar to the situation in other mammals which have been studied: monkeys (Rakic 1977), cats (Kalil 1978; Hickey and Cox 1979; Shatz 1983), ferrets (Linden et al 1981), tree shrews (Brunso-Bechtold and Casagrande 1982) rats (Altman and Bayer 1979; Land and Lund 1979; Bunt et al 1983), hamsters (So et al 1978), squirrels (Cusick and Kaas 1982), opossums (Cavalcante and Rocha-Miranda 1978b) and humans (Hitchcock and Hickey 1980). In cats, ferrets, tree shrews and squirrels it was possible to show that the segregation of ipsilateral and contralateral retinogeniculate fibres begins before the formation of cell laminae in the LGd. This does not appear to be the case in the brushtailed possum, but observations in this species are more difficult because most of the cell laminae are not separated by interlaminar spaces.

Our limited ^{3}H thymidine experiments and histological series of the developing superior colliculus suggest that cell formation in the superior colliculus is at a peak in the first two weeks after birth and has fallen to a very low level by about the 40th postnatal day. It seems likely that retinal fibres invade the superior colliculus while there is still some cell formation and migration occurring since the retinal fibres fill the contralateral superior colliculus at least by the 24th postnatal day. Initially there is considerable overlap of the ipsilateral and contralateral retinotectal fibres but by about the 50th postnatal day the ipsilateral retinal input is confined to the deeper layers of the tectum, in the form of a number of separate patches as occurs in the adult (Sanderson et al 1982). Similar processes of maturation have been observed in other mammals, in which the ipsilateral retinotectal projection is restricted during development and segregation of ipsilateral and contralateral retinotectal fibre occurs: rats (Land and Lund 1979) hamsters (Jen et al 1980), squirrels (Cusick and Kaas 1982) cats (Williams and Chalupa 1982), monkeys (Rakic 1977) and opossums (Cavalcante and Rocha-Miranda 1978b). Where data on cell birth in the superior colliculus are also available (Altman and Bayer 1981) it can be concluded that retinal fibres first reach the tectum towards the end of the period of cell birth in the tectum.

ACKNOWLEDGEMENTS

This work has been made possible by grants from the Flinders University and the Australian Research Grants Schemes and the excellent assistance of Pam Stainer and Lyn Pearson. Johanna Hill typed the manuscript.

REFERENCES

Altman J, Bayer SA (1979) Development of the diencephalon in the rat. IV. Quantitative study of the time of origin of neurons and the internuclear chronological gradients in the thalamus. J Comp Neurol 188: 455.

Altman J, Bayer SA (1981) Time of origin of neurons of the rat superior colliculus in relation to other components of the visual and visuomotor pathways. Exp Brain Res 42: 424.

Beazley LD, Dunlop SA (1983) The evolution of an area centralis and visual streak in the marsupial *Setonix brachyurus*. J Comp Neurol 216: 211.

Brunso-Bechtold JK, Casagrande VA (1982) Early postnatal development of laminar characteristics in the dorsal lateral geniculate nucleus of the tree shrew. J Neurosci 2: 589.

Bunt SM, Lund RD, Land PW (1983) Prenatal development of the optic projection in albino and hooded rats. Dev Brain Res 6: 149.

Cavalcante LA, Rocha-Miranda CE (1978a) Development of retino hypothalamic and accessory optic projections in the opossum. Brain Res 144: 378.

Cavalcante LA, Rocha-Miranda CE (1978b) Postnatal development of retinogeniculate, retinopretectal and retinotectal projections in the opossum. Brain Res 146: 231.

Cavalcante LA, Rocha-Miranda CE (1978c) Development of the dorsal thalamus and the superior colliculus in the opossum, with special reference to optic projection areas. In Rocha-Miranda CE, Lent R (eds): "Opossum Neurobiology", Rio de Janeiro: Academia Brasileira de Ciencias, p 193.

Cusick CG, Kaas JH (1982) Retinal projections in adult and newborn grey squirrels. Dev Brain Res 4: 275.

Hickey TL (1975) Translaminar growth of axons in the kitten dorsal lateral geniculate nucleus following removal of one eye. J Comp Neurol 161: 359.

Hickey TL, Cox NR (1979) Cell birth in the lateral geniculate nucleus of the cat: a ^{3}H-thymidine study. Neurosci Abstr 5: 788.

Hitchcock PF, Hickey TL (1980) Prenatal development of the human lateral geniculate nucleus. J Comp Neurol 194: 395.

Jen LS, Woo HH, So K-F (1980) The uncrossed retino-tectal pathways in golden hamsters - an orthograde HRP study. Neurosci Abstr 6: 207.

Kalil RE (1978) Development of the dorsal lateral geniculate nucleus in the cat. J Comp Neurol 82: 265.

Land PW, Lund RD (1979) Development of the rat's uncrossed retinotectal pathway and its relation to plasticity studies. Science 205: 698.

Lent R (1981) The brain of baby opossums. A model for development and plasticity studies. Trends in Neurosci 4: 84.

Lent R, Mendez-Otero R (1980) Plasticity of the ipsilateral retinotectal projection in early enucleated opossums: changes in retinotopy and magnification factors. Neurosci Lett 18: 37.

Lent R, Rocha-Miranda CE (1978) Aberrant retinofugal projections in the opossum after eye enucleation and tectal lesion. In Rocha-Miranda CE, Lent R (eds): "Opossum Neurobiology", Rio de Janeiro: Academia Brasileira de Ciencias, p 217.

Linden DC, Guillery RW, Cucchiaro J (1981) The dorsal lateral geniculate of the normal ferret and its postnatal development. J Comp Neurol 203: 189.

Lyne AG (1967) "Marsupials and Monotremes of Australia". Sydney: Angus and Robertson, p 45.

Lyne AG, Verhagen AMW (1957) Growth of the marsupial, *Trichosurus vulpecula* and a comparison with some higher mammals. Growth 21: 167.

Rakic P (1977) Prenatal development of the visual system in the rhesus monkey. Phil Trans R Soc Lond B 278: 245.

Robb AM, Crewther DP, Crewther SG, Nelson JE (1982) Visual acuity in the northern native cat, *Dasyurus hallucatus*. Proc Aust Physiol and Pharmacol Soc 13: 197.

Sanderson KJ, Dixon PG, Pearson LP (1982) Postnatal development of retinal projections in the brushtailed possum, *Trichosurus vulpecula*. Dev Brain Res 5: 161.

Sanderson KJ, Pearson LP, Dixon PG (1978) Altered retinal projections in brushtailed possum, *Trichosurus vulpecula*. J Comp Neurol 180: 841.

Shatz CJ (1983) The prenatal development of the cat's retinogeniculate pathway. J Neurosci 3: 482.

So K-F, Schneider GE, Frost DO (1978) Postnatal development of retinal projections to the lateral geniculate body in Syrian hamsters. Brain Res 142: 343.

Warner FJ (1969) The development of the diencephalon in *Trichosurus vulpecula*. Okajimas Fol anat jap 46: 265.

Warner FJ (1970) The development of the pretectal nuclei in *Trichosurus vulpecula*. Okajimas Fol anat jap 47: 73.

Williams RW, Chalupa LM (1982) Prenatal development of retinocollicular projections in the cat: an anterograde tracer transport study. J Neurosci 2: 604.

Wye-Dvorak J (1982) The development of visual pathways in the Tammar Wallaby (*Macropus eugenii*). Proc Aust Physiol and Pharmacol Soc 13: 2.

Development of Visual Pathways in Mammals, pages 155–173

THE POSTNATAL DEVELOPMENT OF THE KITTEN DORSAL LATERAL GENICULATE NUCLEUS

Michael J. Friedlander

Dept. of Physiology and Biophysics
University of Alabama in Birmingham
Birmingham, Alabama, 35294, USA

INTRODUCTION

The dorsal lateral geniculate nucleus (LGN_d) of the kitten undergoes considerable growth and development, postnatally. The nucleus changes its orientation and increases in volume (Kalil, 1978). The individual cells of the LGN_d increase their soma size rapidly over the first 6 postnatal weeks (Garey, et. al., 1973; Cragg et. al., 1976; Kalil, 1978; Hickey, 1980; Kalil, 1980). This time period coincides with the growth and separation of ocular dominance columns in the visual cortex (Levay et. al., 1978). This growth presumably reflects the development of the geniculocortical axon arborizations in visual cortex which has been suggested to relate to somatic growth (Guillery, 1972; Levay et. al., 1980) (although to date, there is no direct evidence on the size of individual geniculocortical axon arborizations in the kitten). Similarly, only preliminary data (Friedlander et. al., 1983a; Friedlander et. al., 1983b) on the growth and development of individual retinogeniculate axon arborizations within the LGN_d has been reported [although autoradiographic transport studies suggest that the monocular retinal innervation to the appropriate laminae is completed by birth in the kitten (Shatz, 1983)].

A large body of evidence (see Hirsch and Leventhal, 1978; Sherman and Spear, 1982 for reviews) suggests that the first 6-8 postnatal weeks of a kitten's life are a "critical period" for the development of the central visual pathways. Modification of the visual environment during this period causes severe deficits to occur in cells of the visual cortex (Wiesel and Hubel, 1963a; Wilson and Sherman, 1977; Shatz and Stryker, 1978) LGN_d (Wiesel and Hubel, 1963b; Sherman et. al., 1972; Kratz et. al., 1978b; Eysel et. al., 1979; Lehmkuhle et. al., 1980; Derrington and Hawken, 1981; Mower et. al., 1981; Zetlan et. al., 1981; Friedlander et. al., 1982; Geisert et. al., 1982; Friedlander and Stanford, 1983) and retinal ganglion cells (Sur et. al., 1982; Leventhal and Hirsch, 1983). It has therefore become a point of major interest to describe the normal developmental processes for the neurones in these pathways. In this chapter, I will consider the physiological and morphological maturation of the retinogeniculate pathway - particularly of the LGN_d cells and the terminal arborizations of the retinal ganglion cell axons which innervate the LGN_d.

ADULT LGN_d FUNCTION

In the adult cat, the retinal ganglion cells and the cells of the LGN_d can be classified into at least 3 major functional classes (W-, X-, and Y- cells -- see Sherman and Spear, 1982 for a recent review). The X- and Y- cells have been studied most extensively. Both of these cell types have concentrically organized receptive field centers and surrounds which give antagonistic responses. These fields may have either "on" or "off" type centers. At any eccentricity within the visual field, X-cells have smaller receptive fields than Y-cells. X-cells also usually respond to a maintained contrast in their receptive field center in a sustained fashion, while Y-cells give more transient responses. X-cells sum visual inputs in a linear fashion while Y-cells are non-linear; X-cells have more slowly conducting axons than Y-cells; X-cells are not as responsive to large high velocity targets than Y-cells and X-cells are more sensitive to finer spatial detail (higher spatial frequencies) than Y-cells. The W-cells have been less completely characterized but generally have the most slowly conducting axons, sluggish responses to visual stimuli, large receptive fields that may show either linear or non-linear spatial summation properties and they have poor contrast sensitivity. In addition, W-cells may respond preferentially to moving visual targets.

The projections of the W-, X- and Y- retinal ganglion

cells to the LGN_d also differ (Hoffmann et.al., 1972; Hoffmann, 1973; Cleland and Levick, 1974; Fukuda and Stone, 1974; Wilson et.al., 1976; Dreher and Sefton, 1979). The X-cells project most heavily to the A-laminae; the Y-cells project to the A-laminae, lamina C and the medial interlaminar nucleus (MIN) of the LGN_d. The W-cells project to the C-laminae and the MIN.

ADULT LGN_d STRUCTURE

Recent work has demonstrated that the individual retinal ganglion cells and the cells of the LGN_d of these 3 different functional classes also have distinct morphological features. The indirect but elegant structure/function correlations of Wassle and his colleagues (Boycott and Wassle, 1974; Wassle et.al., 1975; Wassle and Illing, 1980; Illing and Wassle, 1981; Wassle et.al., 1981) have indicated that retinal ganglion cells with alpha type soma - dendritic structure are functionally Y-cells and most cells with beta-type structure are probably X-cells. The morphology of retinal W-cells is not clear, as yet (but see Stanford and Sherman [1983]). The direct study of Stanford and Sherman (1983) has generally supported these findings. In addition, the terminal axonal arborizations of these retinal ganglion cells within the LGN_d have a characteristic structure. Both Bowling and Michael (1980) and Sur and Sherman (1982) have found that the Y-axons innervate a wider area (within the LGN_d) than X-axons. There is no data on the structure of the terminal arborizations of W-cells within the LGN_d.

The cells of the LGN_d also have characteristic morphological features for each of the 3 functional groups (Friedlander et.al., 1979; Friedlander et.al., 1981; Stanford et.al., 1981; Stanford et.al., 1983). LGN_d X-cells have small somata, thin dendrites, and complex dendritic appendages. The dendritic tree of X-cells is vertically oriented along the projection lines in the LGN_d and restricted to the lamina of origin. Y-cells have large somata, thick straight dendrites that are arranged in a spherical fashion. The Y-cell dendrites cross laminar boundaries. The W-cells have small somata (as do X-cells) but they only occur in the C-laminae. Their morphological signature is a horizontally oriented dendritic tree that runs orthogonal to the projection lines of the LGN_d.

The axons of the 3 classes of LGN_d relay cells also have different structural properties. The axons of W-, X-, and Y- cells have collateral branches (Friedlander et.al., 1981; Stanford et.al., 1983) that innervate the perigeniculate nucleus (PGN) with the Y-cells having the heaviest projection. In addition, over 25% of the X-cells and less than 5% of the Y-cells have intrageniculate axon collaterals in the adult cat LGN_d (Friedlander et.al., 1981). The Y-geniculocortical axons primarily innervate cortical lamina IVa, IVb and VI (with a heavier innervation of IVa than IVb - Humphrey and Sherman, 1982) but see Gilbert and Wiesel, (1979). The X- geniculocortical axons innervate these same laminae but have a heavier projection to IVb than IVa. To date, there have been no studies of the projection of individual W- geniculocortical axons.

NEONATAL LGN_d FUNCTION

The receptive field properties of many retinal ganglion cells and LGN_d cells in the kitten are immature for several weeks after birth. The vascular network around the kitten lens causes poor optics for 3 weeks after birth (2 weeks after eye opening - Thorn et.al., 1976; Bonds and Freeman, 1978; Freeman and Lai, 1978). However, after 3 postnatal weeks of age, this is not an important factor. Most retinal ganglion cells and LGN_d cells (particularly outside the area centralis) have larger receptive field centers in the 3-5 week old kitten than in the adult. This has been suggested to be due to immature properties of the inhibitory surround mechanisms of the receptive fields (Hamasaki and Flynn, 1978; Rusoff and Dubin, 1977; Hamasaki and Sutija, 1979). Poor optics are probably not the cause of these large fields as occassionally a small receptive field is mapped for neighboring cells in the same kitten. When the size of the receptive fields for retinal ganglion cells are normalized for the growth of the eye, the actual area of retina (in um^2) that influences the cells doesn't change (Tucker, 1978; Rusoff and Dubin, 1977; Olson and Freeman, 1980) - [due to a decreased area of visual field subtended by each um of retinal extent with growth of the eye]. The smaller receptive field centers that develop with age are probably therefore due to increased development of the inhibitory surround mechanisms or less sensitive centers. Immature inhibitory surrounds have been reported in retinal ganglion cells (Rusoff and Dubin, 1977) and LGN_d cells (Daniels et. al., 1978; Friedlander, 1982). Additionally, the retinal circuitry that is responsible for receptive field surround mechanisms is thought to play a major role in the non-linear spatial summation properties of Y-retinal ganglion cells (Hochstein and Shapley, 1976). This property is then relayed to LGN_d Y-cells. Since the receptive field surrounds are late to mature, this characteristic non-linear feature of Y-cells may not be apparent in the neonate.

NEONATAL LGNd STRUCTURE/FUNCTION

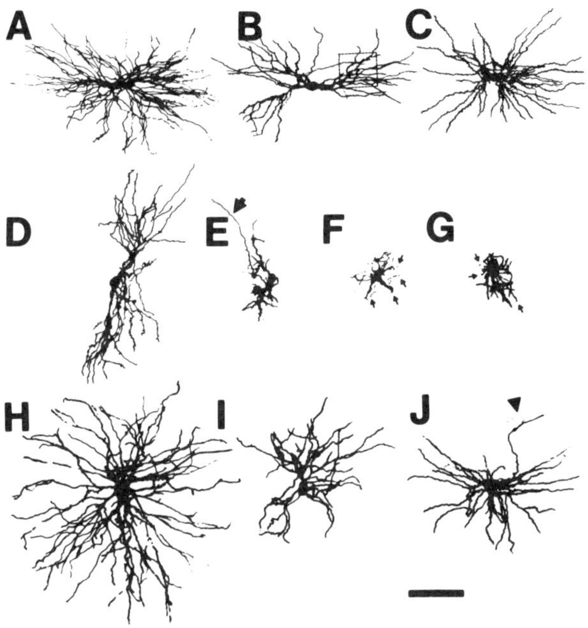

Fig. 1. Drawings of W (A-C), X (D-G) and Y (H-J) LGN_d cells that have been physiologically characterized and intracellularly injected with HRP. First column (A, D and H) are cells from adults. All other cells are from 3-4 week old kittens. Scale = 100 um and is drawn parallel to the LGN_d lamination. Arrows point to axons in E and J and to terminal swellings on immature dendrites in F and G. (Figure reprinted from Friedlander, MJ [1982] Nature 300:180).

Our laboratory has been concerned with the morphological development of the functional types of LGN_d cells and the axonal arborizations of the retinal ganglion cells which innervate them. We have been using the direct method of intracellular injection of the enyzme marker substance, horseradish peroxidase (HRP) into individual, physiologically characterized LGN_d cells and retinal ganglion cell axons to compare the postnatal maturation of their physiology and morphology. We have also used this same technique in the LGN_d in kittens reared with monocular visual deprivation in order to determine if this causes arrest of neural development or degenerative changes.

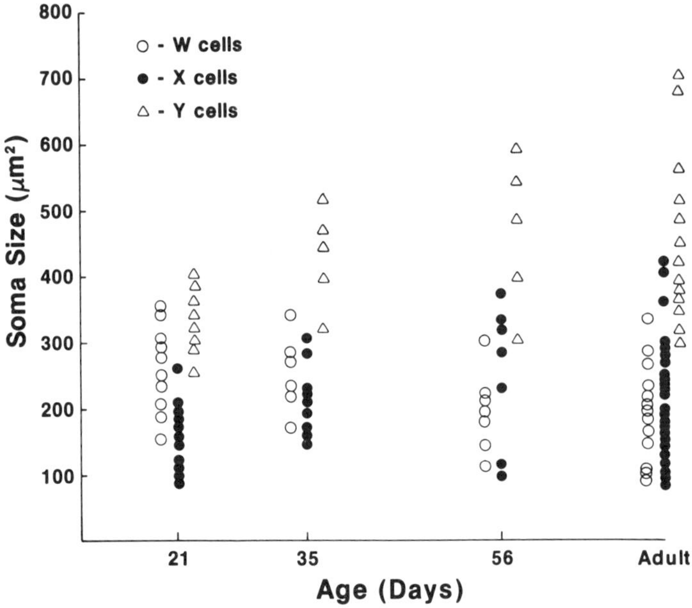

Fig. 2. Distribution of soma sizes (ordinate) measured by planimetry vs. age (abcissa) for physiologically characterized LGN_d cells that were injected intracellularly with HRP.

Figure 1 illustrates examples of W-, X-, and Y-cells from adult cats and 3-4 week old kittens. All of these cells were physiologically characterized and injected intracellularly with HRP. The W-cells were very responsive to visual stimuli, particularly to moving targets. They showed little fatigue to repeated stimulus presentation at 2 Hz. Note that their morphological properties are similar to adult W-cells. The X-like cells may be regularly responsive and have relatively small receptive fields (as did the cell illustrated in figure 1E) or have large receptive fields and sluggish responses that fatigue readily (as did the cells illustrated in figures 1F and 1G). Note that the physiologically early maturing X-like cells (1E) have a soma dendritic structure similar to but smaller than adult X-cells. The physiologically immature

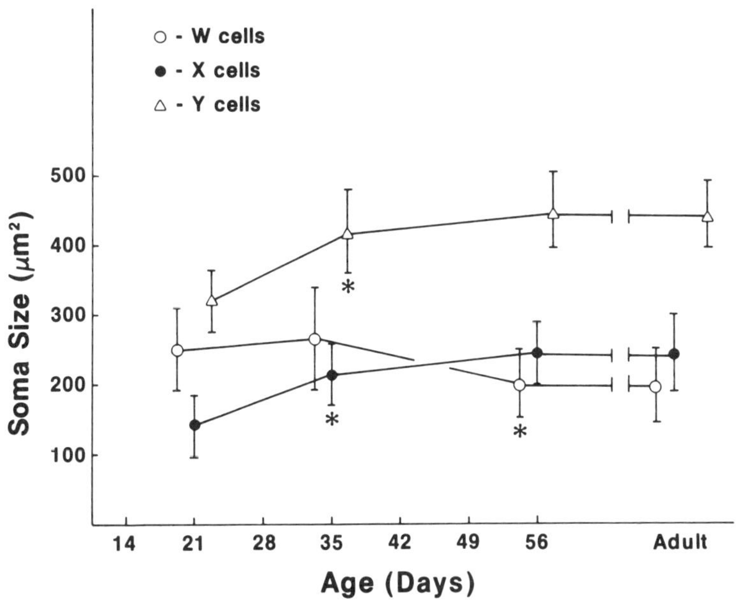

Fig. 3. Mean soma sizes (ordinate) ±S.D. vs. age (abcissa) for the functionally identified LGN_d cells whose sizes are shown in figure 2. The asterisks mark ages where the distribution is significantly different ($p<.05$) than the next earlier age studied.

X-cells are also structurally immature (very restricted

dendritic arborizations with terminal appendages). However, some Y-cells already have adult like physiological properties at this age (non-linear spatial summation, strong receptive field surround responses, brisk responses to high velocity visual stimuli). Examples of these are illustrated in figure 1I and 1J. These cells have similar dendritic structure and shape to adult Y-cells but have a smaller dendritic extent.

The development of the somata at different postnatal ages of these physiologically identified cells are illustrated in figures 2 and 3. Note that both X and Y-cells show a significant growth from 3 to 5 postnatal weeks. It is interesting to note that the W-cells actually show a significant decline in size from 5 to 8 postnatal weeks. The W-cell somatic growth may occur prior to the earliest experimental age (3 weeks).

The axonal arborizations in the LGN_d of retinal ganglion cells also show a significant postnatal growth (Friedlander et.al., 1983a; Friedlander et.al., 1983b). The early maturing Y-axons (which already have very brisk responses and strong non-linear spatial summation within their receptive fields) are readily classifiable at 4-5 weeks of age. An example of one such axon that has been intra-axonally injected with HRP is shown in figure 4. The pattern of branching and innervation within the LGN_d is very similar to that reported for Y-axons in the adult cat (Bowling and Michael, 1980; Sur and Sherman, 1982). However, when the individual terminal boutons are viewed under high power microscopy, they are often covered by filamentous processes and filopodia (indicative of active growth). These types of processes have not been reported on adult retinal axon terminal boutons. In addition, the number of boutons per Y-axon (within the A-lamina) of the LGN_d is significantly less in the kitten than in the adult as illustrated in figure 5. There is a marked postnatal growth (in the number of boutons) of the retinal axons within the LGN_d. In addition, the width of the terminal fields of individual Y-retinal axon arborizations within the LGN_d A-laminae increases between 5 weeks of age and adulthood but this effect is largely compensated for by increasing LGN_d volume (Kalil, 1978; Friedlander et.al., 1983a; Friedlander et.al., 1983b). It has been suggested

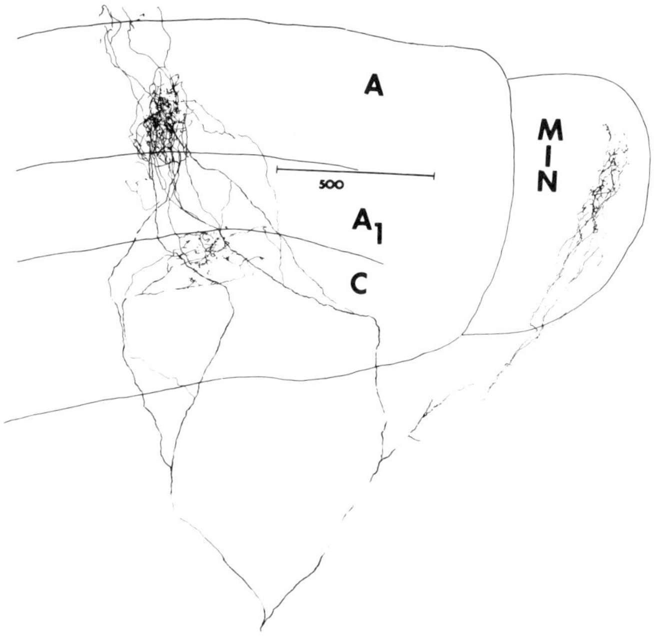

Fig. 4. Drawing of a Y-retinal axon's terminal arborization within the LGN_d of a 4 1/2 week old kitten. Scale bar = 500 um. The axon was injected with HRP in the optic tract below the C complex. The receptive field properties included phase independent non-linear spatial summation within the receptive field, brisk responsiveness to a high velocity (>200°/sec) light target (off center receptive field), an inhibitory on surround and a latency to stimulation of the optic chiasm of 0.9 msec. The axon was activated through stimulation of the contralateral eye. The branching pattern is similar to that reported for adult retinal Y-axons. However, fewer terminal boutons occur than in the adult. Also, some spillover of boutons (<2%) occurs into lamina A_1.

that prenatal and neonatal axons in the CNS show an exuberance of connections that require specific "pruning" during development (Rakic, 1977; Shatz, 1983). While such processes may occur early in the retinogeniculate pathway,

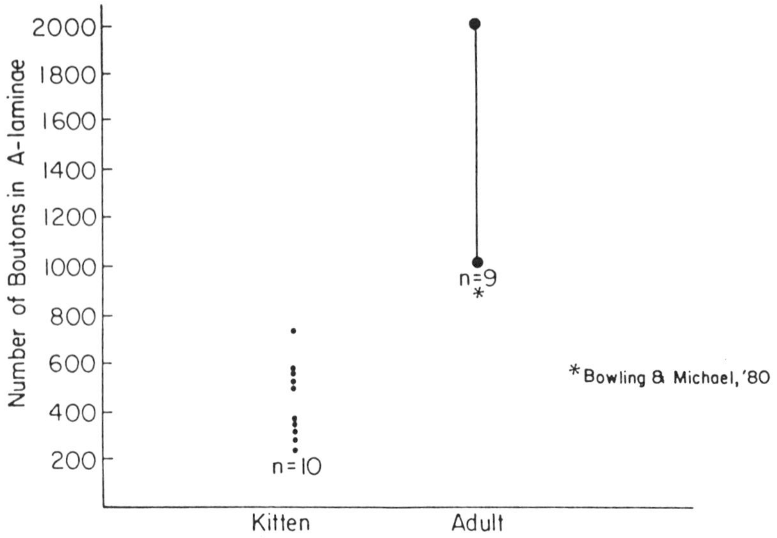

Fig. 5. Numbers of terminal boutons (ordinate) for 5 week old kitten vs. adult (abcissa) retinal Y-ganglion cell axon arborizations within the A-lamina of the LGN_d are compared. The adult values are taken from the study of Bowling and Michael, (1980). The kitten values are from the studies of Friedlander et.al., (1983a; 1983b). These populations are significantly different ($p<.001$ on a Mann-Whitney U-test). There is therefore, a significant postnatal increase in the number of terminal boutons on individual Y-retinal axons.

an increase in synaptic connections must also occur. This is not necessarily contradictory to the pruning hypothesis. The increase in terminal boutons on retinal Y-axon terminals in the LGN_d may reflect later formation of appropriate connections. This could be coupled with a concomitant reduction in the earlier formed inappropriate connections. (An important question for future investigation is how many synapses are made per retinal terminal bouton in the neonate vs. the adult).

MONOCULAR DEPRIVATION EFFECTS ON LGN_d STRUCTURE/FUNCTION

During conditions of rearing kittens in an abnormal visual environment (monocular visual deprivation-MD) striking effects are seen on the cells of the LGN_d (Sherman et.al., 1972; Friedlander et.al., 1982; Friedlander and Stanford, 1983) and their innervating retinal ganglion cell axonal arborizations (Sur et. al., 1982). The X-retinal ganglion cells' terminal arborizations occupy a greater volume of the LGN_d A-laminae innervated by the deprived eye than in normal cats. Concomitantly the Y-retinal ganglion cells' terminal arborizations have a very limited projection to the A-laminae. The LGN_d cells also show marked changes in their structure/function relationships after monocular deprivation (Friedlander et.al., 1982 and figure 6 of this chapter). While many X-cells in A-laminae innervated by the deprived eye in MD animals have normal structure and function (figure 6-middle panel-first row), some X-cells have the morphology of Y-cells (figure 6-end panel-first row). This may result from a functional reinnervation by X-retinal ganglion cells or a masking of Y-input to an LGN_d cell that was genetically Y. These X-cells may be compared to the X-cell from the normal animals LGN_d illustrated in figure 6-first panel- first row. (It is of interest to note that Friedlander et. al. (1981) observed that cells in normal adult cat LGN_d with some X- and some Y- properties had this same type of morphology). It has been suggested by Friedlander et. al. (1982) that a competition between X- and Y- cell optic tract terminals for postsynaptic targets on LGN_d cells may occur. These cells with X- physiological properties but the structure of Y-cells may represent such cells where visual deprivation has resulted in a maintenance of the inappropriate retinal input that would be functionally eliminated under conditions of normal postnatal development.

The effects of monocular deprivation on LGN_d Y-cells are summarized in the bottom row of drawings in figure 6. The Y-cell in the first panel-bottom row is from a normal animal's LGN_d. As illustrated by the example in figure 6-bottom row-second panel, some Y-cells in LGN_d laminae innervated by the deprived eye have normal dendritic structure but abnormally small somata. These Y-cells display normal Y-receptive field properties, as well.

They, therefore represent a group of Y-cells that survive the effects of visual deprivation for the most part.

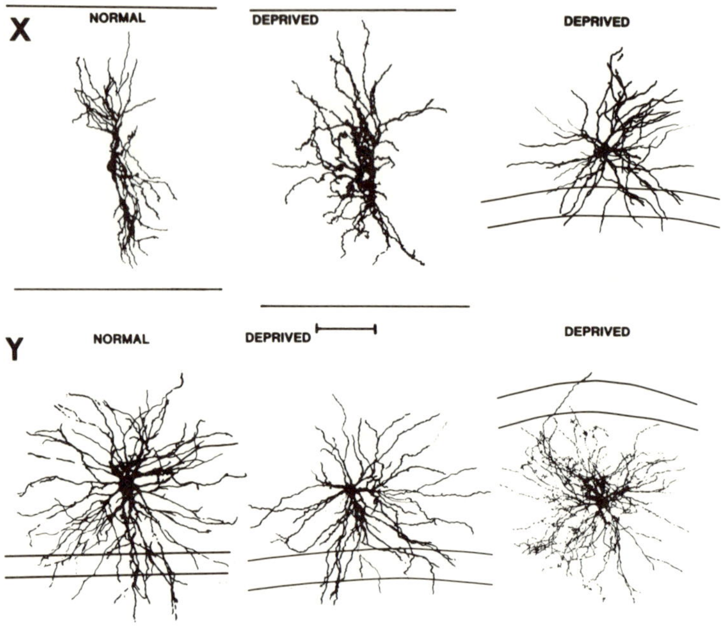

Fig. 6. Drawings of LGN_d cells that have been physiologically characterized and injected intracellularly with HRP in normal cats and from A-laminae innervated by the deprived eye of cats reared with monocular eyelid suture. Scale bar = 100 um. Solid lines represent laminar borders of the LGN_d. (Spaces between pairs of lines represent interlaminar zones). Normal X- and Y- cell are shown in first panel in each row. X-cells in laminae innervated by deprived eye in MD cats were either normal (middle panel- first row) or had Y-type morphology (end panel first row). Y-cells in laminae innervated by the deprived eye in MD cats were either normal except for small somata (middle panel- bottom row) or physiologically and morphologically abnormal (end panel-bottom row).

Alternatively, the cell illustrated in the end panel-lower row of figure 6 had abnormal physiological properties (irregular and weak responsiveness to visual stimulation within its receptive field and to electrical stimulation of its optic tract input). When it did respond, its properties were like those of a Y-cell. However, the dendritic structure is quite abnormal (particularly for a Y-cell - note the fine dendrites covered in swellings). Therefore at least 2 mechanisms may explain the reported loss in recorded Y-cells (Friedlander et.al., 1982) in the LGN_d of monocularly deprived cats - 1) cells that should have received Y-optic tract input are functionally reinnervated by X-optic tract axons and 2) some Y-cells develop abnormal responses (due to pathological changes in the postsynaptic surface such as dendritic varicosities and changes in diameter which may result in poorer retinogeniculate coupling).

CONCLUSIONS

The neurones of the adult cat LGN_d have characteristic morphological features for each functional group (Friedlander et.al., 1981; Stanford et.al., 1983). This difference is already present in the 3 week old kitten (Friedlander, 1982). This implies that there is further separate processing that occurs central to the retinal W-, X-, and Y- circuitry. The postnatal maturation sequence of these groups differ. The early developing Y-cell group (both structurally and functionally) may represent a population of Y-cells that are the least susceptible to postnatal environmental modification. I suggest that these cells may be coincident with those Y-cells found to have no physiological and minimal structural change during monocular deprivation. The postnatal expansion of the Y optic axon terminal arborizations may reflect an increased convergence of retinal terminals on these Y-cells or an additional contacting and recruitment of other as yet immature LGN_d cells into the Y-population. The very immature cells in the kitten LGN_d (both physiologically and morphologically) behave in an X-like fashion. However, as already discussed in this chapter, the late maturation of the receptive field surrounds may mask Y-like (non-linear spatial summation) features. Therefore, some of these cells could be late developing Y-cells and may be highly susceptible to postnatal environmental modification (reflected by the population of cells in the monocularly deprived cat's LGN_d that develop abnormal physiology and dendritic structure). The cells of the monocularly deprived cat's LGN_d that develop intrinsic Y-structural properties but receive a functional X- retinal innervation may reflect an X- vs. Y- competition at the level of the retinal axon terminal boutons. The Y-terminal arborizations have a significant postnatal growth (Friedlander et.al., 1983a; Friedlander et.al., 1983b). In another study (Sur et. al., 1982) it was demonstrated that monocular deprivation caused a reduction (or failure to grow) in these Y-terminal arborizations. Appropriate signalling over the retinogeniculate pathway may be necessary during development in order for these LGN_d cells to attract and/or "harden" these Y-inputs.

The W-cells of the C-laminae are very early to mature. Their soma sizes are adult-like at 3-4 postnatal weeks and then show a reduction. These cells have a low contrast sensitivity in the adult (Sur and Sherman, 1982a). In the neonate (Levay et.al., 1978) it has been reported that while the development of ocular dominace columns in cortical layer IV is incomplete in the young kitten, layer I (where C-laminae W-cells project) columns are more distinct. The W-cell pathway therefore, is in a position to be an important factor in the neonates visual system. We know very little about the effects of visual deprivation on these cells. However, their precociousness may allow this pathway to avoid the deleterious effects of visual deprivation.

Supported by USPH Grant EY03805 and NATO Research Grant 0497/82. Technical assistance of Ms. Loretta Hawkins and the typing skills of Ms. Carol Hutcherson are gratefully acknowledged.

Bonds AB, Freeman RD (1978). Development of optical quality in the kitten eye. Vis Res 18:391.
Bowling DB, Michael CR (1980). Projection patterns of single physiologically characterized optic tract fibers in the cat. Nature (Lond) 286:899.
Boycott, BB, Wassle H (1974). The morphological types of ganglion cells of the domestic cat's retina. J Physiol (Lond) 240:397.
Cleland BG, Levick WR (1974). Properties of rarely encountered types of ganglion cells in the cat's retina and an overall classification. J Physiol (Lond) 240:457.
Cragg B, Anker R, Wan YK (1976). The effect of age on the reversibility of cellular atrophy in the LGN of the cat

following monocular deprivation: A test of two hypotheses about cell growth. J Comp Neurol 168:345.

Daniels JD, Pettigrew JD, Norman JL (1978). Development of single-neuron responses in kitten's lateral geniculate nucleus. J Neurophysiol 41:1373.

Derrington M, Hawken MJ (1981) Spatial and temporal properties of cat geniculate neurons after prolonged deprivation. J Physiol (Lond) 314:107.

Dreher B, Sefton AJ (1979). Properties of neurones in cat's dorsal lateral geniculate nucleus: a comparison between medial interlaminar and laminated parts of the nucleus. J Comp Neurol 183:47.

Eysel VT, Grusser OJ, Hoffmann K-P (1979). Monocular deprivation and signal transmission by X- and Y- neurons of the cat lateral geniculate nucleus. Exp Brain Res 34:521.

Freeman RD, Lai CE (1978). Development of the optical surfaces of the kitten's eye. Vis Res 18:409.

Friedlander MJ (1982). Structure of physiologically classified neurones in the kitten dorsal lateral geniculate nucleus. Nature 300:180.

Friedlander MJ, Lin C-S, Sherman SM (1979). Structure of physiologically identified X- and Y- cells in the cat's lateral geniculate nucleus. Science 204:1114.

Friedlander MJ, Lin C-S, Stanford LR, Sherman SM (1981). Morphology of functionally identified neurons in the lateral geniculate nucleus of the cat. J Neurophysiol 46:80.

Friedlander MJ, Martin KAC, Vahle-Hinz C (1983a). The postnatal development of structure of physiologically identified retinal ganglion cell (rgc) axons in the kitten. J Physiol 336:28P.

Friedlander MJ, Stanford LR (1983). The effect of monocular deprivation on the distribution of cell types in the cat's dorsal lateral geniculate nucleus - a sampling study with fine-tipped micropipettes. Exp Brain Res (in press).

Friedlander MJ, Stanford LR, Sherman SM (1982). Effects of monocular deprivation on the structure/function relationship of individual neurons in the cat's lateral geniculate nucleus. J Neurosci 2:321.

Friedlander MJ, Vahle-Hinz C, Martin KAC (1983b). The structure of the terminal arborizations of physiologically classified optic tract axons in the 3 to 5 week old kitten. Invest Ophthalmol Vis Sci, (Suppl) 24:138.

Fukuda Y, Stone J (1974). Retinal distribution and central projections of Y-, X- and W- cells of the cat's retina. J Neurophysiol 37:749.

Garey LJ, Fisken RA, Powell TPS (1973). Observations on the growth of cells in the lateral geniculate nucleus of the cat. Brain Res 52:359.

Geisert EE, Spear PD, Zetlan S, Langsetmo A (1982). Recovery of Y-cells in the lateral geniculate nucleus of monocularly deprived cats. J Neurosci 2:577.

Gilbert CD, Wiesel TN (1979). Morphology and intracortical projections of functionally characterized neurones in the cat visual cortex. Nature (Lond) 280:120.

Guillery RW (1972). Binocular competition in the control of geniculate cell growth. J Comp Neurol 144:117.

Hamasaki DI, Flynn JT (1977). Physiological properties of retinal ganglion cells of 3 week old kittens. Vis Res 17:275.

Hamasaki DI, Sutija VG (1979). Development of X- and Y-cells in kittens. Exp Brain Res 35:9.

Hickey TL (1980). Development of the dorsal lateral geniculate nucleus in normal and visually deprived cats. J Comp Neurol 189:467.

Hirsch HVB, Leventhal AG (1978). Functional modification of the developing visual system. In Jacobson M (ed): "Handbook of Sensory Physiology. Development of Sensory Systems", Vol 9, Berlin: Springer-Verlag, p 279.

Hochstein S, Shapley RM (1976). Linear and non-linear subunits in Y cat retinal ganglion cells. J Physiol (Lond) 262:265.

Hoffmann K-P (1973). Conduction velocity in pathways from retina to superior colliculus in the cat: a correlation with receptive field properties. J. Neurophysiol 36:409.

Hoffmann K-P, Stone J, Sherman SM (1972). Relay of receptive-field properties in dorsal lateral geniculate nucleus of the cat. J Neurophysiol 35:518.

Humphrey AL, Sur M, Sherman SM (1982). Cortical axon terminal arborizations and soma location of single functionally identified lateral geniculate nucleus neurons. Soc Neuro Abstr 8:2.

Illing R-B, Wassle H (1981). The retinal projections to the thalamus in the cat: A quantitative investigation and a comparison with the retinotectal pathway. J Comp Neurol 202:265.

Kalil R (1978). Development of the dorsal lateral geniculate nucleus in the cat. J Comp Neurol 182:265.

Kalil R (1980). A quantitative study of the effects of

monocular enucleation and deprivation on cell growth in the dorsal lateral geniculate nucleus of the cat. J Comp Neurol 189:483.
Kratz KE, Webb SV, Sherman SM (1978a). Study of the cat's medial interlaminar nucleus: a subdivision of the dorsal lateral geniculate nucleus. J Comp Neurol 180:601.
Kratz KE, Webb SV, Sherman SM (1978b). Effects of early monocular eyelid suture upon neurons in the cat's medial interlaminar nucleus. J Comp Neurol 181:615.
Lehmkuhle S, Kratz KE, Mangel SC, Sherman SM (1980). Effects of early monocular lid suture on spatial and temporal sensitivity of neurons in dorsal lateral geniculate nucleus of the cat. J Neurophysiol 43:542.
Levay S, Stryker MP, Shatz CJ (1978). Ocular dominance columns and their development in layer IV of the cat's visual cortex: a quantitative study. J Comp Neurol 179:223.
Levay S, Wiesel TN, Hubel DH (1980). The development of ocular dominance columns in normal and visually deprived monkeys. J Comp Neurol 191:1.
Leventhal AG, Hirsch HVB (1983). Effects of visual deprivation upon the morphology of retinal ganglion cells projecting to the dorsal lateral geniculate nucleus of the cat. J Neurosci 3:332.
Mower GD, Burchfiel JL, Duffy FH (1981). The effects of dark-rearing on development and plasticity of the lateral geniculate nucleus. Dev Brain Res 1:418.
Olson CR, Freeman RD (1980). Rescaling of the retinal map of visual space during growth of the kitten's eye. Brain Res 186:55.
Rakic P (1977). Prenatal development of the visual system in rhesus monkey. Philos Trans R Soc Lond, Biol 278::245.
Rusoff AC, Dubin, MW (1977). Development of receptive-field properties of retinal ganglion cells in kittens. J Neurophysiol 40:1188.
Shatz CJ (1983). The prenatal development of the cat's retinogeniculate pathway. J Neurosci 3:482.
Shatz CJ, Stryker MP (1978). Ocular dominance in layer IV of the cat's visual cortex and the effects of monocular deprivation. J Physiol, Lond 281:267.
Sherman SM, Hoffmann K-P, Stone J (1972). Loss of a specific cell type from the dorsal lateral geniculate nucleus in visually deprived cats. J Neurophysiol 35:532.
Sherman SM, Spear PD (1982). Organization of visual

pathways in normal and visually deprived cats. Physiol Rev 62:738.

Stanford LR, Friedlander MJ, Sherman SM (1981). Morphology of physiologically identified W-cells in the C-laminae of the cat's lateral geniculate nucleus. J Neurosci 1:578.

Stanford LR, Friedlander MJ, Sherman SM (1983). Morphological and physiological properties of geniculate W-cells of the cat: A comparison with X- and Y- cells. J Neurophysiol 50:582.

Stanford LR, Sherman SM (1983). Morphology of physiologically identified ganglion cells in the cat retina. Soc Neuro Abstr 9:894.

Sur M, Sherman SM (1982). The structure of the terminal arborizations of physiologically identified X- and Y- optic tract axons in the LGN of the cat. Science 218:389.

Sur M, Humphrey AL, Sherman SM (1982). Effects of monocular deprivation on the terminal arborization of functionally identified retinal ganglion cell axons. Nature 300:183.

Thorn F, Gollender M, Erickson P (1976). The development of the kitten's visual optics. Vis Res 16:1145.

Tucker GS (1978). Light microscopic analysis of the kitten retina: postnatal development in the area centralis. J Comp Neurol 180:489.

Wassle H, Levick WR, Cleland BG (1975). The distribution of the alpha type of ganglion cells in the cat's retina. J Comp Neurol 159:419.

Wassle H, Illing R-B (1980). The retinal projection to the superior colliculus in the cat: a quantitative study with HRP. J Comp Neurol 190:333.

Wassle H, Boycott BB, Illing R-B (1981). Morphology and mosaic of on- and off- beta cells in the cat retina and some functional considerations. Proc R Soc Lond, Biol 212:177.

Wiesel TN, Hubel DH (1963a). Single cell responses in striate cortex of kittens deprived of vision in one eye. J Neurophysiol 26:1003.

Wiesel TN, Hubel DH (1963b). Effects of visual deprivation on morphology and physiology of cells in the cat's lateral geniculate body. J Neurophysiol 26:978.

Wilson PD, Rowe MH, Stone J (1976). Properties of relay cells in the cat's lateral geniculate nucleus. A comparison of W-cells with X- and Y-cells. J Neurophysiol 39:1193.

Wilson JR, Sherman SM (1977). Differential effects of

early monocular eyelid deprivation on binocular and monocular segments of cat striate cortex. J Neurophysiol 40:892.

Zetlan, SR, Spear PD, Geisert EE (1981). The role of cortico-geniculate projections in the loss of Y-cells in monocularly deprived cats. Vis Res 21:1035.

Development of Visual Pathways in Mammals, pages 175–178

EARLY POSTNATAL DEVELOPMENT OF THE MONKEY LATERAL GENICULATE NUCLEUS

Pedro Pasik, M.D. and Tauba Pasik, M.D.

Department of Neurology
Mount Sinai School of Medicine, C.U.N.Y.
New York, N.Y. 10029, U.S.A.

The parvocellular laminae of the monkey dorsal lateral geniculate nucleus contain an almost pure population of neurons characterized physiologically as X-cells (Dreher, Fukada & Rodieck, 1976; Schiller & Malpeli, 1978). Although the magnocellular laminae are of mixed composition (Kaplan & Shapley, 1982; Marrocco, McClurkin & Young, 1982), Y-cells are found only in this division. This relative segregation allows the study of differential developmental features of the two subsystems. We have examined this issue in specimens from rhesus monkeys from birth to 16 weeks of age, at the light and electron microscopic levels, under strictly controlled conditions for changes in the tissue during processing of the material. Extensive use of computer-assisted stereologic calculations (Coupland, 1968; Anker & Cragg, 1974) yielded realistic estimates of total numbers of the elements under study.

Results show that the total volume of the geniculate cell laminae increases about 17% between 2 and 4 weeks, and stabilizes thereafter. The changes in the parvocellular division (Pv) parallel those of the total laminar volume, whereas those of the magnocellular component (Mg) exhibit little change during the sampled period. The mean cross-sectional area of cell somata reaches a peak by the 4th week in Pv, whereas the main increase in Mg occurs after the 8th week and this value fails to reach an asymptote at 16 weeks. The neurons in Pv develop at similar rates regardless of their reported physiologic differences, namely cells receiving crossed vs uncrossed retinal afferents, cells with on-center vs off-center receptive fields, and cells from zones representing central vs peripheral retina.

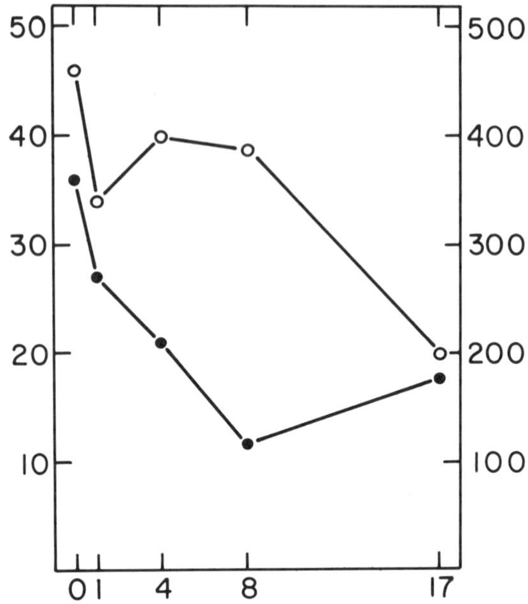

Figure 1. Data of magnocellular laminae represented by open circles and the ordinate on the left side. Data of parvocellular laminae indicated by solid circles and the ordinate on the right side. Ordinates values x10^7. Abscissa in postnatal weeks.

At the ultrastructural level, the weighted mean diameter of retinal boutons, obtained from stereologically reconstructed frequency distributions, tends to increase during maturation in both Pv and Mg, and is consistently larger in Mg. The estimated total number of boutons, derived from the stereologic density and the total laminar volume, show an initial proliferation, which is more marked in Mg, followed by a decrease in both laminar types, reaching approximately 50% of the values at birth by the 16th week. The most dramatic change occurs in the estimated number of synapses made by retinal terminals, and in the cummulative area of synaptic contacts. The number of synapses, obtained with similar computations as for the retinal boutons, decreases markedly during the first 8 weeks in Pv, and after this period in Mg (Fig. 1). This pattern is primarily due to the elimination of retinal to principal cell synapses. Retinal to interneuron synapses remain stable, and are consistently more numerous in Mg than in Pv, which is in accord with the higher proportion of interneurons found in this geniculate division (Hámori, Pasik & Pasik, 1983). The percentage

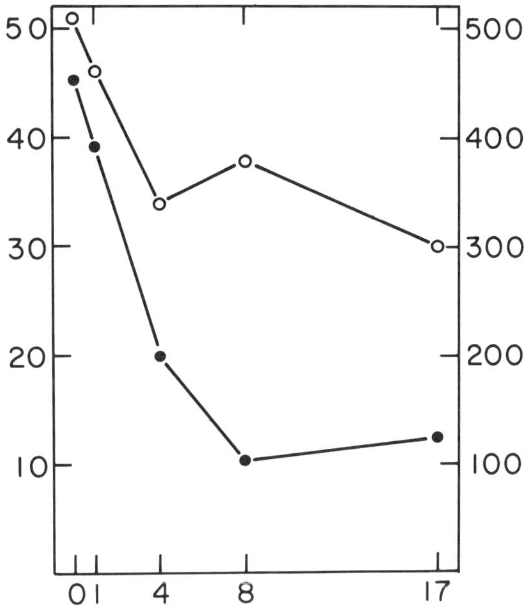

Figure 2. Cummulative Area of Synaptic Contacts. Notations as in Fig. 1, except that the ordinate values are mm^2.

of decrease is similar to that observed in the number of retinal boutons and, therefore, could be attributed to bouton elimination. An additional reduction in the cummulative area of synaptic contacts is present in Pv (Fig. 2) where the value at 16 weeks is only 25% of that at birth.

The above findings indicate the occurrence of substantial changes in the monkey dorsal lateral geniculate nucleus during the first four postnatal months. They are characterized by growth of the cell laminae, both in their cellular and neuropil components, a considerable loss of retinal boutons and synapses, and an even greater restriction in the total area of synaptic contacts. The recently reported elimination of optic axons extending into the early postnatal period (Rakic & Riley, 1983) can account only in part for the present results according to our analysis of the available data. The dynamics of these changes suggests that most of the features stabilize earlier in parvocellular than in magnocellular laminae which would imply a longer period of plasticity in the latter components.

ACKNOWLEDGEMENTS

We wish to acknowledge the collaboration of Drs. G. Holstein, J. Hamori, and M. Gottlieb in various phases of this work. Aided in part by NIH Grants #EY-01926, TR32EY-01867, NS-18657, and the American Parkinson's Disease Association.

REFERENCES

Anker R, Cragg B (1974). Estimation of the number of synapses in a volume of nervous tissue from counts in thin sections by electron microscopy. J Neurocytol 3:525.

Coupland R (1968). Determining sizes and distribution of sizes of spherical bodies such as chromaffin granules in tissue sections. Nature 217:384.

Dreher B, Fukada Y, Rodieck RW (1976). Identification, classification and anatomical segregation of cells with X-like and Y-like properties in the lateral geniculate nucleus of old-world primates. J Physiol (London) 258:433.

Hámori J, Pasik P, Pasik T (1983). Differential frequency of P-cells and I-cells in magnocellular and parvocellular laminae of monkey lateral geniculate nucleus. An ultrastructural study. Exp Brain Res 52:57.

Kaplan E, Shapley RM (1982). X and Y cells in the lateral geniculate nucleus of macaque monkeys. J Physiol (London) 330:125.

Marrocco RT, McClurkin JW, Young RA (1982). Spatial summation and conduction latency classification of cells of the lateral geniculate nucleus of macaques. J Neurosci 2:1275.

Rakic P, Riley KP (1983). Overproduction and elimination of retinal axons in the fetal rhesus monkey. Science 219:1441.

Schiller PH, Malpeli JG (1978) Functional specificity of lateral geniculate nucleus laminae of the rhesus monkey. J Neurophysiol 41:788.

Development of Visual Pathways in Mammals, pages 179–197

RETINOCOLLICULAR PLASTICITY IN THE OPOSSUM REVISITED

C.E. Rocha-Miranda; R. Méndez-Otero; A.S. Ramôa; E. Volchan; L.G. Gawryszewski

Instituto de Biofísica da UFRJ
Rio de Janeiro, R.J. 21941, Brasil

Studies in neuroplasticity have often used as a model the orderly and congruent direct connections of the two retinae with the superior colliculus (SC). An easily reproducible paradigm used to examine the rearrangement of the retinocollicular system has been the observation in the adult animal of the early uniocular enucleation, whereby the optic layers ipsilateral to the remaining eye are made available for aberrant connections by the loss or the non-acquisition of the connections with the contralateral retina. The results of early uniocular enucleation upon this system have been studied in the rat (see Lund, Land, Boles 1980) the rabbit (Chow, Mathers, Spear 1973), the hamster (see Chalupa 1981); the brushtailed possum (Sanderson, Pearson, Dixon 1978) and an American opossum (Lent, Rocha-Miranda 1978; Lent, Méndez-Otero 1980). In all these species there is an expanded projection of the ipsilateral retina on the SC but conflicting results have been reported for the visuotopic representation both among and within species. The anatomical rearrangement has been explained by sprouting, preservation of juvenile projections and changes in the cell death ratio, the emphasis on any one of them appearing to reflect more the current knowledge of intermediate developmental stages rather than a direct evidence. On the other hand, the insufficient knowledge of the properties of the ipsilateral pathway due to weaker and less frequent responses to visual stimuli when compared to the contralateral has hampered physiological studies. For instance, in the opossum Lent and Méndez-Otero (1978) reported that early uniocular enucleation leads to the establishment of a well-ordered, albeit altered, retinotopy at the SC on the side of the remaining eye in conjunction with an expanded projection of the temporal retina. They advanced an interpretation based on the variation of the magnification factor (MF) along the colliculus, suggesting that the ganglion cells at the borders of the temporal retina displayed the most intense plastic response. However, in a recent study of the visual field properties of the opossum's SC (Ramôa, Rocha-Miranda, Méndez-Otero, Josuá 1983) distinct representations for each eye were reported. If the eye ipsilateral to the SC has a different strategy to represent the visual field, then the evaluation of the effects of enucleation ought to take this fact into account. In this work some distinctions in the visuotopy of the contra- and ipsilateral eyes on the SC will be examined and then the effects of early enucleation matched against the projection of the corresponding pathway in the normal animal.

RETINOCOLLICULAR PROJECTIONS IN THE NORMAL AND THE ENUCLEATED OPOSSUM

The SC of the opossum can be subdivided in three regions based on the pattern of its retinal endings: a rostral pole (RP), a direct binocular region (DBR) and a caudal region (Rocha-Miranda, Cavalcante, Gawryszewski, Linden, Volchan 1978). The RP is a partition of the optic layers that receives direct retinal afferents exclusively from the contralateral temporal retina. The DBR is an intermediate region that receives direct projections from both retinae: a crossed component from the central portion of the nasal retina and a smaller uncrossed projection from the temporal retina. Here the crossed terminals are distributed mainly to the lower two optic layers, while the uncrossed form a dense patch in *stratum zonale* and are only sparsely distributed to the other two layers. The caudal region has retinal projections exclusively from the peripheral portion of the contralateral nasal retina. Study of the visuotopy of the contralateral eye (Volchan, Gawryszewski, Rocha-Miranda 1982) showed that at the level of the horizontal meridian at least 150 degrees of the uniocular visual field are represented at the SC, about 40 degrees of which impinge on the ipsilateral visual field. Since the representation of the presumptive vertical meridian was found to occur close to the anterior border of the patch of uncrossed retinal terminals the RP contains the representation of the ipsilateral portion of the binocular visual field. The meridian of longitude 45 degrees in the contralateral field has its representation about the caudal border of the DBR, close to the temporal limit of binocular responses. Although the RP does not receive direct projections from the ipsilateral eye most recording sites therein yield binocular responses. Thus a sector of the visual field of at least 80 degrees can be considered binocular and is represented at the RP and the DBR, the remainder of the visual field being represented at the caudal region and responding exclusively to the stimulation of the contralateral eye.

Early enucleation modifies the pattern of retinal endings mainly in the SC ipsilateral to the remaining eye. A dense uncrossed projection extends from the rostral limits of all optic layers to their caudal reaches (Lent, Rocha-Miranda 1978). Only that sector of the field seen by the temporal retina is represented in the ipsilateral SC (Lent, Méndez-Otero 1980) consistent with the exclusive labelling of temporal ganglion cells after HRP injections in the upper layers (Volchan, Gawryszewski, Rocha-Miranda 1982). Furthermore, in the former investigation a parabola-like function was described by the loci of the MF when plotted against the eccentricity or the rostro-caudal collicular axis. The highest values were derived from stimulation of the nasal and the temporal borders of the visual field, now represented respectively at the rostral and caudal collicular borders. Since this appeared to be quite unlike the variation of the MF at the SC ipsilateral to the remaining eye in acutely enucleated animals or at the contralateral SC in normals (op. cit.), the results were interpreted as reflecting a rearrangement at sites close to the representation of the two borders of the temporal retina rendered vacant by removal of the crossed projection.

EXPERIMENTAL CONDITIONS AND DATA ANALYSIS

Several projects from this Laboratory have contributed to the results herein described. The data analysis, however, has been restricted to a group of fifteen adult opossums (Didelphis marsupialis aurita, Wied 1826) studied under comparable conditions. Eight were normal animals and seven had been enucleated uniocularly at one to three weeks of age. All were subject to a midpontine pretrigeminal section prior to recording. The other conditions are described in Ramôa, Rocha-Miranda, Méndez-Otero and Josuá (1983). To summmarize, while recording multi-unit activity from the upper two layers of the SC, response fields were mapped at the centre of a tangent screen by back-projecting luminous discs on a low adapting background, under pancuronium bromide. The head was positioned in a head holder which allowed for rotation about a vertical axis passing through the centre of an eye. Two imaginary orthogonal planes, one vertical parallel to the sagital plane and one horizontal, contained this centre and their intersection determined the reference meridians on the tangent screen 57 cm away. By monitoring the fundus image it was possible to place the blind spot's projection at 4.3 degrees above the horizon, i.e. its mean elevation at normal head posture. Likewise, its azimuth was corrected to the mean deviation from the vertical reference meridian of 26.5 degrees (Oswaldo-Cruz, Hokoç, Sousa 1979). An equatorial azimuthal representation of the visual field was used, with each eye at the centre of its map. Thus, there is a vertical reference meridian for the contralateral (Vo) and another (Vo') for the ipsilateral eye. The identification of penetration sites and the reconstruction of the optic layers was performed in Nissl stained parassagital sections. The data from normal animals have been normalized to a standard SC and presented in a dorsal view where the antero-posterior (AP) axis has been flattened and the medio-lateral (ML) projected on the horizontal stereotaxic plane. For the reconstructions, the parassagital sections were aligned by the caudal limit of the optic layers which in the opossum approaches a straight line normal to the sagital plane. Normalized data from two other animals which received eye injection of a marker (tritiated proline or HRP) have been used for an estimate of the dimensions of the three subdivisions of the optic layers. Another normalized representation had to be drawn for the SC ipsilateral to the remaining eye in view of the shrinkage along the ML axis.

VISUOTOPIC REPRESENTATION CONVEYED BY EACH EYE

The visuotopic representation in the SC of normal opossums was reconstructed correlating recording sites with response field centres. The sample consisted of 86 penetrations, all of which yielded responses from the contralateral eye and 73 from the ipsilateral. Binocular responses were found at the rostral 1.8 mm. There were 29 binocular sites at the RP and 44 at the DBR. At a binocular site stimulation of the ipsilateral eye usually evoked a hefty response, though often weaker than that evoked by the other. Differences in representation became apparent by alternately mapping the response fields of each eye along an AP row of penetrations. When the contralateral eye was estimulated a caudo-rostral progression of the electrode site resulted in a predictable shift of the representation from temporal to nasal in the contralateral hemifield and across the binocular part of the ipsilateral hemifield. For the ipsilateral eye a similar shift of field centres occurred when recording from the first caudal binocular sites onwards. But upon reaching the neighbourhood of the region representing the central visual field the following alternatives were observed in order of frequency: (a) the response field centres persisted close to

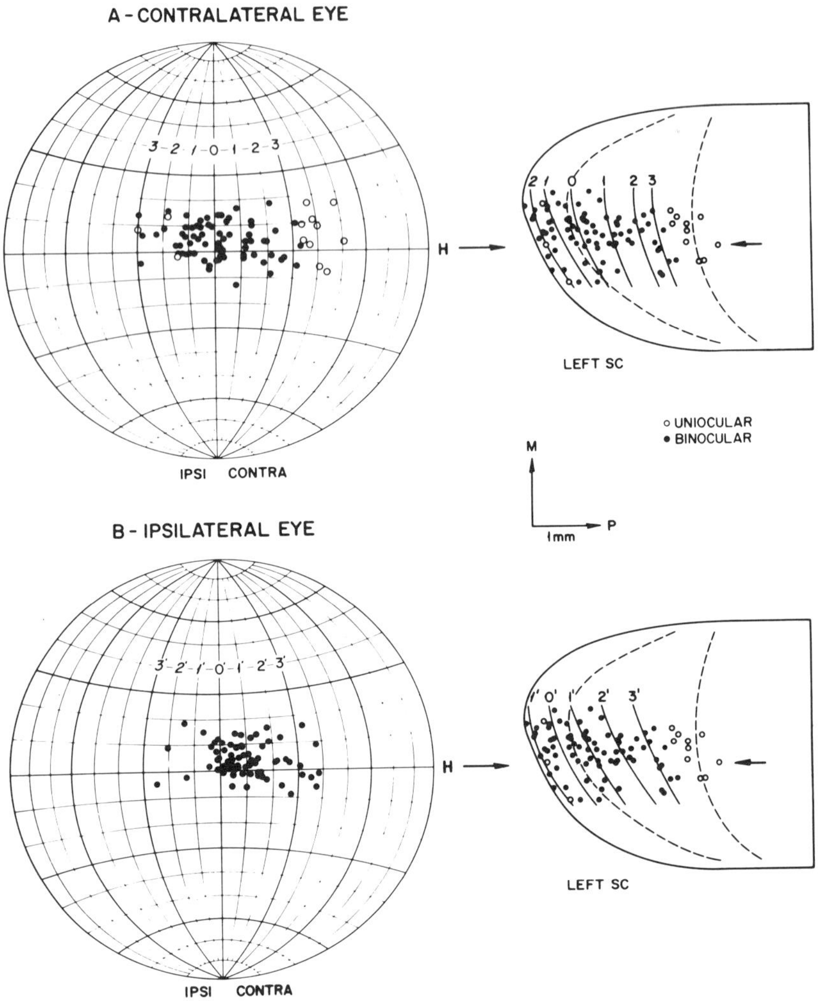

Fig. 1. Distribution of response field centres from the SC on visual field maps centred at each eye (left). The isoazimuth lines have been drawn in intervals of one decade of visual angle on a dorsal view of the left optic layers by relating position in the field with penetration site (right). Dark circles: binocular responses. Light circles: responses to the contralateral eye. Dashed lines: limits of the DBR. H = horizontal meridian.

Vo' (b) the response field centres invaded the ipsilateral hemifield, albeit trailing behind those of the other eye or (c) the site became monocular. Pooling of this data from several animals yields the distinct maps in figure 1 where isoazimuth lines have been drawn for each eye on the colliculus. For their determination penetration sites were represented on the standard SC and labeled according to the longitude of their response field centres in 10 degrees steps on either side of Vo or Vo' and boundaries estimated by eye on the pooled data. The most salient distinctions which can be appreciated in figure 1 are a rostral shift of the isolongitude lines for the ipsilateral eye with respect to the contralalteral, a scanty representation of the ipsilateral visual field by the ipsilateral eye and a clumping of field centres about Vo'. Observe that Vo and Vo' are represented at about 750 μm and 450 μm from the rostral end, respectively, a point to which we shall return later. Due to clustering of the penetrations about the representation of the horizontal meridian the same procedure yielded few isoelevation lines. Observe that in figure 1 the horizontal meridians are in register for both eyes.

The presence of two visuotopies stands out in the analysis of other parameters of visual field representation as well. A sensitive one is the MF which in this case yields two values, one for each eye, from a pair of penetrations. A statistical treatment of the pooled data for each eye indicated that the MF's for the ipsilateral eye (MFi) were different and higher than those for the contralateral (MFc) (Wilcoxon signed ranks test, $.01 < p < .02$). Since this could result from a sampling bias the variation of the MF was analized along one of its relevant dimensions, the AP axis, and the differences between the MF's for each eye along this axis plotted in figure 2. Considering the pattern of isolongitude lines and in order to diminish the mingling of MF values from different longitudes the position of a penetration along the AP axis was determined, in this case, with respect to the anterior limit of the SC. Previous work predicted that the MFc is not homogeneously distributed along this dimension (Volchan, Gawryszewski, Rocha-Miranda 1982). Likewise, the difference between pairs of MF's is not constant. Conspicuously, at the rostral SC MFi is often much higher than MFc (Fig. 2). If this last statement were true it would be of considerable heuristic value to determine where in the colliculus are there regions displaying a distinct change of MF for the two eyes. A search for homogeneous (minimum s.d.)

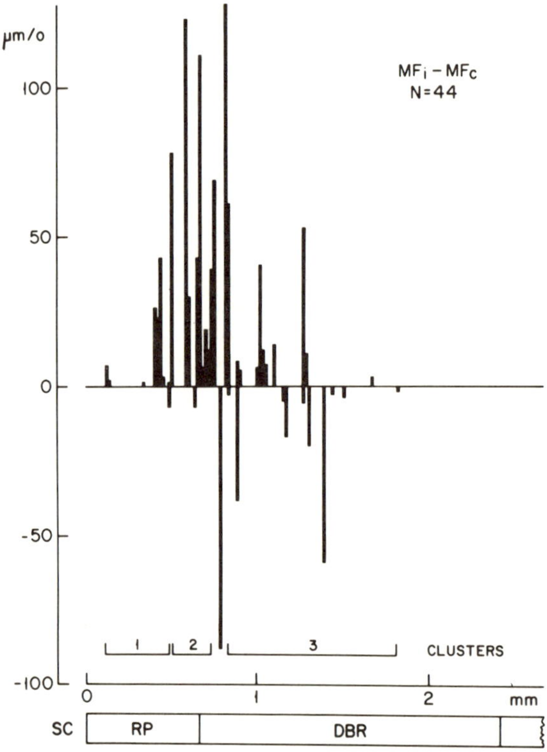

Fig. 2. Differences of the magnification factors for the ipsilateral (MFi) and contralateral (MFc) eyes (ordinates) along the AP axis of the SC (abcissae). Observe that MFi is greater than MFc mainly at the interface RP/DBR.

clusters was applied to the data in figure 2 by using Fisher's algorithm for partition by exact optimization (Hartingam 1975). By this method a rostral cluster with values of MFi only slightly higher than MFc is delineated at the rostral 500 μm of the AP axis, followed by a second cluster with very high values of MFi located within the next 250-300 μm segment of the AP axis and a third cluster beyond that with no appreciable difference between the MF's of the two eyes. This method suggests, then, that large differences between the MFi and the MFc, favouring the first, should be expected at the SC about the boundary of RP with the DBR.

DISPARITIES AND THE REFERENCE MERIDIANS

An incidental observation during the course of these experiments was that a convergent horizontal disparity was present in 67 out of 73 response field pairs as estimated from the azimuths of their centres. Table 1 shows some statistics for this variable in segments (bins), 333 μm long, of the AP axis. For the horizontal disparities the following points should be emphasized: the medians are relatively constant within the DBR, about 5.3 degrees of convergence, and RP disparities are higher and have a greater variability. Since these results show the same general tendency observed in a previous single unit study (Ramôa, Rocha-Miranda, Méndez-Otero, Josuá 1983), it is reasonable to suppose that disparities measured from multi-unit

Table 1

Horizontal disparities
(+ = convergent; - = divergent)

Bin	1st quartile	median	3rd quartile	range
1st RP	6.8	11.7	13.0	-3.1 to 25.6
2nd RP	6.7	8.9	12.3	3.1 to 16.2
1st DBR	2.2	5.3	9.1	-1.6 to 14.8
2nd DBR	-0.2	5.3	8.0	-2.2 to 10.4
3rd DBR	2.5	5.0	7.2	0.6 to 7.3
4th DBR	4.5	5.6	9.8	4.3 to 11.4

field centres reflect this variable's central

tendency for that cluster of binocular units at a particular site. For the vertical disparities it was observed that median values along the AP segments were either zero or departed by no more than 1.5 degree from zero, lending, therefore, no reason to evoke ciclorotation of the eyes as an explanation for the horizontal disparities.

The convergent disparities of pairs of response fields have been translated on figure 3 into points in space at which the field centres would be in register. For 67 pairs with convergent disparities 59 would superimpose before the screen. Furthermore, the points of RP fields (light squares) tend to be closer to the animal than those of DBR ones (dark squares). Similar findings in the single unit study (op. cit.) led to the suggestion of distinct functions for these subdivisions in the processing of the third dimension. More relevant to the present objectives is the fact that Vo and Vo' were thought to contain the projection of the central axes of binocular correspondence under the assumption that the latter were parallel to the sagital plane. If, as these results suggest, these axes are convergent their representation on the SC will not coincide with that of Vo and Vo'. The consequences of this assumption shall now be examined.

The projection on the screen of convergent central binocular axes falls towards the hemifield of laterality opposite to that of the eye. Thus, when recording from the left SC (Fig. 3) the projection of the central binocular axis of the right eye (Vc) would fall somewhere on the left hemifield, which in this case is the ipsilateral hemifield, and would be represented rostrally to the projection of Vo. For the left eye, the central binocular axis (Vi) would fall on the right hemifield, which is contralateral to the left colliculus, and would be represented caudally to Vo'. In the opossum, as in the cat (Harting, Guillery 1976), the rostral border of the uncrossed retinocollicular projection is likely to represent the naso-temporal border. Median disparities along the DBR are around 5.3 degrees (Table 1) so one may assume this values for disparities occuring at the region representing Vc and Vi. Assuming symmetry about the midline, the angle formed by the plane that contains the centre of the eye and Vo (or Vo') and Vc (or Vi) will then have 2.65 degrees for each eye (Fig. 3). The representation of Vc and Vi on

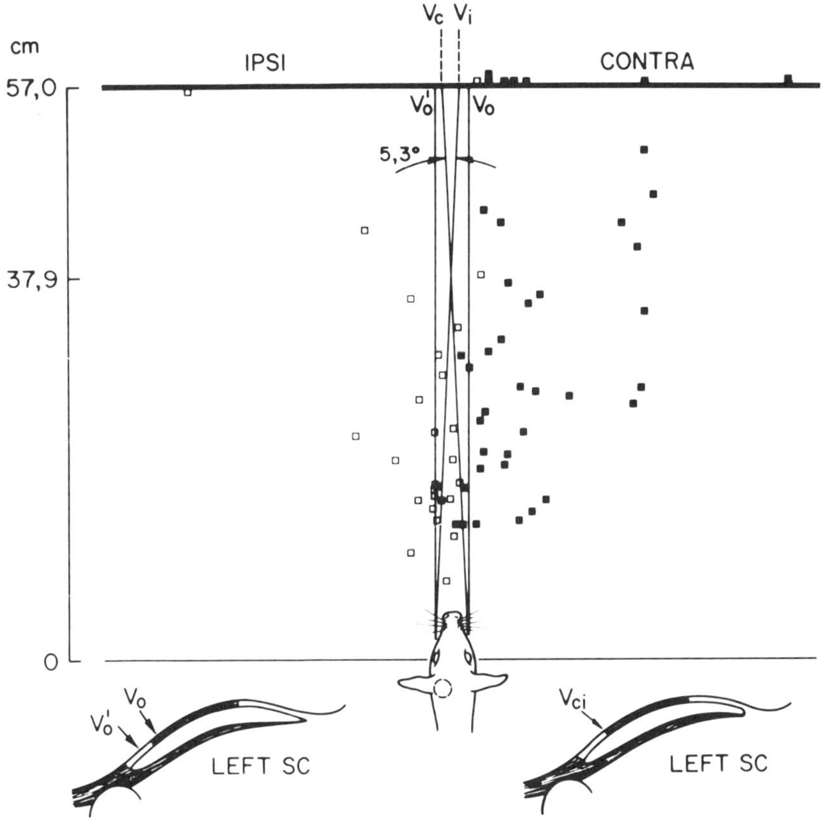

Fig. 3. Points of overlap of field centres with convergent horizontal disparity. Convergence beyond 57 cm is represented on the back of the screen. Vo and Vo' are the vertical reference meridians. Vc and Vi represent the central binocular axes derived from the median DBR disparity (5.3^{o}) which yields an overlap at 37.9 cm. Filled squares: DBR fields; empty squares: RP fields. Insets: parassagital section of the left SC with the representation of Vo and Vo' and Vc and Vi (as a function of MFc and MFi).

the colliculus can then be estimated using the average MF for each eye within a region bounded by Vo and Vo' as a first approximation. A more precise determination of the representations of Vo and Vo' than that used in figure 1 was achieved by calculating the "X" intercept on a plot of the azimuths versus the AP axis from recording sites not including the RP. The correlation coefficients of these variables were 0.978 and 0.936 for the contralateral and ipsilateral eyes, respectively. This method placed the representation of Vo at 777 μm (95% confidence interval: 737 to 812 μm) and Vo' at 527 μm (95% confidence interval: 435 to 599 μm) on the AP axis. Sixteen pairs of penetrations had their mid-intervals within the above intervals and their average MF were 43 μm/deg for the contralateral and 72 μm/deg for the ipsilateral eye. It is possible, then, to propose tentatively that the representation of Vc might have its average representation on the colliculus at the level of the H meridian displaced rostralward from Vo by 114 μm (43 x 2.65) which places it at 777 - 114 = 663 μm on the AP axis. Likewise, that of Vi might have to be displaced caudalward from Vo' by 191 (72 x 2.65) μm, placing it at 527 + 191 = 718 μm on the AP axis. These results indicate that by assuming this convergence of the central binocular axes their representation on the SC would be centred at the rostral edge of the DBR and probably overlap considerably (inset Fig. 3). Since these experiments were not designed primarily to quantify disparities the above corrections have not been introduced in the visuotopic maps. It is important, however, to realize that they would reduce considerably the shift between collicular maps for the two eyes (Fig. 1) but would not significantly alter the distinct pattern of representation at the rostral end of the SC for the ipsilateral eye (Fig. 1) which results in the high values of MFi.

VISUOTOPIC REPRESENTATION IN THE EARLY ENUCLEATED OPOSSUM

The description of the visuotopic representation of early enucleated opossums has been based on 63 and 140 penetrations on the SC contra- and ipsilateral to the remaining eye, respectively, of 7 adult animals enucleated at 7, 10, 15 and 22 days after birth. All members of this group yielded similar patterns. As with normals, the SC was explored by parassagital rows of penetrations directed to the representation of the H meridian in both colliculi, supplemented by other rows representing higher or lower elevations, in some cases. The SC contralateral to this eye was of normal dimensions and the contra and ipsilateral hemifields are represented as in normal animals (Fig. 4). On the ipsilateral SC, however, there was considerable shrinkage along the ML but not along the AP axes, as shown by the normalized outline of the optic layers in figure 4. Furthermore, the visual representation was altered on this side. First, field centres were all within the domain of the temporal retina, although the recording sites were distributed from the very rostral end to close to the caudal border. Penetrations sites having no visual representation were restricted to the caudal 8 to 16% of the AP axis. This represents a two fold increase in length of the collicular region at the level of the H meridian dedicated to the representation of this sector of the visual field. Second, the response field centres from rostral collicular sites appear to display only the most frequent alternative found in normals: after approaching to within a few degrees of Vo', more in keeping with a distribution about the hypothetical Vi, response field centres tend to represent the same region in space. Besides these two clear points figure 4 hints at other changes which are hard to distinguish qualitatively from the normal representation: occasional reversals at the rostral end and some degree of re-representation at the caudal one.

A measure of this rearrangement was achieved by the determination of the MF along the AP axis of the SC. Figure 5 allows a comparison between the normal and the early enucleated preparation, care being taken to use only those values close to the representation of the H meridian. As expected, the distribution of the MFc's in the early enucleated animals was similar to that obtained in normals. On the other hand the MFi's show differences, reproducing the pattern found by Lent and Méndez-Otero (1980). While one can confidently say that at the caudal collicular half the two groups are quite distinct, since the ipsilateral eye does not evoke a response at this region in normal animals, the situation at the rostral half is not

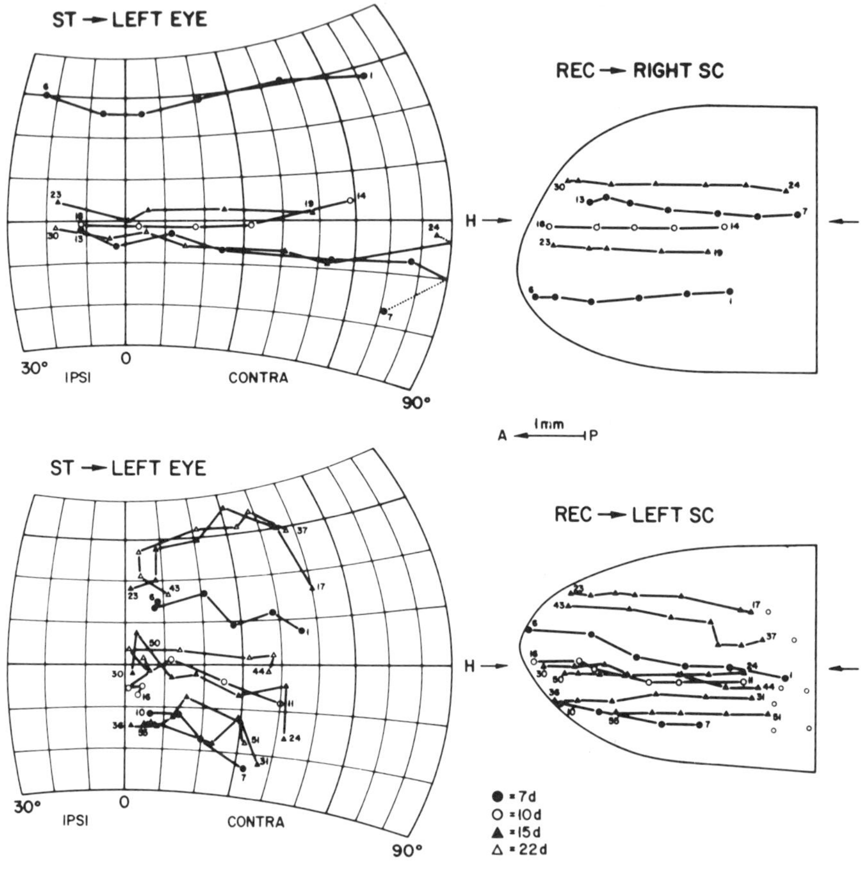

Fig. 4. Visual representation on both SC of the early enucleated opossum with a remaining left eye. Only a sector of its visual field is shown. Data pooled from 5 adult animals enucleated between 7 and 22 days (symbols). Penetration sites indicated on standard dorsal views of the contralateral (normal) and ipsilateral (shrunken) SC.

so clear. Although obviously tempting to be atributed to a rearrangement of inputs, the rise of the MFi from lower values at an intermediate region in the colliculus to higher ones at the rostral end cannot by itself be accepted as evidence since this is also the trend observed in the normal preparation. Nevertheless, statistical analysis (Kruskal-Wallis and Mann-Whitney) of the MFi's of the normal and early-enucleated animals at comparable levels on the AP axis, i.e. the rostral 1.8 mm, rejected the null hypothesis at $p < .01$. This, however, could be due to the absence or scarcity in the enucleated animals of the low values of MFi present at the rostral end of the normal SC.

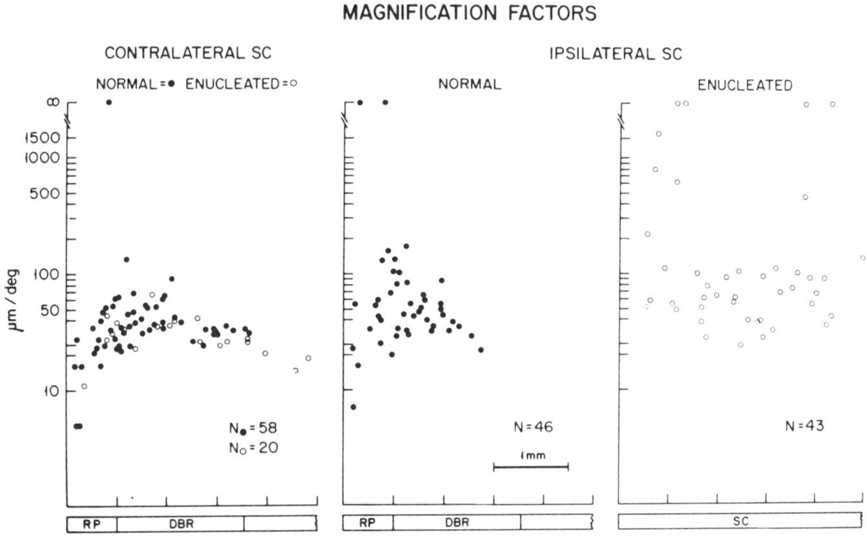

Fig. 5. Semi log plot of the magnification factors (MF) along the AP axis of the SC contra- and ipsilateral to the stimulated eye in normal (dark circles) and early enucleated opossums (light circles).

Which changes are attributed to plastic rearrangements become even more difficult to determine with the present tools when preliminary results in two animals enucleated when adults and examined several months later are taken into account. The representation of the temporal sector appears to extend farther back on the AP axis and occasional high values for the MFi were recorded at this level. As in the normal animal there is a discrete representation of the ipsilateral visual field.

CONCLUSION

This investigation set out to examine the visual representation in the SC of the opossum with an eye to defining a proper control for the evaluation of the neuroplastic changes subsequent to early enucleation using relatively simple tools. Maybe influenced by the considerable anatomical re-arrangement, large qualitative differences were expected between the various preparations. Instead, most of the differences encountered could only be described quantitatively. Notwithstanding this work has uncovered some aspects of the collicular organization that are not only relevant for enucleation as a model of neuroplasticity but also for the understanding of binocular organization in this animal. Some general considerations can now be made.

This work has shown that the pattern of representation of the ipsilateral retinocollicular projection is sufficiently distinct from that of the contralateral to render hazardous the use of the latter as a control for functional rearrangements in early enucleation. At the interface between the RP and the DBR, where central visual space is represented, there's a tendency for the ipsilateral eye to represent considerably smaller increments of space than the contralateral. As a result the MFi will have much larger values than MFc at this region. Thus, while from the point of view of neuroplasticity a rise of the MFi at the rostral end of the colliculus cannot be taken *per se* as indicative of rearrangement, it reveals the end-result of interactions which shaped the adult pattern. How is this achieved is at present an open question.

On the other hand the evidence from response field correspondences has argued against the shift in representation of longitudes between the two eyes as a real difference in organization, at least for the central meridians. Rather it appears to stem from the choice of the reference meridians. If these include the central binocular axes many isolongitude lines will come virtually into register. Thus, by adjusting their representations proportionally to MFc and MFi, Vc and Vi become localized at the rostral border of the uncrossed retino-collicular pathway, as predicted by the anatomy (Harting & Guillery, 1976; Rocha-Miranda, Cavalcante, Gawryszewski, Linden, Volchan 1978). Although it may result from an artifact of paralysis, the convergence of the central binocular axes might be a physiological arrangement for a nocturnal animal that probably uses its vision to behave towards nearby objects. Furthermore, Picanço-Diniz, Silveira and Oswaldo-Cruz (1983) have provided evidence which indicates that opossums are on the average myopic (2.3 D). It is not known what effect if any removal of an eye may have upon the development of the other, so the adoption of the hypothesis of a convergent central axis should rely on further experimental evidence. The average azimuth of the blind spots did not differ in the two groups, lending no reason to suppose that development in the absence of binocularity leads to an aberrant posture of the remaining eye. This, however, needs not apply to the central connections.

Although this work has shown that high values of MF at the borders of the visual field representation as well as expansions of representation are not *per se* good evidence for rearrangement, it should be stressed that the early enucleated preparation has not been found identical to either the normal or the late enucleated controls. Particular to early enucleation seems to be a representation with no field centres in the ipsilateral hemifield, i.e., an absence of the physiological (and anatomical) RP, accompanied by a paucity of low MFi rostrally. Also characteristic is the increase in the rostro-caudal extent of the collicular region responsive to the ipsilateral eye, coherent with the expanded territory of uncrossed terminals. The extent within which the temporal retina is represented in the ipsilateral SC of early-enucleates is significantly larger, but here again the magnitude of the changes will be a function of the adopted control. The preliminary results in late enucleates indicate that the removal of an eye unaccompanied by any detectable anatomical rearrangement will increase the AP extent of the collicular segment responsive to the ipsilateral eye. A caudal rise of the MFi may also accompany this change. Furthermore, the evaluation of the increase in the area of influence of the ipsilateral eye at the caudal level will depend on one's interpretation of what has happened to the RP of early-enucleates. Have its cells been innervated by uncrossed terminals or have they died during maturation? This and many other questions remain unanswered in this deceptively simple model of neuroplastic rearrangement of the nervous system.

Acknowledgements: DR. Claudio Struchiner oriented the statistical analysis. Elisa Carvalho Dias and Lidia Kős participated in various phases of this work and Maria Luiza da Silva was responsible for the lay-out of the publication. Thanks are also extended to all members of the Department of Neurobiology who helped through their criticism.

Chalupa LM (1981). Some observations on the functional organization of the golden hamster's visual system. Behav Brain Res 3:189.

Chow K-L, Mathers LH, Spear PD (1973). Spreading of uncrossed retinal projection in superior colliculus of neonatally enucleated rabbits. J Comp Neurol 151:307.

Harting JK, Guillery RW (1976). Organization of retinocollicular pathways in the cat. J comp Neurol 166:133.

Hartingam JA (1975). "Clustering Algorithms." New York: John Wiley & Sons, p 351.

Lent R, Méndez-Otero R (1980). Plasticity of the ipsilateral retinotectal projection in early enucleated opossums: changes in retinotopy and magnification factors. Neuroscience Letters 18:37.

Lent R, Rocha-Miranda CE (1978). Aberrant retinofugal projections in the opossum after eye enucleation and tectal lesion. In Rocha-Miranda CE, Lent R (eds): "Opossum Neurobiology," Rio de Janeiro: Academia Brasileira de Ciências, p 217.

Lund RD, Land PW, Boles J (1980). Normal and abnormal uncrossed retinotectal pathways in rats: an HRP study in adults. J comp Neurol 189:711.

Oswaldo-Cruz E, Hokoç JN, Sousa APB (1979). A schematic eye for the opossum. Vision Res 19:263.

Picanço-Diniz CW, Silveira LCL, Oswaldo-Cruz E (1983). Electrophysiological determination of the refractive state of the eye of the opossum. Vision Res (accepted for publication).

Ramôa AS, Rocha-Miranda CE, Méndez-Otero R, Josuá KM (1983). Visual receptive fields in the superficial layers of the opossum's superior colliculus: representation of the ipsi and contralateral hemifields by each eye. Exp Brain Res 49:373.

Rocha-Miranda CE, Cavalcante LA, Gawryszewski LG, Linden R, Volchan E (1978). The vertical meridian representation and the pattern of retinotectal projections in the opossum. In Rocha-Miranda CE, Lent R (eds): "Opossum Neurobiology," Rio de Janeiro: Academia Brasileira de Ciências, p 113.

Sanderson KJ, Pearson LJ, Dixon PG (1978). Altered retinal projections in brushtailed possum Trichosurus vulpecula following removal of one eye. J comp Neurol 180:841.

Volchan E, Gawryszewski LG, Rocha-Miranda CE (1982). Visuotopic organization of the superior colliculus of the opossum. Exp Brain Res 46:263.

Development of Visual Pathways in Mammals, pages 199–215

The Development of Visual Cells in Cat Superior Colliculus

Barry E. Stein

Department of Physiology & Biophysics
Medical College of Virginia
Richmond, Virginia 23298

A great deal of attention was focused on the visual role of the superior colliculus when its ablation in cat was shown to produce profound visual neglect despite an intact geniculostriate system (Sprague & Meikle 1965). Since removal of the superior colliculus and ablation of visual cortex appeared to produce qualitatively different visual dysfunctions, it was postulated, or intimated, that there were "two visual systems" (Sprague 1966, Diamond & Hall 1969, Schneider 1969). We now know that these "systems" are interrelated components of a complex whole, and that removal of visual cortex alters the physiological profiles of superior colliculus cells. While a complete functional segregation of visual cortex and superior colliculus does not exist, their different contributions to specific aspects of visual behavior are apparent: the superior colliculus is involved primarily in attending, orienting and following behaviors, whereas the visual cortex plays a principal (though not exclusive) role in the analysis of stimulus detail.

The heightened interest in the superior colliculus continued unabated, even when the original "two visual system" hypothesis was no longer tenable, and has resulted in a wealth of anatomical and physiological information about this structure. The magnitude of interest is illustrated by the number of review articles appearing since 1970 (e.g. Kruger 1970, McIlwain 1972, Palmer et al 1972, Schiller 1972, Sprague 1972, Gordon 1975, Schneider 1975, Sparks & Pollack 1977, Goldberg & Robinson 1978,

Stein 1978a, Coulter et al 1979, Wurtz & Albano 1980, Huerta & Harting 1982, Stein 1984).

This chapter focuses on the development of the visual receptive field properties of superior collicular cells, which is an integral part of the maturation of the remarkably accurate visually-guided behaviors of adult animals (the properties and development of non-visual superior colliculus cells are dealt with elsewhere - see Stein 1978a,1984). Although there are some significant variations in the characteristics of cells across species and structures, the developmental features of the cat superior colliculus reflect general principles by which the mammalian brain constructs complex physiological properties in individual neurons. And, since much of the developmental work on the superior colliculus has been done in the cat, it will be the principal species discussed.

In order to understand better the direction of its developmental changes, we will first review some of the characteristic properties of the mature superior colliculus.

SENSORY AND MOTOR ORGANIZATION

While the superior colliculus consists of at least seven laminae (Kanaseki & Sprague 1974), it is often operationally separated into only two divisions: superficial (layers I-III) and deep (layers IV-VII), based on differences between them in morphology, afferent-efferent connections, physiology and behavioral involve ments (see Edwards 1980, Stein & Gordon 1981, Stein 1984).

Visual cells are concentrated most heavily in superficial laminae and their incidence decreases progressively with depth. Despite the fact that deeper lamina visual cells may have larger receptive fields (McIlwain 1975) and receive corticotectal influences from different areas (Ogasawara et al 1983) than do their superficial counterparts, they share many of the same features, including the same topography: cells with nasal receptive fields are located rostral in the superior colliculus, whereas those with temporal receptive fields are located caudal. Similarly, superior visual space is represented medially in the superior colliculus, and inferior visual space is represented laterally.

Cells of the deep laminae send descending efferents to a variety of extraoculomotor-related structures (Graham 1977, Edwards & Henkel 1978), and these "motor" cells are in topographic register with the visuotopy. Thus, focal electrical stimulation of the superior colliculus produces an eye movement that centers the area centralis on the same point in space represented at the stimulation site in the visuotopic map (Stein et al 1976, Syka & Radil-Weiss 1971, Harris, 1980, Straschill & Rieger 1973, Roucoux & Crommelinck 1976). During "normal" behavior a localized visual stimulus produces a restricted focus of sensory activity in a given locus, which in turn is converted into an output signal that moves the eyes accordingly. Apparently, the topographic nature of the sensory and motor maps is integral to the production of accurate orientation behavior: it is found in the sensory representations and motor output systems of many species, and studies show that early disruption of a map by surgical intervention alters later orientation responses (see Stein 1984 for relevant references). Although the mechanisms by which this critical sensorimotor transduction takes place in the superior colliculus are not yet clear, the receptive field properties of its cells determine which stimuli have the greatest access to the circuitry involved.

RECEPTIVE FIELD PROPERTIES

The receptive field properties of visual cells in cat superior colliculus have been detailed in a number of laboratories under a variety of conditions (e.g. Marchiafava & Pepeu 1966, Straschill & Taghavy 1967, McIlwain & Buser 1968, Sprague et al 1968, Sterling & Wickelgren 1969, Berman & Cynader 1972, Stein & Arigbede 1972a, Dreher & Hoffmann 1973, Harutiunian-Kozak et al 1973, Pinter & Harris 1982), and there is good agreement that the most obvious characteristics of superior colliculus cells are that they:

1. prefer moving to stationary flashed stimuli;
2. prefer stimuli moved in specific directions across the receptive field (directionally selective) within a specific velocity range;

3. receive inputs from both eyes (binocular)
4. exhibit receptive fields that are not systematically divided into separate "on" and "off" regions;
5. have larger receptive fields than generally found in the geniculostriate system;
6. have suppressive flanks bordering the excitatory receptive field;
7. are excited optimally by stimuli far smaller than the borders of their receptive fields;
8. exhibit transient responses even to maintained stimuli;
9. have a propensity for habituation when the same stimulus is repeated, even at relatively long interstimulus intervals.

Of course, not every cell exhibits each of these characteristics, and some properties are more likely to be exhibited by cells in one location than another. For example, deep layer cells habituate more readily than superficial layer cells, and monocular cells become increasingly more frequent at increasingly temporal eccentricities.

DEVELOPMENT OF THE SUPERIOR COLLICULUS

Sensory and Motor Topographies

Presumably, the characteristics and topographic arrangement of cells described above are crucial elements in how the superior colliculus performs its role in visually-guided behavior. Since newborn kittens are virtually blind, and only gradually develop accurate visuomotor skills, we suspected that many, if not all, of these properties of superior colliculus cells would also require a protracted period of development. We were correct physiologically, but not anatomically.

The anatomical pattern of visual inputs to the superior colliculus appears to be adult-like before birth (and has been described at prenatal or early postnatal stages by a number of investigators - Anker 1977, Williams & Chalupa 1982, Graybiel 1975, Stein & Edwards 1979); however, these inputs are not functionally mature at this time (see below). At present the functional maturation of the visuotopic map still has not been described adequately using physiological techniques, and our information is derived from behavioral experiments. These experiments show that the earliest visual orientation is to stimuli falling on the central retina (see Sireteanu & Maurer 1982), presumably because the central representation in the superior colliculus is developing first. It is not clear, however, whether these early orientation responses are limited by the maturation of sensory aspects of the superior colliculus or by the maturation of its efferents. In fact, the question of the maturation of superior colliculus efferents has also received very little attention.

To determine when the various efferent projections to extraoculomotor-related structures are present in young kittens, my colleagues and I (Stein et al 1982) used axonal tracing techniques in kittens six hours to five weeks of age. The major medial (predorsal bundle) and lateral projections from the superior colliculus were already present even in the youngest animal studied, and the various efferent targets thought to be involved in eye movements contained label. These included the central gray matter overlying the oculomotor nucleus (where the axons of superior colliculus cells can contact the dendrites of oculomotor neurons - Edwards & Henkel 1978), and the portions of the pontine and medullary reticular formation that send inputs to the abducens nucleus; in addition, we have now observed label in newborn kittens in the medial accessory nucleus of the superior olive - a relay to the posterior vermis of the cerebellum involved in shifts of gaze (Stein, Spencer & Edwards, in preparation). Between one and six days of age, some postnatal maturation was evident in the labelling density of these target structures, but did not approach the dramatic functional maturation that appears to take place.

Because we were able to evoke eye movements in two-day-old kittens by electrical stimulation of the superior colliculus, it was apparent that at least some of these efferents are functional many days before visually active cells can be recorded (Stein et al 1980). However, even though some topography in the eye movement map was apparent, many stimulation loci within the neonatal superior colliculus were ineffective. Several factors may have accounted for this observation, including maturational changes in the functional state of the extraocular muscles (see Lennerstrand & Hanson, 1978a,b). Furthermore, it is unlikely that the pattern of functional contacts is directly reflected in the labelling pattern exhibited using axonal tracing techniques.

The presence of a motor map in neonatal animals, however incomplete, has significant implications for theories of visuomotor development. Self-generated eye movements appear to be necessary for an animal to "make sense" of visual input (see Hein et al 1979), making them a prerequisite of visuomotor development. Thus, we would expect eye movements to precede the development of sensory responses to visual stimuli, and the presence of eye movement-evoked responses from the superior colliculus before its cells respond to sensory input is consistent with this hypothesis. Still to be determined is which natural stimuli evoke eye movements from the superior colliculus in a normal animal. It is possible, but clearly speculative, that nonvisual inputs that can activate superior colliculus cells in neonates serve this purpose (see Stein 1984).

Development of Visual Cells

The first cells to respond to visual stimuli appear in the superior colliculus at seven days of age (Stein et al 1973a), and for many days these "immature" cells exhibit properties very different from those of the adult cat (Stein et al 1973a, b, Norton 1974, Stein & Gallagher 1981). Characteristically, they

1. exhibit a preference for stationary rather than moving stimuli;
2. often require stimuli to be of long duration (e.g. 400 msec) and high intensity;
3. have exceedingly long latencies - sometimes longer than 1 sec.;
4. respond with fewer impulses/response and lower reliability than in adults;
5. have optimal stimuli that are larger than those found in adults;
6. have larger receptive fields than do adults;
7. show more pronounced response habituation than adults;
8. are monocular (contralateral) only;
9. have receptive fields that are not systematically divided into "on" and "off" regions;

10. do not exhibit suppressive areas bordering the receptive fields;
11. require lower velocities of moving stimuli to elicit a response than do movement-sensitive cells in the adult;
12. do not exhibit directional selectivity even when responsive to moving stimuli;
13. display transient responses even when maintained stimuli are presented.

With the exception of numbers 9 and 13, the properties of kitten superior colliculus visual cells are strikingly different from those found in adult animals.

This physiological profile changes dramatically over a 6-8 week period as the properties of neonatal superior colliculus cells mature. The first binocular cell recorded by Stein et al (1973a) appeared in a ten-day-old kitten; this was followed by a rapid increase in the proportion of binocular cells. Similarly, as moving stimuli became more effective, the effectiveness of stationary flashed stimuli waned. Moreover, with the increased effectiveness of moving stimuli, the adult-like discrimination of the parameters of movement became evident. Directionally selective cells first appeared toward the end of the second postnatal week of development and increased rapidly thereafter. Toward the end of the third week response latencies began to decrease, presumably because of increased myelination in the visual pathways (see Moore et al 1976). In addition, during the 6-8 week maturational period, response reliability increased, susceptibility to habituation decreased, and suppressive surrounds appeared.

At the same time that the properties of superior colliculus cells are maturing, there is a concomitant change in visually-guided behavior. It is difficult to relate this behavioral development to specific changes in the superior colliculus because many areas of the visual system are developing at the same time; however, destroying the superior colliculus in developing animals does disrupt visually-guided behavior, and there is correlational evidence for parallel behavioral and physiological development (see Norton 1981).

Influences of Retinal Development

Because an admixture of cells with mature and immature response properties has been observed in neonatal animals at the same retinal eccentricities (Stein et al 1973a, b, Daniels et al 1978), it is unlikely that the poor optical quality of the neonatal eye (see Thorn et al 1976, Bonds & Freeman 1978) can account for estimates of the maturation of these properties. However, the immaturity of retinal ganglion cells at this time undoubtedly does influence the development of mature response characteristics of superior colliculus cells. Large receptive fields without surround inhibition and with the inconsistent, weak and slow responses characteristic of an immature system have been observed in the retinas of kittens 3-4 weeks of age (Rusoff & Dubin 1977, Hamasaki & Flynn 1977).

A long maturational sequence for some of the properties of Y-cells in the retinogeniculate system has been noted by Daniels et al (1978) and (in the medial intralaminar nucleus) by Wilson et al (1982). This is particularly important for the superior colliculus since the Y-indirect system (retina-lateral geniculate-visual cortex-superior colliculus) forms the corticotectal pathway (Hoffmann 1973).

Influences of Corticotectal Development

When Stein & Magalhaes-Castro (1975) ablated visual cortex in neonates, thereby removing the Y-indirect pathway to the superior colliculus, the maturation of the complex properties of superior colliculus cells was permanently arrested. When these same animals had matured, recordings in their superior colliculi showed a much lower percentage of directionally selective, binocular and movement-sensitive cells than found in normal adult cats. These cells, then, had properties similar to those found in neonates, and these observations have been confirmed by others (Berman & Cynader 1976, Mize & Murphy 1976, Flandrin & Jeannerod 1977). Ablation of visual cortex in adult cats produces the same kind of "regressive" effect on the receptive field properties of superior colliculus cells (Wickelgren & Sterling 1969, Palmer et al 1972). The control exerted by visual cortex over superior colliculus cells is particularly evident in studies using cortical cooling to produce a reversible deactivation of cortex. These studies demonstrate that severe response depression occurs in most superior colliculus cells when the cortical influence is eliminated (Wickelgren & Sterling 1969, Stein & Arigbede 1927b, Stein 1978b, Ogasawara et al 1983).

The observations just described indicate that the maturation of many of the properties of superior colliculus cells may depend on the maturation of the corticotectal pathway. To study the maturation of this pathway directly, Stein & Edwards (1979) used axonal tracing methods in kittens 6 hours to 12 days of age, and demonstrated that the corticotectal projection from 17-18 already appears adult-like in newborn animals. Corticotectal projections had been demonstrated in prenatal animals using degeneration techniques (Anker 1977), but the apparent completeness of this projection in newborns shown by axonal tracing methods was surprising, given the total absence of visually responsive superior colliculus cells.

As noted earlier, the presence of a projection early in development does not necessarily reflect functionally mature connections. Therefore, Stein & Gallagher (1981) used the method of reversible cortical deactivation (cooling) to study the functional maturation of the corticotectal pathway. They found that despite the presence of a well-developed corticotectal projection in newborn animals, the cooling of visual cortex rarely influenced the activity of visual cells in the superior colliculus before 13 days of age. However, at the end of the second week of postnatal development and thereafter, increasing proportions of superior colliculus cells were affected by cortical cooling (Fig. 1a) as the proportion of binocular and directionally selective cells increased (Fig. 1b). These parallels in the development of receptive field properties and cooling-induced depression were not coincidental, for the presence of binocularity and directional selectivity in a cell proved to be a reliable predictor that it would be influenced by cortical cooling. This occurred regardless of the animal's age (Fig. 1c), so that a binocular, directionally selective cell in a 13-day-old animal was as likely to be influenced by cortical cooling as a cell with these properties in an adult cat.

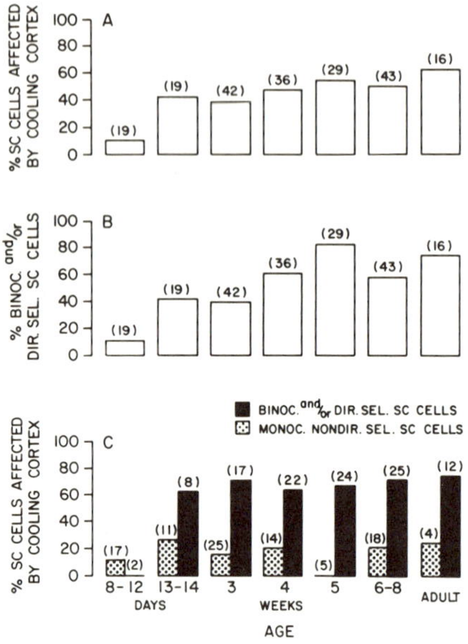

Fig. 1: The development of corticotectal influence on superior colliculus cells. In each graph, the number in parenthesis above a bar represents the sample size in each case. In A, the percentage of cells affected by cortical cooling is plotted as a function of age. Note the dramatic increase in the percentage of cells affected by cortical cooling at 13-14 days of age and more gradual and irregular changes thereafter. In B, the percentage of binocular and directionally selective cells shows a similar developmental course. These changes in A and B are not independent. As shown in C, the presence of binocularity or directional selectivity is a reliable predictor that a given superior colliculus cell will be affected by cortical cooling regardless of age. Thus, at least 62% of the cells with these properties were depressed by cortical cooling at 13-14 days of age (when these properties were first evident) and this percentage changed very little during later maturation. (From Stein & Gallagher 1981).

On the basis of these data, it is apparent that the corticotectal system does not begin to be functional until almost two weeks of age. Curiously, we obtained no evidence for a gradual increase in cortical control over individual superior colliculus cells; rather, the data were more indicative of a sudden maturational change, as if a "gate" opened allowing cortex to influence its target cells in an "all-or-none" fashion. Since this "gate" opened at different times for different superior colliculus cells, the corticotectal system as a whole appeared to have a protracted developmental period.

At present we do not know where the critical postnatal maturational changes takes place in the corticotectal system to allow it to control visual cells in the superior colliculus. Two obvious loci to consider are the cortical cells themselves and the corticotectal synapse. It is possible that the adult-like cortical cells observed in 8-day-old kittens by Hubel & Wiesel (1963) are not the specialized lamina V corticotectal cells described by Palmer & Rosenquist (1974). The presence of a population of immature cortical cells in young kittens (Pettigrew 1974, Blakemore & Van Sluyters 1975) is consistent with the description of a long period of cortical synaptogenesis (Cragg 1975) and the comparatively long postnatal maturation of the indirect Y-cell system (see Influence of Retinal Development). It is also consistent with the possibility that corticotectal cells mature more slowly than other cortical cells, and an abstract by Wolf & Albus (1981) supports this hypothesis. They reported that the properties of lamina V cells (presumably corticotectal cells) develop later than other visual cortical cells. However, the conflicting data of Tsumoto & Suda (1982) led them to just the opposite conclusion. Yet, whether or not corticotectal cells develop mature receptive field properties before or after other visual cells in cortex, their synapses in the superior colliculus may require a protracted period of postnatal development. Thus, both the maturation of corticotectal cells and their functional contacts in the superior colliculus require further study before we can determine the critical postnatal events that allow the cortex to profoundly influence visual cells in the superior colliculus.

REFERENCES

Anker R (1977). The prenatal development of some of the visual pathways in the cat. J Comp Neurol 173:185.

Berman N, Cynader M, (1972). Comparison of receptive-field organization of the superior colliculus in Siamese and normal cats. J Physiol London 224:363.

Berman N, Cynader M (1976). Early versus late visual cortex lesions: Effects on receptive fields in cat superior colliculus. Exp Brain Res 25:131.

Blakemore C, Van Sluyters RC (1975). Innate and environmental factors in the development of the kitten's visual cortex. J Physiol London 248:663.

Bonds AB, Freeman RD (1978). Development of optical quality in the kitten eye. Vision Res 18:391.

Coulter JD, Bowker RM, Wise SP, Murray EA, Castiglioni AJ, Westlund KN (1979). Cortical, tectal and medullary descending pathways to the cervical spinal cord. Prog Brain Res 50:263.

Cragg BG (1975). The development of synapses in the visual system of the cat. J Comp Neurol 160:147.

Daniels JD, Pettigrew JD, Norman JL (1978). Development of single-neuron responses in kitten's lateral geniculate nucleus. J Neurophysiol 41:1373.

Diamond IT, Hall WC (1969). Evolution of neocortex. Science 164:251.

Dreher B, Hoffmann K-P (1973). Properties of excitatory and inhibitory regions in the receptive fields of single units in the cat's superior colliculus. Exp Brain Res 16:333.

Edwards SB (1980). The deep cell layers of the superior colliculus. Their reticular characteristics and structural organization. In Hobson A, Brazier M (eds): "The Reticular Formation Revisited," New York: Raven, p 193.

Edwards SB, Henkel CK (1978). Superior colliculus connections with the extraocular motor nuclei in the cat. J Comp Neurol 179:451.

Flandrin JM, Jeannerod M (1977). Lack of recovery in collicular neurons from the effects of early deprivation or neonatal cortical lesion in the kitten. Brain Res 120:362.

Goldberg ME, Robinson DL (1978). Visual system: Superior colliculus. In Masterton RB (ed): "Handbook of Behavioral Neurobiology," New York: Plenum, p 119.

Gordon BG (1975). Superior colliculus: Structure, physiology and possible functions. MTP Int Rev Sci 3:185.

Graham J (1977). An autoradiographic study of the efferent connections of the superior colliculus in the cat. J Comp Neurol 173:629.

Graybiel AM (1975). Anatomical organization of retinotectal afferents in the cat: An autoradiographic study. Brain Res 96:1.

Hamasaki DI, Flynn JT (1977). Physiological properties of retinal ganglion cells of 3-week-old kittens. Vision Res 17:275.

Harris LR (1980). The superior colliculus and movements of the head and eyes in cats. J Physiol 300:367.

Harutiunian-Kozak B, Dec K, Wrobel A (1973). The organization of visual receptive fields of neurons in the cat colliculus superior. Acta Neurobiol Expl Warsaw 33:563.

Hein A, Vital-Durand F, Salinger W, Diamond R (1979). Eye movements initiate visual-motor development in the cat. Science 204:1321.

Hoffmann K-P (1973). Conduction velocity in pathways from retina to superior colliculus in the cat: A correlation with receptive-field properties. J Neurophysiol 36:409.

Hubel DH, Wiesel TN (1963). Receptive fields of cells in striate cortex of very young, visually inexperienced kittens. J Neurophysiol 206:419.

Huerta MF, Harting JK (1982). Tectal control of spinal cord activity: Neuroanatomical demonstration of pathways connecting the superior colliculus with the cervical spinal cord grey. Prog Brain Res 51:293.

Kanaseki T, Sprague JM (1974). Anatomical organization of prectectal nuclei and tectal laminae in the cat. J Comp Neurol 158:319.

Kruger L (1970). The topography of the visual projection to the mesencephalon: A comparative survey. Brain Behav Evol 3:169.

Lennerstrand C, Hanson J (1978a). The postnatal development of the inferior oblique muscle of the cat. I. Isometric twitch and tetanic properties. Acta Physiol Scand 103:132.

Lennerstrand C, Hanson J (1978b). The postnatal development of the inferior oblique muscle of the cat. II. Effects of repetitive stimulation on isometric tension responses. Acta Physiol Scand 103:144.

Marchiafava PL, Pepeu GC (1966). Electrophysiological study of tectal responses to optic nerve volley. Arch Ital Biol 104:406.

McIlwain JT (1972). Central vision: Visual cortex and superior colliculus. Ann Rev Physiol 34:291.

McIlwain JT (1975). Visual receptive fields and their images in superior colliculus of the cat. J Neurophysiol 38:219.

McIlwain JT, Buser P (1968). Receptive fields of single cells in cat's superior colliculus. Exp Brain Res 5:314.

Mize RR, Murphy EH (1976). Alterations in receptive field properties of superior colliculus cells produced by visual cortex ablation in infant and adult cats. J Comp Neurol 168:393.

Moore CL, Kalil R, Richards W (1976). Development of myelination in optic tract of the cat. J Comp Neurol 165:125.

Norton TT (1974). Receptive-field properties of superior colliculus cells and development of visual behavior in kittens. J Neurophysiol 37:674.

Norton TT (1981). Development of the visual system and visually guided behavior. In Aslin R, Alberts J, Petersen MR (eds): "Development of Perception," New York: Academic, p 133.

Ogasawara K, McHaffie JG, Stein BE (1983). Differential influences of striate and extrastriate cortex on the response of visual cells in cat superior colliculus. Invest Ophthalmol Suppl 24:224.

Palmer LA, Rosenquist AC (1974). Visual receptive fields of single striate cortical units projecting to the superior colliculus in the cat. Brain Res 67:27.

Palmer LA, Rosenquist AC, Sprague JM (1972). Corticotectal systems in the cat: Their structure and function. In Frigyesi T, Rinvik E, Yahr MD (eds): "Corticothalamic Projections and Sensorimotor Activities," New York: Raven, p 491.

Pettigrew JD (1974). The effect of visual experience on the development of stimulus specificity by kitten cortical neurones. J Physiol London 237:49.

Pinter RB, Harris LR (1981). Temporal and spatial response characteristics of the cat superior colliculus. Brain Res 207:73.

Roucoux A, Crommelinck M (1976). Eye movements evoked by superior colliculus stimulation in the alert cat. Brain Res 106:349.

Rusoff AC, Dubin MW. Development of receptive-field properties of retinal ganglion cells in kittens. J Neurophysiol 40:1188.

Schiller PH (1972). The role of the monkey superior colliculus in eye movement and vision. Invest Ophthalmol 11:451.

Schneider GE (1969). Two visual systems: Brain mechanisms for localization and discrimination are dissociated by tectal and cortical lesions. Science 163:895.

Schneider GE (1975). Two visuomotor systems in the Syrian hamster. Neurosci Res Prog Bull 13:255.

Sireteanu R, Maurer D (1982). The development of the kitten's visual field. Vision Res 22:1105.

Sparks DL, Pollack JG (1977). The neural control of saccadic eye movements: The role of the superior colliculus. In Brooks BA, Bajandas FJ (eds): "Eye Movements," New York: Plenum, p 179.

Sprague JM (1966). Visual, acoustic, and somesthetic deficits in the cat after cortical and midbrain lesions. In Purpura DD, Yahr M (eds): "The Thalamus," New York: Columbia Univ. Press, p 391.

Sprague JM (1972). The superior colliculus and pretectum in visual behavior. Invest Ophthalmol 11:473.

Sprague JM, Marchiafava PL, Rizzolatti G (1968). Unit responses to visual stimuli in the superior colliculus of the unanesthetized mid-pontine cat. Arch Ital Biol 106:169.

Sprague JM, Meikle TH Jr (1965). The role of the superior colliculus in visually guided behavior. Exp Neurol 11:115.

Stein BE (1984). Development of the superior colliculus. Ann Rev Neurosci 7:95.

Stein BE (1978a). Development and organization of multimodal representation in cat superior colliculus. Fed Proc 37:2240.

Stein BE (1978b). Nonequivalent visual, auditory and somatic corticotectal influences in cat. J Neurophysiol 41:55.

Stein BE, Arigbede MO (1972a). A parametric study of movement detection properties of neurons in the cat's superior colliculus. Brain Res 45:437.

Stein BE, Arigbede MO (1972b). Unimodal and multimodal response properties of neurons in the cat's superior colliculus. Exp Neurol 36:179.

Stein BE, Clamann HP, Goldberg SJ (1980). Superior colliculus: Control of eye movements in neonatal kittens. Science 210:78.

Stein BE, Edwards SB (1979). Corticotectal and other corticofugal projections in neonatal cat. Brain Res 161:399.
Stein BE, Gallagher H (1981). Maturation of cortical control over superior colliculus cells in cat. Brain Res 223:429.
Stein BE, Goldberg SJ, Clamann HP (1976). The control of eye movements by the superior colliculus in the alert cat. Brain Res 118:469.
Stein BE, Gordon BG (1981). Maturation of the superior colliculus. In Aslin RN, Alberts JR, Petersen MR (eds): "The Development of Perception: Psychobiological Perspectives," Vol 2, New York: Academic, p 157.
Stein BE, Labos E, Kruger L (1973a). Sequence of changes in properties of neurons of superior colliculus of the kitten during maturation. J Neurophysiol 36:667.
Stein BE, Labos E, Kruger L (1973b). Determinants of response latency in neurons of superior colliculus in kittens. J Neurophysiol 36:680.
Stein BE, Magalhaes-Castro B (1975). Effects of neonatal cortical lesions upon the cat superior colliculus. Brain Res 83:480.
Stein BE, Spencer RF, Edwards SB (1982). Efferent projections of the neonatal superior colliculus: Extraoculomotor-related brain stem structures. Brain Res 239:17.
Sterling P, Wickelgren BG (1969). Visual receptive fields in the superior colliculus of the cat. J Neurophysiol 32:1.
Straschill M, Rieger P (1973). Eye movements evoked by focal stimulation of the cat's superior colliculus. Brain Res 59:211.
Syka S, Radil-Weiss T (1971). Electrical stimulation of the tectum in freely moving cats. Brain Res 28:567.
Thorn F, Gollender M, Erickson P (1976). The development of the kitten's visual optics. Vision Res 16:1145.
Tsumoto T, Suda K (1982). Laminar differences in development of afferent innervation to striate cortex neurones in kittens. Exp Brain Res 45:433.
Wickelgren BG, Sterling P (1969). Influence of visual cortex on receptive fields in the superior colliculus of the cat. J Neurophysiol 32:16.
Williams RW, Chalupa LM (1982). Prenatal development of retinocollicular projections in the cat: An anterograde tracer transport study. J Neurosci 2:604:22.

Wilson JR, Tessing DE, Sherman, M (1982). Development of the electrophysiological properties of Y-cells in the kitten's medial interlaminar nucleus. J Neurosci 2:562.

Wolf W, Albus K (1981). Postnatal development of receptive-field properties in the kitten's visual cortex. Neurosci Lett Suppl 7:S202.

Wurtz RH, Albano JE (1980). Visual-motor function of the primate superior colliculus. Ann Rev Neurosci 3:189.

Development of Visual Pathways in Mammals, pages 217–230

Postnatal Development of Corticotectal Neurons in the Kitten Striate Cortex

T. Tsumoto, K. Suda and H. Sato

Department of Physiology,
Kanazawa University Medical School,
Takaramachi, Kanazawa, 920 Japan

Recent studies on the striate cortex of the adult cat have demonstrated that cortical neurons are segregated into layers which differ in the inputs they receive and in the sites to which their neurons project (Holländer 1974; Palmer and Rosenquist 1974; Toyama et al. 1974; Gilbert and Kelly 1975; Leventhal 1979; Lund et al. 1979). Neurons projecting to the superior colliculus (CT cells) are located in layer V of the cortex and neurons projecting through the corpus callosum to the other hemisphere (callosal cells) are mostly in layer III. In newborn kittens, however, Innocenti and his associates reported that laminar and tangential distributions of callosal cells are not as restricted as in the adult, and suggested that these cells acquire their adult distribution postnatally by eliminating their callosal connections and decreasing in number in layers in which mature cells are not located (Innocenti et al. 1977; Innocenti 1981). Is such an aberrant laminar and tangential cell localization in the neonatal cortex also the case for CT cells? If it is the case, when they establish their mature distribution pattern in the striate cortex? To answer these questions is one of the aims of this study. In Part I of this paper, the postnatal development of CT cells was quantitatively studied by using the method of retrograde transport of horseradish peroxidase (HRP).

In the adult cat, CT cells are known to exert a powerful influence on visual response properties, particularly selectivity for the direction of stimulus movement and binocular responsiveness, of neurons in the superior colliculus (SC) (Wickelgren and Sterling 1969; McIlwain and Fields 1970; Hoffmann and Straschill 1971; Berman and Cynader 1975; Mize and Murphy 1976). In young Kittens, Stein and Gallagher (1981) reported that activities of SC cells with the selective properties were often depressed by cooling inactivation of the cortex while those of cells without the selectivities were not. These results suggest that the maturation of visual responsiveness of SC neurons may depend on postnatal development of the corticotectal projection system. Until now, however, there is no direct evidence showing that the rate of functional maturation of CT cells and an age of their full maturation match with those of SC neurons. In Part II of this study, therefore, the postnatal functional maturation of CT cells has been studied and the results were compared with the previous data on functional development of SC neurons.

PART I MORPHOLOGICAL DEVELOPMENT OF CT CELLS

Thirty-seven kittens, ranging in age postnatally from the first day to the 12th week, and four adult cats were used for this study. Experimental procedures for preparing animals and injecting HRP into the SC were given elsewhere (Tsumoto et al. 1983). A single injection of 0.2-1.0 µl of 25 % HRP (Sigma type VI) was made unilaterally into the SC. Subsequent treatments of the brain for HRP histochemistry were done according to the methods of Mesulam (1978), with tetramethyl benzidine as the chromogen. All the injection sites, identified as a core of dense HRP precipitate, were found in the SC. In very young kittens, the region of dense HRP precipitate tended to be slightly more diffuse than in the adult even when a smaller amount of HRP was used. In such cases, therefore, the HRP precipitate sometimes spread to the central gray and in an extreme case to the mesencephalic reticular formation (Fig. 1A). It did not, however, reach the thalamus rostrally and the middle of the inferior colliculus caudally.

HRP-labelled Cells in the Neonatal Cortex

In neonates and kittens younger than the 8th day, CT cells appeared more lightly labelled than in older animals. At birth, HRP-labelled cells could be found throughout the cortex between the splenial and suprasylvian sulci, although their distribution was discontinuous in a few regions (Fig. 1B). As to laminar distribution, however, HRP-labelled cells were obviously confined to layer V of the cortex. The labelled cells were very densely packed in layer V and their somas

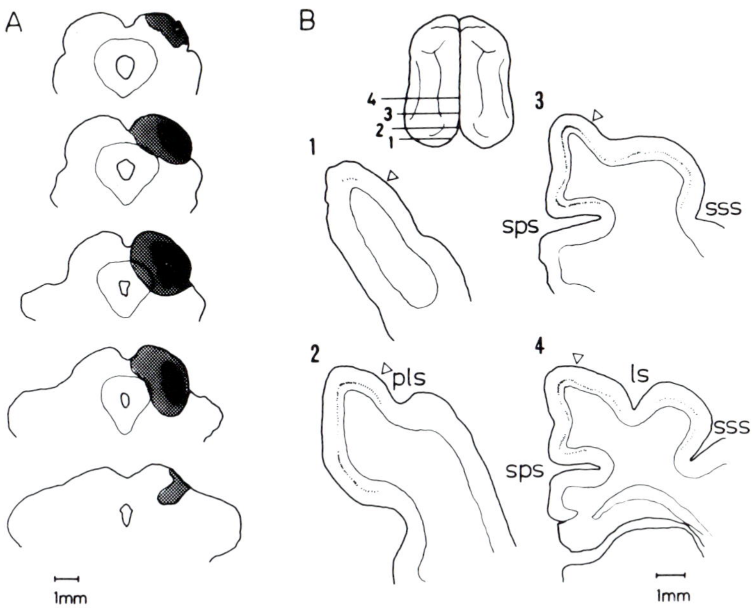

Fig. 1. Distribution of HRP precipitate at the injection site (A) and labelled cells in the ipsilateral visual cortex (B) of a newborn kitten. A, coronal sections through the SC. Filled and crosshatched zones indicate the dense core of HRP precipitate and the surrounding region containing lighter precipitate, respectively. B, coronal sections of the cortex. The level of the sections is indicated by the same number in inset at the top. Each dot represents a cell labelled with HRP. Triangles demarcate the border between areas 17 and 18. pls, sps, sss, and ls indicate postlateral, splenial, suprasylvian, and lateral sulci, respectively. (From Tsumoto et al. 1983)

were already triangular or pyramidal in shape although very slender (Fig. 2A). No HRP granules were found to fill the apical or basal dendrites in the newborn kittens although a Golgi-like image of the dendritic tree was seen in many labelled neurons of kittens older than the fourth week (see Fig. 3). Camera lucida drawings of cell bodies in sections counterstained with cresylviolet demonstrated that the labelled cells in the newborn kittens were relatively large in comparison with non-labelled cells in layer V and adjacent layers (Fig. 2). A very rough topographic correspondence between the striate cortex and SC seemed to exist even in neonatal kittens, in which the intracortical distribution of the labelled cells was much wider than in the adult. This was seen in Fig. 1 which shows a distribution of the labelled cells in the visual cortex of a newborn kitten following an injection of a relatively large amount of HRP (0.5 µl) into the SC. In this kitten the center of HRP injection was located in a slightly lateral portion of the SC, although the visible reaction product spread throughout the mediolateral extent. In the adult cat, it is known that the lateral portion of SC represents the lower visual field (Feldon et al. 1970). In very rough accordance with this, a heavy and continuous labelling

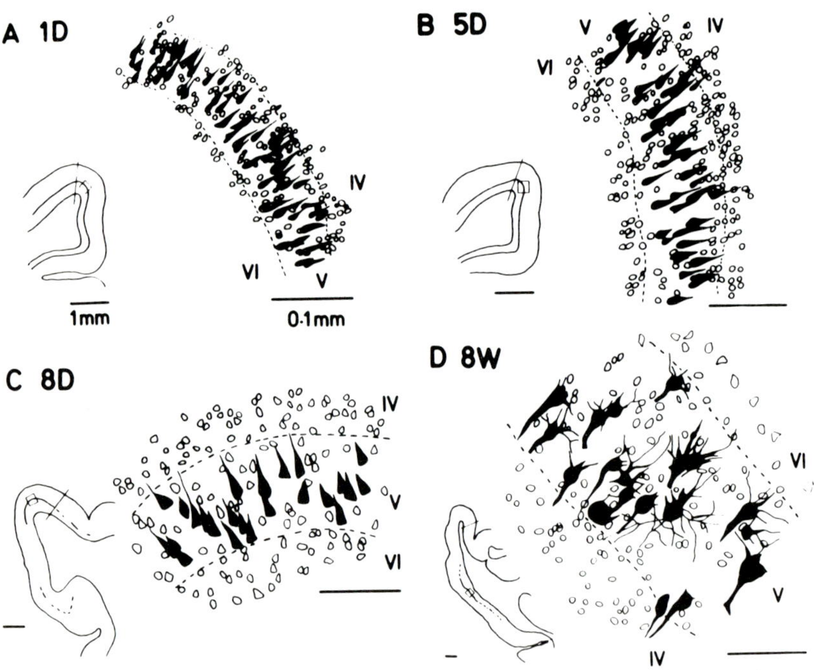

Fig. 2. Camera lucida outlines of HRP-labelled (filled) and unlabelled (open) neuronal somata in and near layer V of the cortex of kittens aged as indicated. Areas roughly corresponding to the drawings are indicated by squares in each inset at the left. In each inset, labelled cells are indicated by solid or broken lines. Thinner lines perpendicular to the cortex indicate the border between areas 17 and 18. (From Tsumoto et al. 1983)

was seen in the relatively rostral striate cortex, which represents the lower visual field (Tusa et al. 1978).

Developmental Changes of HRP-labelled Cells

The high packing density and relatively wide tangential distribution of the HRP-labelled cells in the striate cortex were still seen at the third to fifth days of age (Fig. 2B). At these ages, however, a few cells were observed to have very fine apical dendrites filled with HRP up to layer I of the cortex. Side branches of the apical dendrite were not recognizable yet. Their somas were still very slender and basal dendrites appeared absent or poorly developed in the HRP staining. During the second week of age, a dramatic developmental change seemed to occur in CT cells. At the 8th day the tangential distribution of the labelled cells was not as widespread as that seen in the newborn kittens (Fig. 2C). The packing density of the labelled cells declined remarkably and the soma size became much larger than that of kittens at the first week. A substantial number of cells already had thick apical dendrites stained with HRP. During the third to fourth weeks, CT cells seemed to mature at a slower rate than during the second week. At the fourth week, the dendritic arborization pattern of most of the HRP-labelled cells became nearly adultlike (Fig. 3). Their apical dendrites were well stained

Fig. 3. Darkfield photomicrograph showing HRP-labelled cells located in the medial wall of the lateral gyrus of a 4 week-old kitten. The pial surface is seen at the left. Coronal section. Scale, 100 μm.

with HRP and were often recognizable up to layer I of the cortex. Also, the basal dendrites seemed to mature substantially during the third to fourth weeks, although the proximal parts were not thick yet. The cross-sectional areas of their somas were still smaller than those of the adult. During the fifth week the cross-sectional area of the labelled cells increased slightly and attained adultlike values (see Fig. 4A). But, the basal dendrites of the labelled cells were not fully developed yet in terms of arborization pattern. An adutlike arborization pattern of the basal as well as apical dendrites was seen at the 8th week (Fig. 2D). Qualitatively, thus, a full maturation of CT cells seemed to occur at the 8th week of age.

Quantitative Analysis of Development of CT Cells

To analyse developmental changes in size of the CT cells, we calculated cross-sectional areas of the HRP-labelled cells. Procedures for this calculation have been detailed elsewhere (Tsumoto et al. 1983). The population mean of the cross-sectional areas for each age group was plotted in Fig. 4A. During the first week the cross-sectional area of the cells increased at a rate of 10-12 μm^2/day on the average. During the second week, the increasing rate became higher (14-15 μm^2/day), and thus the population mean increased by 61 % from the 5th to 11-12th days. Thereafter, such a dramatic change was not seen. The increasing rate was 3.2-4.9 μm^2/day until 5th week, and it became further lower (1.0-0.7 μm^2/day) until the 8th week when it was completely adultlike. In parallel with the increase in cell size, the packing density of HRP-labelled cells in layer V was also markedly changed with age. The density of HRP-labelled cells in layer V of the cortical region where labelling was heaviest was calculated as the number of the labelled cells divided by the corresponding area of layer V. The mean density for each age group is shown in Fig. 4B (open circles and broken lines). At birth, it was 16.3 cells/$10^4 \mu m^2$ of layer V and this value declined steadily at a rate of 0.5 cells/$10^4 \mu m^2$/day during the first week. A dramatic decrease in density took place during the second week when the decreasing rate was 1.5 cells/$10^4 \mu m^2$/day on the average. After the third week the decreasing rate became very low. At the 8th week the mean density became approximately equal to that of the adult. To see whether these changes occur in parallel with a decrease in the density of non-labelled cells in layer V, the ratio of the labelled to

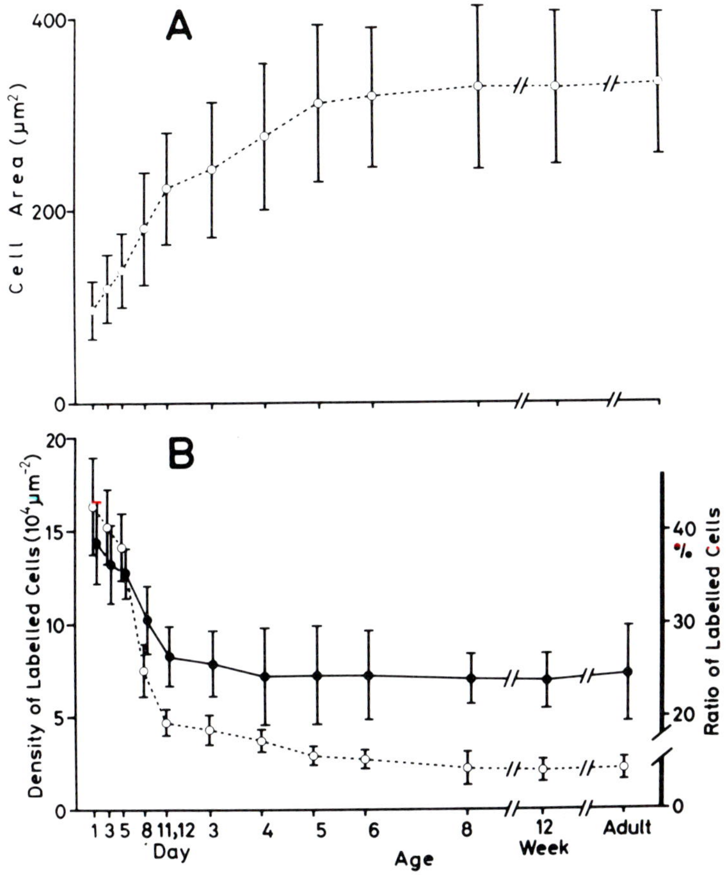

Fig. 4. Developmental changes in the mean cross-sectional areas of somas of HRP-labelled cells (A), and in their mean densities and the mean ratios of labelled to non-labelled cells in layer V (B). In B, the former values are represented by open and the latter by filled circles. Vertical bars, twice the standard deviation. (From Tsumoto et al. 1983)

non-labelled cells in layer V was calculated for each of the camera lucida drawings. The mean for each age group is plotted in Fig. 4B (filled circles and solid lines). It is evident that a significant decrease in the ratio occurred only during

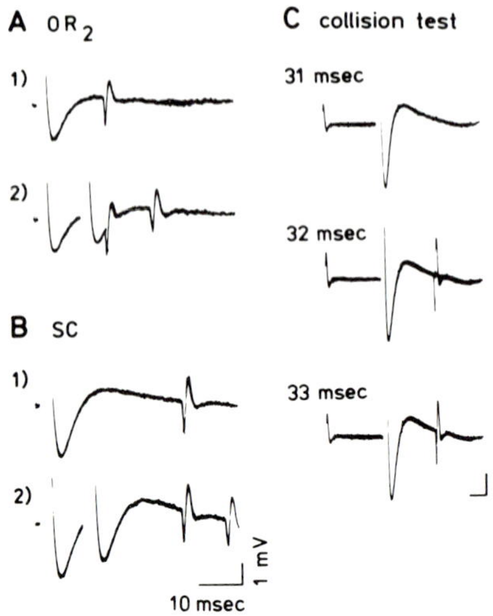

Fig. 5. Antidromic responses of a CT cell recorded from a 9-day-old kitten. A and B: Responses to single (1) and double (2) shocks applied to OR_2 and SC, respectively. C: Collision test. SC shocks were delivered at progressively longer intervals following spontaneous spike discharges. When the interval was 31 msec or shorter, the responses were blocked by the orthodromic spikes. When the interval was lengthened to 33 msec the responses were constantly evoked by SC stimulation. In A-C, five sweeps were superimposed. (From Tsumoto and Suda 1983)

the first and second weeks and the most remarkable change took place in the latter period. The ratio attained an adultlike value (about 25 %) already at the fourth week of age.

PART II FUNCTIONAL DEVELOPMENT OF CT CELLS

Forty-two kittens, ranging in age from the 8th day to 14th week, and six adult cats were used for the electrophysiological study. The experimental procedures for preparing and maintaining the animals, electrical and visual stimulations, and single-unit recordings were detailed elsewhere (Tsumoto and Suda 1983). In kittens younger than 9 weeks, the trachea and femoral vein were cannulated under ketamine anesthesia (i.m., 30-40 mg/kg) after which they were anesthetized with pentobarbitone (Nembutal) (i.v., 5-10 mg/kg). In the older animals, anaesthesia was initiated with Nembutal (i.p., 30-35 mg/kg). A bipolar, stimulating electrode was inserted stereotaxically into the SC. Two other stimulating electrodes of the same type were placed in the optic radiation (OR); 4-5 mm below the cortical surface and just above the dorsal lateral geniculate nucleus. These stimulating sites will be called OR_1 and OR_2, respectively. A double-barrelled glass pipette was inserted into cortex through a closed chamber. One chnnel was filled with 2 M NaCl solution for recording, the other with a solution of Fast Green dye for histological identification of the recording sites.

Antidromic Activation of CT Cells

Of the 1143 cells studied in area 17 of the cortex, 157 were excited antidromically by SC stimulation and thus were identified as CT cells. The criteria for antidromic activation (Fuller and Schlag 1976; Tsumoto and Suda 1982) were slightly modified as follows: (1) constant latency of the responses (Fig. 5A-1 and B-1); (2) the responses had to follow a rate of stimulation higher than 100 Hz (A-2 and B-2); and (3) block of the supposed antidromic spike by an orthodromic spike for a given period (collision test, C). There is a possibility that the SC electrode might concomitantly activate fibers projecting to the subcortical structures other than the SC. Arguments that this possibility is unlikely have been detailed elsewhere (Tsumoto and Suda 1983).

Changes in the Axonal Conduction Velocity with Age

In the adult CT cells, latencies of the antidromic responses to SC stimulation (SC-latencies) were very short (1.4-7.3 msec, mostly shorter than 4.0 msec). During the second and third weeks of age, however, the SC-latencies were very long (26-40 msec at the 2nd and 15-32 msec at the 3rd week). A CT cell with an adultlike latency as short as 6.0 msec was first recorded at the fourth week. During development, the conduction distance from the SC to the cortex must lengthen and consequently, the SC-latency might not be a good indication of axonal maturation. To exclude such a factor, the axonal conduction velocity of each CT cell was calculated as the conduction distance from OR_1 to OR_2 divided by the difference in response latency between the two sites of stimulation. In Fig. 6 the distributions of the axonal conduction velocities of all the CT cells are shown as frequency histograms for all the age groups. It is seen that all the cells sampled in kittens younger than the third week had conduction velocities slower than 2 m/sec, regardless of visual responsiveness. About one-third of CT cells attained the conduction velocities suggesting myelination of their axons at the fourth week, and about 40 percent of CT cells had adultlike values at the fifth week.

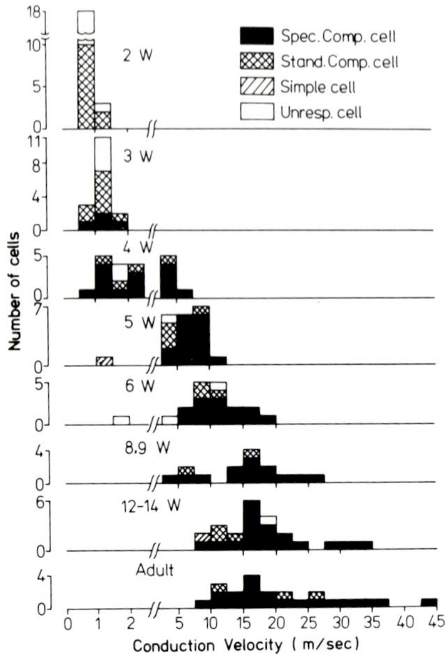

Fig. 6. Frequency histogram of axonal conduction velocities of CT cells in each age group. Filled, crosshatched, hatched and open columns represent special complex, standard complex, simple and visually unresponsive cells, respectively.

Changes in Visual Response Properties with Age

In the adult cortex, the majority of CT cells have a high frequency of spontaneous activity and complex type receptive fields characterized by vigorous responses to a fast moving slit or spot presented to either eye and by an absence of summation of responses for increased slit length (Palmer and Rosenquist 1974). Cells with these properties are called "special complex" cells (Gilbert 1977). In the present experiments, 86 percent of the CT cells recorded in the adult were of special complex type (Fig. 6). During the second week of age, 57 percent of the CT cells had complex properties but could not be classified as special complex (Fig. 6). The other CT cells showed no or rather vague visual responses. During the third week, four of the 12 visually responsive cells had properties of special complex type (Fig. 6). During the fourth week, the properties of a great majority of the CT cells were adutlike and they could be classified as special complex. The difference in the proportion of the special complex cells between the 3rd and 4th weeks was statistically significant (x^2-test, $p < 0.01$).

DISCUSSION

The results presented in Part I have demonstrated that the corticotectal projection is already established at or before birth in the cat, as reported previously (Anker and Cragg 1974; Stein and Edward 1979). However, the cells of origin of this projection (CT cells) are immature at birth in that they have very slender small somata and their dendrites are not yet recognizable in the HRP staining. This does not necessarily mean that the CT cells have no dendrites at birth. Since clearly impregnated apical dendrites of pyramidal neurons in layer V were observed with the Golgi method in neonatal or even fetal monkeys (Lund et al. 1977) and newborn rats (Juraska and Fifkova 1979), it is likely that dendrites of the HRP-labelled cells were too thin to be filled with HRP or beyond the resolution of our microscopes even if they were filled with HRP. The most rapid developmental changes occurred during the second week of age, and then the dendritic arborization pattern and the soma area became virtually adultlike at the fourth and fifth weeks, respectively, although their full maturation was not seen until the 8th week of age.

The results presented in Part II indicate that CT cells as a whole are immature in terms of visual responsiveness and axonal conduction velocity at the second week when the eyelids of the kitten open. The visual response properties develop rapidly during the third and fourth weeks, and become virtually adultlike during the fourth and fifth weeks, although a full maturation of the axonal conduction velocity does not occur until the 8th to 9th weeks. Thus, the time course of development of visual responsiveness seems to coincide with that of dendritic development following growth of the soma. This suggests that maturation of visual responses of cortical cells may be closely related to the maturation of synapses, since dendrites are known to be main post-synaptic sites in the cortex (Colonnier 1968). This suggestion is consistent with the previous findings that a dramatic change in the density of intracortical synapses and in the synaptic delay in the cortex occurs during the fourth and fifth weeks (Cragg 1975; Tsumoto and Suda 1982).

In the SC, the specific response properties of visual neurons, such as direction selectivity and binocularity, first appeared in some cells at the end of the second week and the proportion of such specific cells became virtually adultlike at the fourth week although full maturation required 7 to 8 weeks (Stein et al. 1973; Norton 1974). This time course of development of the SC cells seems to coincide very well with that of functional development of CT cells. Recently Stein and Gallagher (1981) reported that in kittens, SC neurons with specific visual responses were often affected by cortical inactivation by cooling while those without the specific responses were not. Taking these findings altogether, it seems reasonable to conclude that the maturation of visual response properties of SC neurons depends on the functional development of the CT cells.

Distribution and Density of CT cells in the Neonatal Cortex

In Part I of the present study, we found that the HRP-labelled cells were confined to layer V of the cortex at birth, as in the adult. Therefore, it is likely that the laminar arrangement of CT cells is already established before or at birth and not substantially changed after birth. In the tangential dimension, however, CT cells are located more diffusely in kittens at the first week of age than in the adult, although a rough retinotopic arrangement appears to be already established in this projection system at birth. The results in Part I have also shown that the packing density of the HRP-labelled cells in layer V of the striate cortex reduced dramatically during the second week of age. We suggest that this rapid change may not be a simple consequence of an expansion of the layer for the following two reasons: First, the rate of increasing volume of the kitten brain is almost constant during postnatal development (Cragg 1975), suggesting that the volume of layer V may increase at a constant rate. However, we observed that the decrease in cell density occurs in a stepwise fashion. Second, the decreasing density of unlabelled cells in layer V did not proceed as rapidly as that of the labelled cells during the second week. Thus, the rapid decrease in the density of CT cells may be due to either death selective for CT cells in layer V or to elimination of axon collaterals of neurons which mainly project to structures other than the SC. The latter possibility may be supported by the recent findings suggesting that exuberant callosal projections may disappear by eliminating axon collaterals of cortical neurons which send axons to the other areas (Innocenti 1981; O'Leary et al. 1981), although the former possibility cannot be excluded.

Anker RL, Cragg BG (1974). Development of the extrinsic connections of the visual cortex in the cat. J Comp Neurol 154:29.

Berman N, Cynader M (1975). Receptive fields in cat superior colliculus after visual cortex lesions. J Physiol 245:261.

Colonnier M (1968). Synaptic patterns of different cell types in the different laminae of the cat visual cortex. An electron microscope study. Brain Res 9:268.

Cragg BG (1975). The development of synapses in the visual system of the cat. J Comp Neurol 160:147.

Feldon S, Feldon P, Kruger L (1970). Topography of the retinal projection upon the superior colliculus. Vision Res 10:135.

Fuller JH, Schlag JD (1976). Determination of antidromic excitation by the collision test: problems of interpretation. Brain Res 112:283.

Gilbert CD (1977). Laminar differences in receptive field properties of cells in cat primary visual cortex. J Physiol 268:391.

Gilbert CD, Kelly JP (1975). The projections of cells in different layers of the cat's visual cortex. J Comp Neurol 163:81.

Hoffmann KP, Straschill M (1971). Influences of corticotectal and intertectal connections on visual responses in the cat's superior colliculus. Exp Brain Res 12:120.

Holländer H (1974). On the origins of corticotectal projections in the cat. Exp Brain Res 21:430.

Innocenti GM (1981). Growth and reshaping of axons in the establishment of visual callosal connections. Science 212:824.

Innocenti GM, Fiore L, Caminiti R (1977). Exuberant projection into the corpus callosum from the visual cortex of newborn cats. Neurosci Lett 4:237.

Juraska JM, Fifkova E (1979). A Golgi study of the early postnatal development of the visual cortex of the hooded rat. J Comp Neurol 183:247.

Leventhal AG (1979). Evidence that the different classes of relay cells of the cat's lateral geniculate nucleus terminate in different layers of the striate cortex. Exp Brain Res 37:349.

Lund JS, Boothe RG, Lund RD (1977). Development of neurons in the visual cortex (area 17) of the monkey (Macaca nemestrina): A Golgi study from fetal day 127 to postnatal maturity. J Comp Neurol 176:149.

Lund JS, Henry GH, Macqueen CL, Harvey AR (1979). Anatomical organization of the primary visual cortex (area 17) of the cat. A comparison with area 17 of the macaque monkey. J Comp Neurol 184:599.

McIlwain JT, Fields HL (1970). Superior colliculus: Single

unit responses to stimulation of visual cortex in the cat. Science 170:1426.

Mesulam M-M (1978). Tetramethyl benzidine for horseradish peroxidase neurochemistry: A non-carcinogenic blue reaction product with superior sensitivity for visualizing neural efferents. J Histochem Cytochem 26:106.

Mize RR, Murphy EH (1976). Alterations in receptive field properties of superior colliculus cells produced by visual cortex ablation in infant and adult cats. J Comp Neurol 168:393.

Norton TT (1974). Receptive-field properties of superior colliculus cells and development of visual behavior in kittens. J Neurophysiol 37:674.

O'Leary DDM, Stanfield BB, Cowan WM (1981). Evidence that the early postnatal restriction of the cells of origin of the callosal projection is due to the elimination of axonal collaterals rather than to the death of neurons. Develop Brain Res 1:607.

Palmer LA, Rosenquist AC (1974). Visual receptive fields of single striate cortical units projecting to the superior colliculus in the cat. Brain Res 67:27.

Stein BE, Edwards SB (1979). Corticotectal and other corticofugal projections in neonatal cat. Brain Res 161:399.

Stein BE, Gallagher HL (1981). Maturation of cortical control over superior colliculus cells in cat. Brain Res 223:429.

Stein BE, Labos E, Kruger L (1973). Sequence of changes in properties of neurons of superior colliculus of the kitten during maturation. J Neurophysiol 36:667.

Toyama K, Matsunami K, Ohno T, Tokashiki S (1974). An intracellular study of neuronal organization in the visual cortex. Exp Brain Res 21:45.

Tsumoto T, Suda K (1982). Laminar differences in development of afferent innervation to striate cortex neurons in kittens. Exp Brain Res 45:433.

Tsumoto T, Suda K, Sato H (1983). Postnatal dvelopment of corticotectal neurons in the kitten striate cortex: A quantitative study with the horseradish peroxidase technique. J Comp Neurol 219:88 .

Tsumoto T, Suda K (1983). Postnatal development of the corticotectal neurons in the kitten striate cortex: An electrophysiological study. Develop Brain Res 11:29.

Tusa RJ, Palmer LA, Rosenquist AC (1978). The retinotopic organization of area 17 (striate cortex) in the cat. J Comp Neurol 177:213.

Wickelgren BG, Sterling P (1969). Influence of visual cortex on receptive field in the superior colliculus of the cat. J Neurophysiol 32:16.

Development of Visual Pathways in Mammals, pages 231–242

AFFERENT INPUT AND DEVELOPMENT OF VISUAL CALLOSAL CONNECTIONS IN HAMSTER

Robert W. Rhoades, Ph.D.
Richard D. Mooney, Ph.D.
Dept. of Anatomy
University of Medicine and Dentistry of New Jersey-School of Osteopathic Medicine and Rutgers Medical School
Piscataway, NJ 08854

Stephen E. Fish, Ph.D.
Neurobiology Program
Northeastern Ohio Universities College of Medicine
Rootstown, OH 44272

In recent years the callosal projections of the mammalian neocortex have become an important model system for studying the normal development and plasticity of corticofugal pathways. While the ontogeny of the callosal connections between many cortical areas has been examined (eg. Wise and Jones, 1976; Ivy, Akers and Killackey, 1979; Feng and Brugge, 1983), those which interconnect the visual cortices have received particular attention. This system is especially appropriate for developmental studies since its sensory input can be readily modified or even completely eliminated.

In normal adult hamsters (Fig. 1), as in most other mammals (for reviews see Berlucchi, 1981; and Elberger, 1982) callosal cells and terminals interconnecting the striate and prestriate visual cortices (area 17 and 18 in cat; areas 17 and 18a in rodents) are located, for the most part, near the border between these two areas.[1]

---

[1] This discussion will be limited to areas 17 and 18(a) since development of callosal connections between these areas has received the greatest attention.

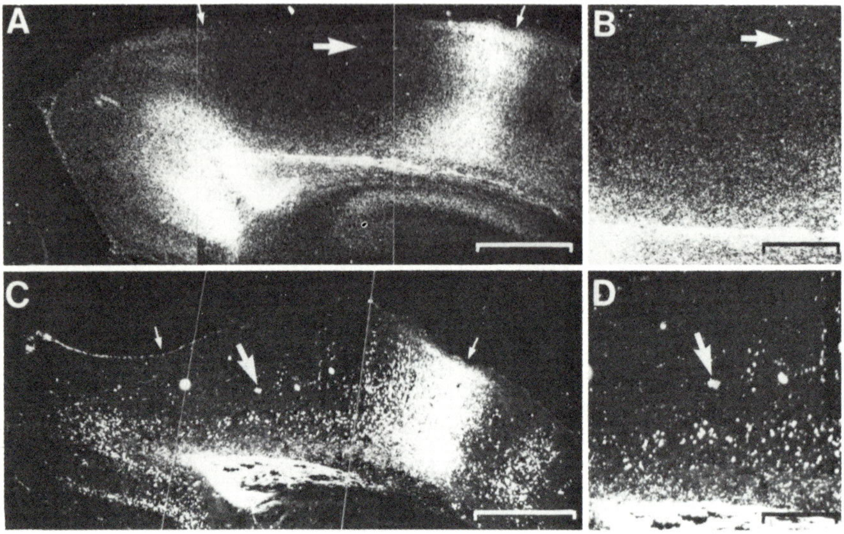

Fig. 1. A Darkfield montage showing callosal labelling as demonstrated by autoradiography, in a normal adult hamster. Label can be seen in all layers near the 17-18a border (right small arrow) and at the medial border of area 17 (left small arrow). At the 17-18a border label is heavy in layers V and VI, light in lamina IV and heavy in layers I-III. Light labelling can also be seen in lamina VI and lowermost lamina V in the medial striate cortex. B is a higher power micrograph which illustrates this medial labelling in greater detail. C is a polarized darkfield montage which illustrates the distribution of visual callosal neurons in a normal adult animal. At the 17-18a border (right small arrow) numerous labelled neurons were present in layers II-III and V. A few labelled cells were also visible in laminae IV and VI. In medial area 17 a number of labelled neurons could also be seen in lower layer V. These are shown in more detail in D. The calibrations for A and C are 1mm; those for B and D are 0.5mm. Larger arrows point to corresponding points in A and B and C and D, respectively.

Small numbers of callosal neurons are also located in medial area 17 (Fig. 1C and D; also see Dursteler, Blakemore and Garey, 1979; Miller and Vogt, 1982) and lamina VI of the entire striate cortex in rodents receives callosal input (Fig. 1C and D; also see Yorke and Caviness, 1975;Cusick and Lund, 1981; Rhoades and Fish, 1983).

The distribution of visual callosal projection neurons is not nearly so restricted in neonatal animals. Innocenti, Fiore and Caminiti (1977) first noted that visual callosal neurons were much more widespread in neonatal than in adult cats and this is now also known to be the case in hamsters (Fig. 2A; Mooney, Rhoades and Fish, 1984), rats (Miller and Vogt, 1982; Olavarria, Malach, Lee and Van Sluyters, 1983), mice (Rhoades, Mooney and Fish, 1984) and rabbits (Chow, Baumbach and Lawson, 1981). The distribution of callosal terminals shortly after birth shows considerably more variation across species. In perinatal hamsters (Fig. 2B; also see Mooney et al. 1984) callosal terminals occupy layers V and VI across the entire dorsal posterior neocortex. Widespread tangential distributions of callosal terminals have also been reported for neonatal rats (Miller and Vogt, 1982) and rabbits (Chow et al., 1981), but not for kittens (Stein and Edwards, 1979; Innocenti, 1981).

The postnatal progression from a diffuse to a restricted pattern of callosal projection neurons has been shown to involve retraction of axon branches (Innocenti, 1981; O'Leary, Stanfield and Cowan, 1981; Ivy and Killackey, 1982) <u>and</u> to be sensitive to a variety of surgical and environmental manipulations (again, see Elberger, 1982 for a review).Experiments in hamster (Fig. 3), rat and cat have demonstrated that visual input is almost certainly <u>not</u> a necessary condition for the achievement of a restricted pattern of visual callosal connectivity (Lund and Mitchell, 1979; Innocenti and Frost, 1980; Cusick and Lund, 1982), but that the presence of the retinae is required for the refinement of this projection (Fig. 4; Rhoades and Fish, 1983; Innocenti and Frost, 1980; Cusick and Lund, 1982; Olavarra et al., 1983;

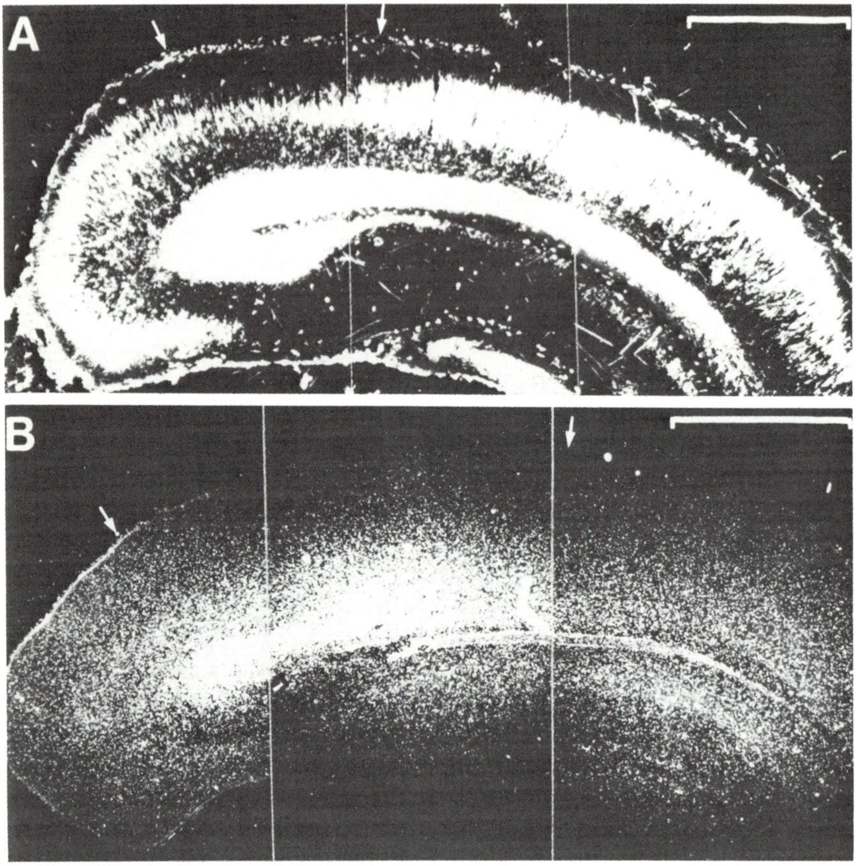

Fig. 2. A is a polarized darkfield montage showing the distribution of visual callosal cells in a 6 day old hamster. Labelled neurons are restricted primarily to the subplate region and they are continuous across the mediolateral extent of the cortex. Outside presumptive area 17 (approximate borders denoted by small arrows) the labelled neurons form a bilaminar pattern. Within area 17 only the more ventral of these two bands contains labelled cells. B is a darkfield montage which shows the visual callosal projection as demonstrated by autoradiography in a 6 day old hamster. Here silver grains are visible in the subplate region throughout presumptive area 17 (arrows). The calibrations in both A and B are 0.5mm.

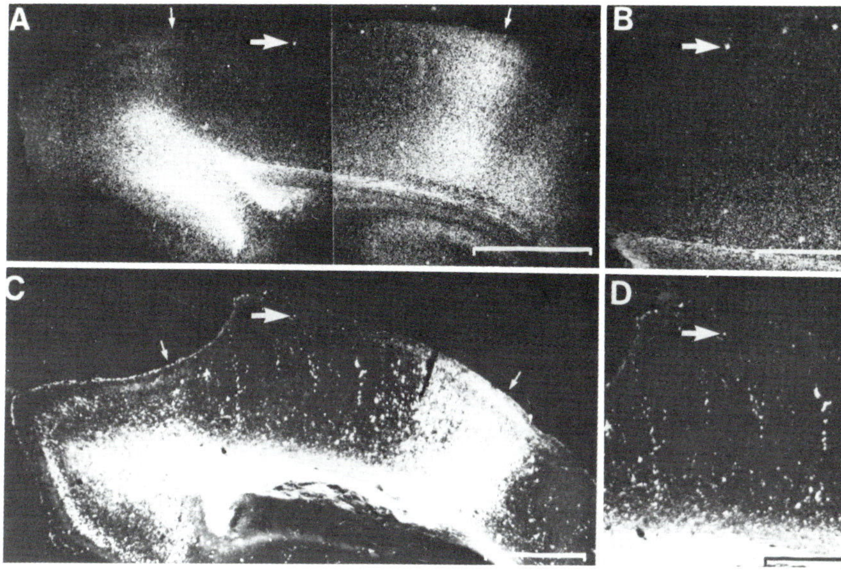

Fig. 3. Visual callosal labelling in hamster reared from birth to adulthood in total darkness. A and B illustrate terminal labelling as demonstrated by autoradiography. Note the similarity to the data from the normal animal (Fig. 1). C and D show the distribution of callosal cells in a dark reared hamster. While the number of labelled cells in medial area 17 may be slightly lower than normal, the overall distribution of labelled neurons is very similar to that in a normal animal. All conventions are as in Figure 1.

Rhoades et al., 1984). These results then indicate that some aspect of the input from the retinae to the thalamus and thence to the visual cortex is necessary for a restricted pattern of callosal connectivity, but that this retinal influence upon cortical development apparently requires no visual stimulation.

While several interpretations of the above findings are possible (see, for example, Innocenti and Frost,

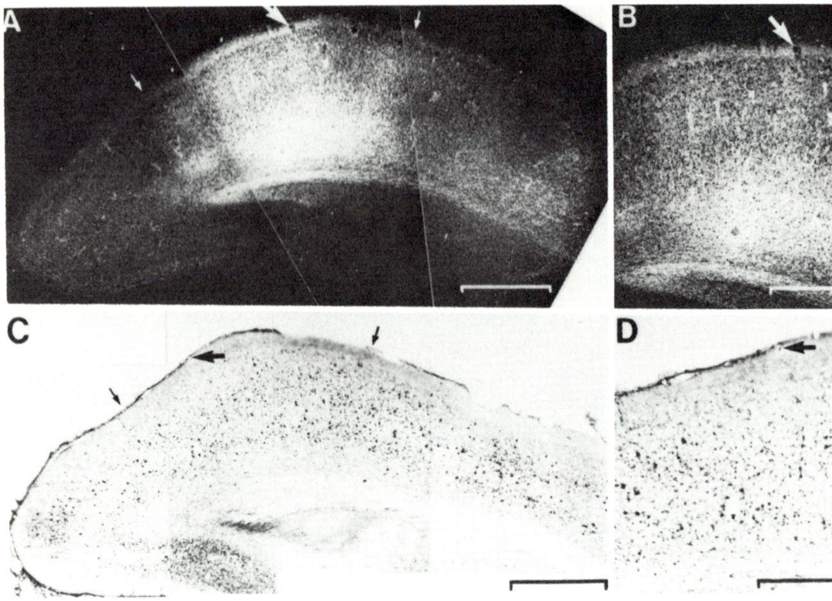

Fig. 4. A and B are darkfield photomicrographs that illustrate the autoradiographic labelling of the visual callosal projection in an adult hamster subjected to removal of both eyes at birth. Here labelling can be seen in all layers throughout area 17. C and D are lightfield photomicrographs which show the distribution of visual callosal neurons in a bilateral enucleate. Labelled cells are scattered fairly uniformly throughout areas 17 and 18a primarily in layers II-III and V. All conventions are the same as those in Figure 1.

1980), one hypothesis which appears extremely attractive is that retinal activity, even the spontaneous activity which occurs in the absence of visual input, may be critical in the focusing of callosal connections. Berlucchi (1981; also see Shatz, 1977) has suggested that a positive correlation between the firing of the thalamic and callosal afferents to a given cortical neuron may be a necessary condition for the stabilization (Changeux and Danchin, 1976) of the latter connection. Insofar as the thalamic input to a given cortical cell determines its activity, this hypothesis necessitates that the firing of the thalamic afferents to a callosal projection neuron and its target in the contralateral hemisphere must be positively correlated if the callosal connection is to be maintained. Thus if correlated activity in the two hemispheres is to play a role in the focusing of callosal connections, the thalamic inputs to the 17-18a border areas, and only these areas, must be positively correlated, even in the dark. The topographic organization of the mammalian visual system and the functional characteristics of retinal ganglion cells are consistent with this possibility.

Retrograde tracing experiments (eg. Cooper and Pettigrew, 1979; Illing and Wassle, 1981; Terao, Inatomi and Toshihiro, 1982) have shown that there is a relatively narrow band along the vertical meridian of the retina where ganglion cells may project to the main laminae of either one or the other dorsal lateral geniculate nuclei (LGNd). The portions of the geniculates innervated by the retinal vertical meridians, of course, project to the 17-18(a) borders in the two hemispheres. Recent experiments in rabbit (Arnett and Spraker, 1981) have shown that the firing of neighboring, and only neighboring, retinal ganglion cells exhibits a high positive correlation, <u>even in the dark</u>. Thus if the firing of retinal ganglion cells determines that of the cortical neurons to which they indirectly project then it is possible that the activity of cells in the 17-18(a) border regions of the two hemispheres may be positively correlated, even in the absence of visual stimulation. Under these conditions, other parts of the striate cortex would receive input from only one retina (monocular regions) or corresponding, but presumably uncorrelated, input from the two retinae.

A direct test of this hypothesis would be to alter retinal activity in a way that should have a readily predictable effect upon callosal connections. To accomplish this, we have reared hamsters from early gestation through maturity in an environment where the only illumination was that provided by brief stroboscopic flashes (see Chalupa and Rhoades, 1978). Such stimulation should synchronize ganglion cell firing throughout all portions of both retinae and thus all portions of the visual cortices in both hemispheres. If callosal connections are stabilized by positively correlated activity in the two hemispheres then one would expect an abnormal retention of such connections in these animals. This is, in fact, what occurred.

Throughout the striate cortex of the strobe reared hamsters many more callosal neurons were present in lamina V than is normally the case (Fig. 5A-D). In rostral

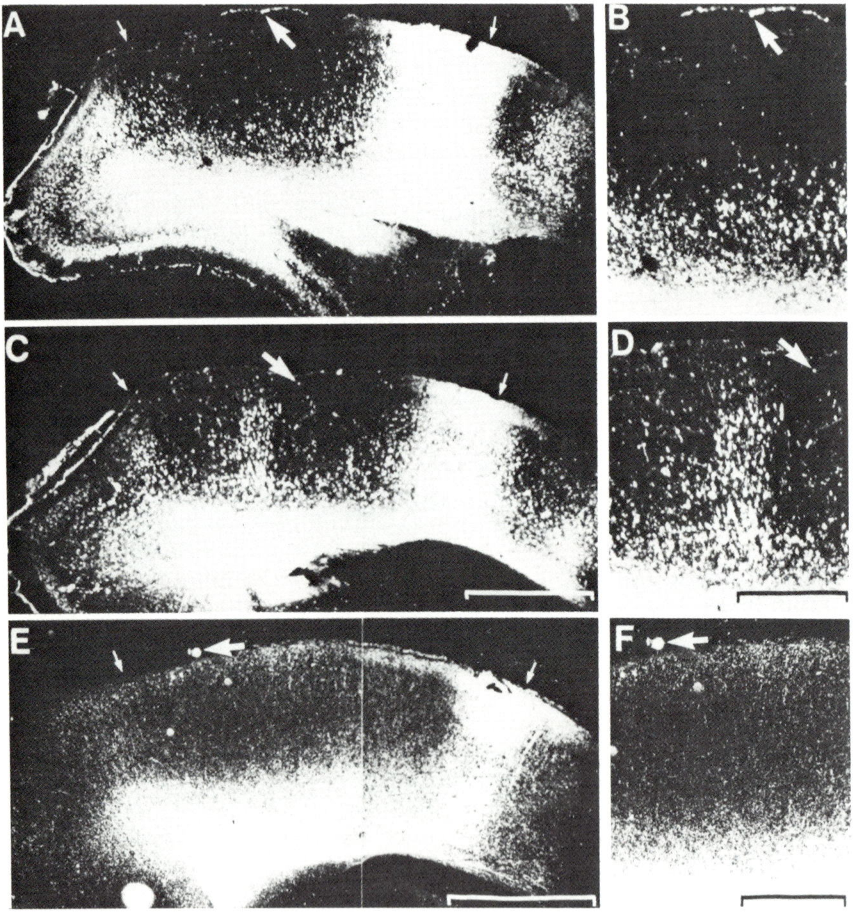

Fig. 5. A and B show the distribution of callosal neurons is a section through the posterior visual cortex in an adult hamster which was strobe-reared from birth. Note the abnormally high density of labelled cells in lamina V. C and D show the distribution of visual callosal cells in a section 250μm rostral to that shown in A and B. In addition to the high number of labelled cells in lamina V, a column of labelled neurons extending dorsally through lamina II-III is present in medial area 17. E and F illustrate visual callosal terminal labelling, as demonstrated by autoradiography, in a strobe reared hamster. Note the extremely dense labelling in lamina V and VI in the medial striate cortex and also the abnormal medial extension of the labelling in layer I. All conventions as in Figure 1. Calibrations in C and D also apply to A and B, respectively.

sections a dense patch or column of labelled cells which included both infra- and supragranular neurons was also visible (Fig. 5C and D). Autoradiogrpahic experiments in strobe reared hamsters (Fig. 5E and F) revealed an abnormally high density of callosal labelling in layers I, V and VI throughout the striate cortex.

It seems unlikely that the changes seen in the callosal projections of the strobe reared hamsters were a function of reduced visual experience since the organization of this pathway in the dark reared animals was essentially normal. Rather it seems more likely that the unusual <u>pattern</u> of cortical activity induced by the stroboscopic environment was responsible for the changes observed. That patterned activity may play an important role in the shaping of neural connections is not a new idea (eg. Changeux and Danchin, 1976; Willshaw and von der Malsburg, 1976), but it is only recently that it has begun to receive substantial experimental support in the central nervous system (eg. Stryker, 1981; Archer, Dubin and Stark, 1982; Meyer, 1983; Schmidt and Edwards, 1983).

One additional aspect of our results from the strobe reared animals that deserves mention is the maintenance of a strong focus of callosal connectivity at the 17-18a border. It may be that some special property of this cortical region serves to stabilize callosal connections (note, however, that this does not appear to be the case in neonatally enucleated hamsters). On the other hand, it may be that correlated spontaneous activity in the dark period between strobe flashes, differentially reinforced callosal connectivity in this region. These possibilities are currently being evaluated in experiments where cortical neuronal firing is silenced with tetrodotoxin.

ACKNOWLEDGEMENT

The technical assistance of Jeff Polcer, Joan Fiore, Michelle Math, Dianne Goodman and Alice Brown is gratefully acknowledged. These experiments were supported in part by NS16001, EY04170, EY03546, BNS8004601, The March of Dimes and the UMDNJ Foundation (RWR) and NS18369 (SEF). Thanks to Eleanor Kells for typing the manuscript.

REFERENCES

Archer SM, Dubin MW, Stark LA (1982). Abnormal development of kitten retinogeniculate connectivity in the absence of action potentials. Science 217:743-745.

Arnett D, Spraker TE (1981). Cross-correlation analysis of the maintained discharge of rabbit retinal ganglion cells. J Physiol 317:29-47.

Berlucchi G (1981). Recent advances in the analysis of the neural substrates of interhemispheric communication. In Pompeiano O, Marsan CA (eds): "Brain Mechanisms and Perceptual Awareness", New York: Raven Press, pp 133-152.

Chalupa LM, Rhoades RW (1978). Modification of visual response properties in the superior colliculus of the golden hamster following stroboscopic rearing. J Physiol 274:571-592.

Chow KL, Baumbach HD, Lawson R (1981). Callosal projections of the striate cortex in the neonatal rabbit. Exp Brain Res 42:122-126.

Cooper ML, Pettigrew JD (1979). The decussation of the retinothalamic pathway in the cat, with a note on the major meridians of the cat's eye. J Comp Neurol 187:285-312.

Cusick CG, Lund RD (1981). The distribution of callosal projections to the occipital visual cortex in rats and mice. Brain Res 214:239-259.

Cusick CG, Lund RD (1982). Modification of visual callosal projections in rats. J Comp Neurol 212: 385-398.

Dursteler MR, Blakemore C, Garey LS (1979). Projections to the visual cortex in the golden hamster. J Comp Neurol 183:185-204.

Elberger AJ (1982). The functional role of the corpus callosum in the developing visual system: a review. Progr Neurobiol 18:15-79.

Feng JZ, Brugge JF (1983). Postnatal development of auditory callosal connections in the kitten. J Comp Neurol 214:416-426.

Illing RB, Wassle H (1981). The retinal projection to the thalamus in the cat: a quantitive investigation and a comparison with the retinotectal pathway. J Comp Neurol 202:265-285.

Innocenti GM (1981). Growth and reshaping of axons in the establishment of visual callosal connections. Science 212:824-827.

Innocenti GM, Fiore L, Caminiti R (1977) Exuberant projection into the callosum from the visual cortex of newborn cats. Neurosci Lett 4:237-242.

Innocenti GM, Frost DO (1980). The postnatal development of visual callosal connections in the absence of visual experience or of the eyes. Exp Brain Res 39:365-375.

Ivy GO, Akers RM, Killacky HP (1979). Differential distribution of callosal projection neurons in the neonatal and adult rat. Brain Res 173:532-537.

Ivy GO, Killackey HP (1982). Ontogenetic changes in the projections of neocortical neurons. J Neurosci 2:735-743.

Lund RD, Mitchell DE (1979). The effects of darkrearing on visual callosal connections of cats. Brain Res 167:172-179.

Meyer RL (1983). Tetrodotoxin inhibits the formation of refined retinotopography in goldfish. Dev Brain Res 6:293-298.

Miller M, Vogt BA (1982). Heterotopic callosal projections and the development of callosal afferents in rat visual cortex. Soc Neurosci Absts 8:674.

Mooney RD, Rhoades RW, Fish SE (1984). Neonatal superior collicular lesions alter visual callosal development in hamster. Exp Brain Res (in press).

Olavarria J, Malach R, Lee P, Van Sluyters RC (1983). Development of visual callosal connections in normal and enucleated rats. Invest Ophthalmol Vis Sci 24:9.

O'Leary DDM, Stanfield BB, Cowan WM (1981). Evidence that the early postnatal restriction of the cells of origin of the callosal projection is due to the elimination of axonal collaterals rather than to the death of neurons. Dev Brain Res 1:607-617.

Rhoades RW, Fish SE (1983). Bilateral enucleation alters visual callosal but not corticotectal or corticogeniculate projections in the hamster. Exp Brain Res 51: 451.

Rhoades RW, Mooney RD, Fish SE. A comparison of visual callosal organization in normal, bilaterally enucleated, and congenitally anophthalamic mice. sub to J Neurosci.

Schmidt JT, Edwards DL (1983). Activity sharpens the map during the regeneration of the retinotectal projection in goldfish. Brain Res 269:29-39.

Shatz CJ (1977). Anatomy of interhemispheric connections in the visual system of Boston Siamese and ordinary cats. J Comp Neurol 173:497-518.

Stein BE, Edwards SB (1979). Corticotectal and other corticofugal projections in neonatal cat. Brain Res 161:399-409.

Stryker MP (1981). Late segregation of geniculate afferents to the cat's visual cortex after recovery from impulse blockade. Soc Neurosci Absts 7:842.

Terao N, Inatomi A, Toshihiro M (1982). Anatomical evidence of the overlapped distribution of ipsilaterally and contralaterally projecting ganglion cells to the lateral geniculate nucleus in the cat retina: a morphologic study with fluorescent tracers. Invest Ophthalmol Vis Sci 23:796-798.

Willshaw DJ, von der Marlsburg C (1976). How patterned neural connections can be set up by self-organization. Proc Roy Soc Lond B 194:431-445.

Wise SP, Jones EG (1976). The organization and postnatal development of the commissural projection of the somatic sensory cortex of the rat. J Comp Neurol 168:313-343.

Yorke CH, Caviness VS (1975). Interhemispheric neocortical connections of the corpus callosum in the normal mouse: a study based on anterograde and retrograde methods. J Comp Neurol 164:233-246.

Development of Visual Pathways in Mammals, pages 243–253

ROLE OF AXON ELIMINATION IN THE DEVELOPMENT OF VISUAL CORTEX

G.M. Innocenti

Institute of Anatomy, University of Lausanne

Rue du Bugnon 9, 1011 Lausanne-CHUV, Switzerland.

Callosal connections are now known to undergo an important postnatal reorganization in several sensory areas and animal species (Innocenti et al., 1977; Ivy et al., 1979; Innocenti and Caminiti, 1980; Chow et al., 1981; O'Leary et al., 1981; Feng and Brugge, 1983). In areas 17 and 18 of newborn kittens, for example, neurons projecting through the corpus callosum have a widespread tangential distribution; in adult cats, on the contrary, they are restricted to near the 17/18 border (Innocenti et al., 1977; Innocenti and Caminiti, 1980).

In the transition from the diffuse to the localized distribution, neurons at specific locations in neocortex eliminate their callosal axons and seem to form (or possibly maintain) connections in the ipsilateral hemisphere (Innocenti, 1981). A similar mechanism is responsible for the reshaping of callosal connections in the parietal cortex of the rat (O'Leary et al., 1981; Ivy and Killackey, 1982).

What is the rôle of this apparent wastefulness? The <u>elimination</u> of transitory callosal projections could achieve one of the following:

i) creation of acallosal sectors in neocortex.

ii) constancy of number of callosal axons in adults; a similar rôle was proposed for neuronal death in other systems (Cowan, 1973; for discussion see Katz and Grenander, 1982).

iii) elimination of transitory axon collaterals (Innocenti, 1979; O'Leary et al., 1981) as a way for individual neurons to chose their final target from a wider lot to which they may have originally projected.

The hypotheses listed above are not mutually exclusive; the question is whether they, alone or in combinations, are sufficient to account for the events involved in the reorganization of callosal connections. Here, I review recent work which suggests that this is not the case.

Although the loss of callosal projections during development has now been reported in a number of studies, the available evidence does not allow to estimate the magnitude of this loss. This estimate is critical to understand the rôle of axon loss in callosal development. The available studies have employed retrograde transport techniques and shown that in certain regions of the neocortex neurons could be labeled through the corpus callosum in newborn, but not in adult animals. Difficulties in producing injections of identical size and location in animals of different ages, and important volumetric changes in the cortex during postnatal maturation are two of the difficulties discouraging the use of retrograde transport techniques to quantify the loss of callosal connections in development. Furthermore, it is possible that maturation may not lead to real axon loss but just impair the ability of some callosal axons to take up or transport label to their cell bodies. Although such an explanation can hardly account for the loss of callosal projections observed in area 17 (for discussion, see Koppel and Innocenti, 1983), even for this region the magnitude of axonal loss in normal development remains impossible to estimate from retrograde transport data.

A quantitative electron microscopic study (Koppel and Innocenti, 1983; Fig. 1) has established a new value of 23 million axons in the corpus callosum of the adult cat. This figure is almost 5 times larger than the currently accepted figure (for references, see Doty, 1973) of 5 million, based on light microscopic data. However, even more striking was our figure of 79 million axons found in the newborn kitten (Fig. 1). More than 70 % of the axons present at birth have been lost by adulthood: at birth, there would be enough axons to constitute 3.3 corpora callosa. This loss is considerably greater than that reported in other fibre tracts (Rager and Rager, 1976; Ng and Stone, 1982; Reh and Kalil, 1982; Rogers and Ehrlich, 1983; Rakic and Riley, 1983; Williams et al., 1983). But the loss may even have been underestimated for the following reasons: i) some axons may have already been lost prenatally, and ii) new axons may still be growing into the corpus callosum during the period of axonal loss.

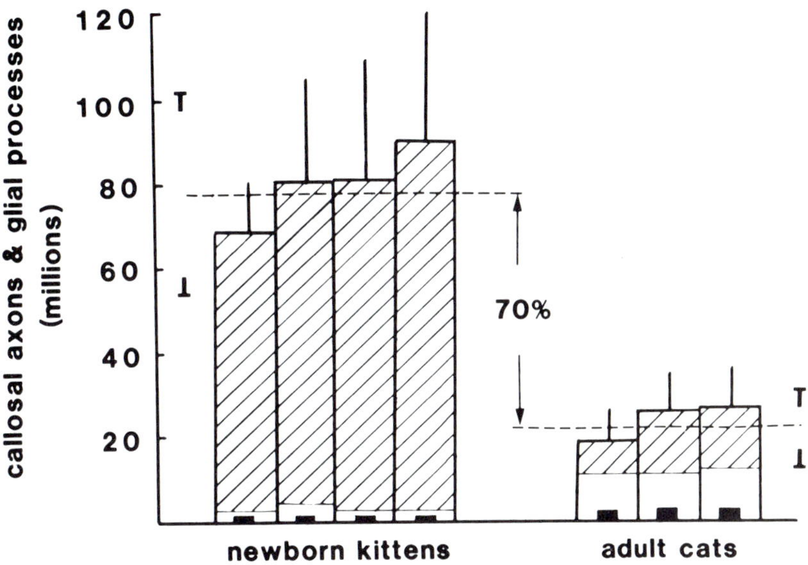

Fig. 1. Histogram showing total numbers of elements per corpus callosum in each of the seven animals analysed. The bars, indicating total axon numbers (mean values per animal; standard deviations represented by vertical lines) have been subdivided; in the newborn animal the large (diameter greater than 0.5 μm) and small axons are represented by open and hatched sectors, respectively, while in the adult the same conventions denote unmyelinated and myelinated axons. The small filled rectangles represent numbers of glial processes. Mean numbers of axons for each age group are indicated by the dashed horizontal lines and their standard deviations by the two pairs of "Ts" on the left and right of the histogram (from Koppel and Innocenti, 1983, with permission).

Even a superficial glance at the available maps of adult callosal connections (Ebner and Myers, 1965; see their Fig. 1) shows that the acallosal portions of an adult neocortex are far less than 70 % and at birth, the density of callosally projecting neurons is not particularly high in the regions destined to become acallosal (Innocenti et al., 1977; Innocenti and Caminiti, 1980). Therefore the developmental elimination of callosal axons is disproportionately large for the task of creating acallosal sectors in neocortex.

An additional although preliminary finding is that of important differences in the number of axons in the corpora callosa of different adults. Individual differences in the number and distribution of callosally projecting neurons in adult cats were also noticed in retrograde transport studies (Innocenti and Frost, in preparation). Thus, apparently, the elimination of transitory axons in development does not achieve absolute constancy of callosal axon number although it may achieve constancy of axon number relative to some other brain parameter.

A hypothesis shared by different groups (Innocenti, 1979; O'Leary et al., 1981; Ivy and Killackey,1982) is that immature cortical neurons, and in particular those which give rise to transitory callosal axons, may possess a number of axon collaterals. Cortical neurons might select their target through an indiscriminate growth of axon collaterals followed by elimination of the "inappropriate" ones. One (or several) collaterals could course through the corpus callosum and this could theoretically account for the high number of callosal axons in newborn kittens. We have tested this possibility by simultaneously injecting the fluorescent tracers diamidino yellow (DY) and fast blue (FB) (Bentivoglio et al., 1980; Kaizer et al., 1983) into different areas of the same or of different hemispheres in newborn kittens (Innocenti and Clarke, 1983b). In some animals, areas 17/18 and 19/21a or PMLS were injected in the same hemisphere; in others PMLS was injected in the two hemispheres. We looked for double-labeled neurons in all visual areas, and most carefully in areas 17, 18 and PMLS.

Previous studies had shown (Segraves and Innocenti, 1982) that in adult cats, different sets of neurons project from areas 17/18 to different contralateral areas as well as to the same or different areas in the two hemispheres.

Here, as in other areas studied thus far (Catsman-Berrevoets et al., 1980; Graziosi et al., 1982; Schwartz and Goldman-Rakic, 1982; Segraves and Innocenti, 1982) an extremely small minority (usually about 1 % in our experiments, but precise figure varies with area) of adult cortical neurons can be shown to send axon collaterals to widely separate targets.

The results obtained in kittens were unexpectedly similar to those in the adults: few neurons with collateralized axons were found; each area seemed to receive from a private set of efferent neurons. Therefore, in cortical regions such as medial area 17, which will become acallosal

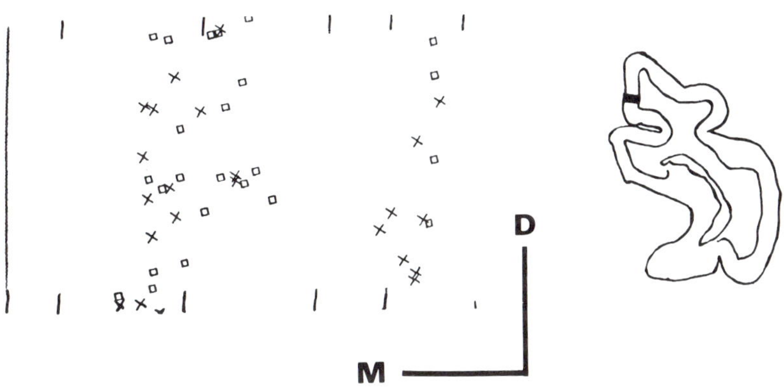

Fig. 2. Distribution of area 17 neurons projecting transitorily through the corpus callosum to different contralateral targets. Crosses and squares denote FB and DY labeled neurons, respectively; segments mark the boundary of layers (I, II + III, IV, V, VI) as determined after counterstaining. Sector of cortex corresponds to the black areas in the inset. Kitten injected on postnatal day 2 in areas 17/18 and in area PMLS of the same hemisphere and sacrificed on postnatal day 11. Crosses and squares, respectively: transitorily-callosal neurons projecting to a broad area including PMLS, 21a and 19, and transitorily-callosal neurons projecting to contralateral 17 and 18. No neurons were double labeled in the regions shown and very few in the other areas. Bars correspond to 300 µm. M: medial; D: dorsal.

in the adult, several different sets of neurons exist in development, each projecting temporarily to a different contralateral area (Fig. 2). Thus, several weeks before the transitory callosal axons are eliminated, single neurons have already sent an axon to one or perhaps few target areas. It appears improbable that the overproduction and elimination of callosal axons should be interpreted as a strategy whereby cortical neurons seek their targets through indiscriminate growth and selective elimination of axon collaterals.

The results of a third series of experiments (Innocenti and Clarke, 1983a; Clarke and Innocenti, 1983) revealed that, unlike we had known thus far, in newborn kittens transitory connections exist between certain parts of cortex which will not become acallosal. This elimination may account for a large portion of the axonal loss, and it also gives some indication as to the general rôle of axon elimination in neocortical development. The work was stimulated by the observation that, while in newborn kittens callosal neurons are diffusely distributed, callosal axons have entered the cortical gray matter at the restricted tangential locations where they will also be found in the adult (Innocenti, 1981). In addition, however, a more widely distributed axonal population exists in the white matter, but apparently either does not enter the gray matter or remains confined to layer VI (Innocenti, 1981). One unanswered question concerned the topographical origin of the axons which enter the gray matter and of those which do not.

We found that injections of retrograde fluorescent tracers restricted to the gray matter labeled neurons in the contralateral hemisphere only when regions corresponding to those entered by the callosal axons were encroached. The neurons labeled by these injections were restricted to the locations (e.g. the 17/18 borders) where they will also be found in the adult; outside these locations (e.g. in most of area 17) labeled neurons are extremely rare.

On the contrary, injections extending into the white matter, or restricted to the latter, labeled callosal neurons not only at the places where they are also found in the adults but also in regions which will become acallosal. In particular, injections into the white matter of area 17/18 labeled a crescent shaped, medio-laterally elongated territory, spanning without interruption areas 17, 18, 19, 21a, and the suprasylvian visual areas. Surprisingly, however, this territory is not restricted to the visual areas. Medially it extends, although inconsistently, over the limbic cortex in the lower bank of the splenial sulcus; laterally, it invariably comprises the auditory fields AI, AII, and DP (Fig. 3).

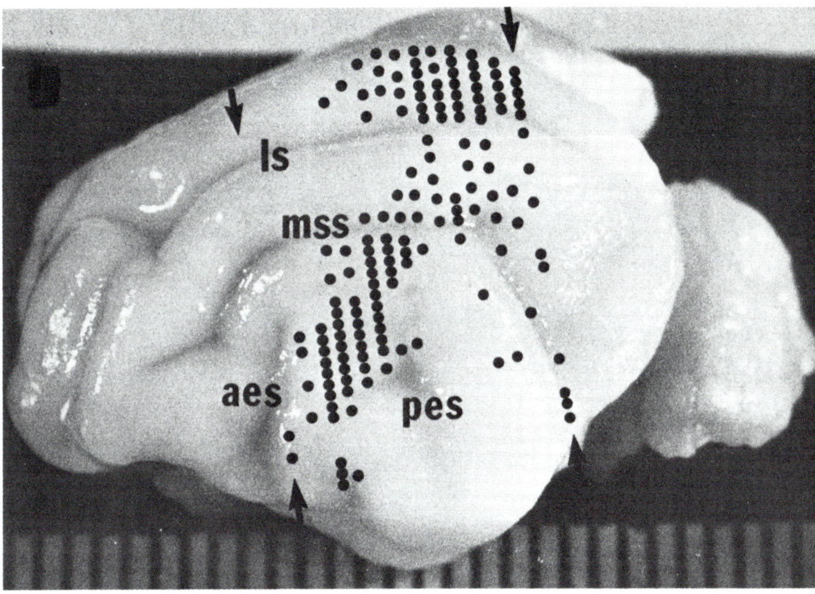

Fig. 3. Lateral view photograph of a kitten brain showing the distribution of neurons retrogradely labeled by FB (filled dots). The hemisphere contralateral to a FB injection (on day 2 1/2, sacrificed on day 9 1/2) in the caudal part of areas 17/18 is shown (the center of injection was 6.58 mm rostral from the occipital pole). In each photograph arrows mark rostral and caudal limits of the reconstructed region. Dots are arrayed parallel to the plane of sectioning; their density along an array gives a rough indication of the density of labeled neurons. The distance between arrays corresponds to that between the sections examined. ls = lateral sulcus; mss = middle suprasylvian sulcus; pes = posterior ectosylvian sulcus; aes = anterior ectosylvian sulcus. The cortex dorsal and rostral to pes contains the A1 and A2 auditory areas. Calibration is in millimeters (small divisions).

The callosal projection from the auditory to the visual cortex is transitory. Only a few neurons can occasionally be labeled in the auditory areas of adult cats, even when much more massive injections of fluorescent tracers or HRP into areas 17 and 18 are used than in the kittens. At what age the elimination occurs is unclear. Preliminary experiments suggest that by postnatal day 38, most of the elimination has taken place. As for the transitory callosal projection from area 17 (Innocenti, 1981), the callosal projection from the auditory to the visual cortex seems to disappear by axonal elimination rather than by neuronal death.

The projection from auditory to visual cortex is bilateral. The ipsilateral (associational) projection is similar in its tangential origin and fate to the callosal one.

Neurons giving rise to the various cortico-cortical projections are radially segregated in the auditory cortex (Clarke and Innocenti, 1983). Neurons projecting transitorily to ipsilateral areas 17/18 are in layer II and in upper part of layer III; callosal neurons projecting transitorily to contralateral area 17/18 are deeper in layer III, and those projecting to contralateral auditory cortex are still deeper in layer III and in layer IV; at least part of this latter projection remains into adulthood (Feng and Brugge, 1983). In addition, a few neurons in layers VI and exceptionally in V participate in all the projections. Similarly to what described above for area 17, very few neurons send bifurcating axons into more than one projection.

These findings indicate that the elimination of transitory axons is not restricted to the corpus callosum but is a more general event in the organization of cortico-cortical connections. The elimination of transitory cortico-cortical axons in development seems to determine i) which areas - and ii) which "points" of each area will be connected. In both cases, a problem of a topographical nature is solved i.e. the creation of selective connections between specific cortical locations.

It is remarkable, however, that even before the elimination of transitory axons, callosal connections are not unselective. Each area, and each "point" within an area, receives from a restricted set of areas and "points" within these areas although less restricted than in the adult (Innocenti and Clarke, 1983a).

This early specificity seems to conform to organizing principles different from those of the adult nervous system; for example, in adult cats primary sensory areas are not directly connected, but some of them are, in newborn kittens. What these "primitive" principles may be and what dictates them is a challenging question for future work.

Bentivoglio M, Kuypers HGJM, Catsman-Berrevoets CE, Loewe H, Dann O (1980). Two new fluorescent retrograde neuronal tracers which are transported over long distance. Neurosci Lett 18:25.

Catsman-Berrevoets CE, Lemon RN, Verburgh CA, Bentivoglio M, Kuypers HGJM (1980). Absence of callosal collaterals derived from rat corticospinal neurons. A study using fluorescent retrograde tracing and electrophysiological techniques. Exp Brain Res 39:433.

Chow KL, Baumbach HD, Lawson R (1981). Callosal projections of the striate cortex in the neonatal rabbit. Exp Brain Res 42: 122.

Clarke S, Innocenti GM (1983). Bilateral transitory projections from auditory to visual areas in kittens. Neurosci Abs in press.

Cowan WM (1973). Neuronal death as a regulative mechanism in the control of cell number in the nervous system. In Rockstein M (ed): "Development and Aging in the Nervous System," New York and London: Academic Press, Inc., p 19.

Doty RW, Negrao N (1973). Forebrain commissures and vision. In Autrum H et al (eds): "Handbook of Sensory Physiology," vol. VII/3, Berlin: Springer Verlag, p 543.

Ebner FF, Myers RE (1965). Distribution of corpus callosum and anterior commissure in cat and raccoon. J Comp Neurol 124:353.

Feng JZ, Brugge JF (1983). Postnatal development of auditory callosal connections in the kitten. J comp Neurol 214:416.

Graziosi ME, Tucci E, Barbaresi P, Ugolini G, Manzoni T (1982). Cortico-cortical neurones of somesthetic area SI as studied in the cat with fluorescent retrograde double-labelling. Neurosci Lett 31:105.

Innocenti GM (1979). Two types of brain plasticity? Progr Brain Res 51:479.

Innocenti GM (1981). Growth and reshaping of axons in the establishment of visual callosal connections. Science 212:824.

Innocenti GM, Caminiti R (1980). Postnatal shaping of callosal connections from sensory areas. Exp Brain Res 38:381.

Innocenti GM, Clarke S (1983a). Transitory callosal projections from the visual cortex of newborn kittens: II. Topographic organization. Neurosci Lett Suppl. 14:S177.

Innocenti GM, Clarke S (1983b). Multiple sets of visual cortical neurons projecting transitorily through the corpus callosum. Neurosci Lett in press.

Innocenti GM, Fiore L, Caminiti R (1977). Exuberant projection into the corpus callosum from the visual cortex of newborn cats. Neurosci Lett 4:237.

Ivy GO, Akers RM, Killackey HP (1979). Differential distribution of callosal projection neurons in the neonatal and adult rat. Brain Res 173:532.

Ivy GO, Killackey HP (1982). Ontogenetic changes in the projections of neocortical neurons. J Neurosci 2:735.

Katz MJ, Grenander U (1982). Developmental matching and the numerical matching hypothesis for neuronal cell death. J theor Biol 98:501.

Keizer K, Kuypers HGJM, Huisman AM, Dann O (1983). Diamidino yellow dihydrochloride (DY-2HCl); a new fluorescent retrograde neuronal tracer, which migrates only very slowly out of the cell. Exp Brain Res 51:179.

Koppel H, Innocenti GM (1983). Is there a genuine exuberancy of callosal projections in development? A quantitative electron microscopic study in the cat. Neurosci. Lett. in press.

Ng AYK, Stone J (1982). The optic nerve of the cat: appearance and loss of axons during normal development. Develop Brain Res 5:263.

O'Leary DDM, Stanfield BB, Cowan WM (1981). Evidence that the early postnatal restriction of the cells of origin of the callosal projection is due to the elimination of axonal collaterals rather than to the death of neurons. Develop Brain Res 1:607.

Rager G, Rager U (1976). Generation and degeneration of retinal ganglion cells in the chicken. Exp Brain Res 25:551.

Rakic P, Riley KP (1983). Overproduction and elimination of retinal axons in the fetal rhesus monkey. Science 219:1441.

Reh T, Kalil K (1982). Development of the pyramidal tract in the hamster. II. An electron microscopic study. J Comp Neurol 205:77.

Rogers LJ, Ehrlich D (1983). Asymmetry in the chicken forebrain during development and a possible involvement of the supraoptic decussation. Neurosci Lett 37:123.

Schwartz ML, Goldman-Rakic PS (1982). Single cortical neurones have axon collaterals to ipsilateral and contralateral cortex in fetal and adult primates. Nature 299:154.

Segraves MA, Innocenti GM (1982). An examination of the projections of cat visual callosal neurons using double retrograde tracers. Neurosci Lett Suppl. 10:S442.

Acknowledgements: Supported by the Swiss National Science Foundation grants nr. 3.628.0.80 and 3.422.0.83.

Reorganization of Visual System Following Damage or Transplantation

Development of Visual Pathways in Mammals, pages 257–274

REMOVAL OF VISUAL CORTEX IN THE CAT: EFFECTS ON THE MORPHOLOGICAL DEVELOPMENT OF THE RETINO-GENICULO-CORTICAL PATHWAY

Ronald E. Kalil

Department of Ophthalmology
University of Wisconsin
Madison, Wisconsin 53706

When visual cortex (areas 17, 18 and 19) is damaged in the newborn kitten, structural changes occur at all levels of the retino-thalamo-cortical pathway. By comparison, when adult cats receive comparable brain damage the anatomical changes that occur in the central visual pathway are limited in extent, and take place chiefly in the thalamus. The differences between the infant and adult visual system in response to similar cortical lesions are of considerable interest because they may help us to understand the morphological basis of functional recovery from brain damage. Since recovery is often superior when brain damage is experienced early in development rather than in adulthood (Spear, '79), a comparison of some of the anatomical consequences of removing visual cortex in newborn and adult cats will be presented in this chapter in order to provide a structural basis for some of the functional mechanisms that may be involved.

METHODS

Most of the observations reported here were made in adult cats. Unless noted otherwise these animals received unilateral lesions of visual cortex on the day of birth or as adults, and they survived for six months to one year before the effects of the cortical lesions were studied anatomically.

Special care was taken in an attempt to remove nearly identical regions of cortex in both newborn and adult cats.

At both ages, the lesions were intended to remove the lateral and postlateral gyri and the splenial cortex on the medial bank of the hemisphere. Details of the surgical procedures used to make the lesions and the histological methods employed to evaluate their extent, along with representative examples of typical lesions in newborn and adult cats are presented elsewhere (Spear et al., '80).

RETINA

To study the effects of visual cortex lesions on retinal ganglion cells, flat mounts of the retina were stained with cresyl violet and three sets of measurements were made on the ganglion cells from each eye. First, the cross-sectional areas of all ganglion cells located within specific areas of each retina were measured. The regions chosen for these measurements were located paracentrally, about 2.5 mm above the area centralis and 1.0 mm nasal or temporal to the zero vertical meridian. Since the area used for each set of cell size measurements was held constant at 0.13 mm^2, the number of cells encountered in each region gave an accurate measure of cell density. However, since the number of large ganglion cells is low, relative to cells that are small or medium in size, too few large cells are sampled in an area of only 0.13 mm^2 to provide reliable information about their size or density. Therefore, two additional sets of measurements were made on large cells only. Cross-sectional areas were determined by measuring the sizes of 100 large cells in the nasal and temporal hemi-retinae of each eye. These measurements were made in two regions that were each 1.2 mm^2, and centered 2.4 mm above and 1.6 mm nasal or temporal to the area centralis. Finally, average large cell density was calculated for upper nasal and temporal quadrants of each eye by counting all of the large cells present. For each set of measurements, comparisons were made between corresponding hemi-retinae of the two eyes.

Frequency histograms show that the size distribution of ganglion cells in the normal adult cat is trimodal (Boycott and Wässle, '74; Stone and Keens, '80; Hughes, '81). In order to make quantitative comparisons in the present experiments, ganglion cells were grouped according to cell body size, measured in terms of cross-sectional area, as follows: small ganglion cells, $50\mu^2$ to $250\mu^2$; medium sized ganglion cells, $251\mu^2$ to $650\mu^2$; and large ganglion cells, $850\mu^2$ to

$1900\mu^2$. Our measurements indicate that the mean cross-sectional area of small ganglion cells is $164\mu^2 \pm 2.0$ (S.E), that for medium sized cells is $394\mu^2 \pm 11.2$ (S.E.), while large cells average $1335\mu^2 \pm 48.8$ (S.E).

In cats with visual cortex damage as adults there is no change in the distribution of ganglion cell size. The relative proportions and absolute numbers of small, medium and large sized ganglion cells appear essentially identical to those of normal animals. In contrast, marked retrograde transneuronal degeneration of retinal ganglion cells is evident in cats with visual cortex damage at birth. Moreover this degeneration is very specific and appears to affect only medium sized ganglion cells (fig. 1). Thus when

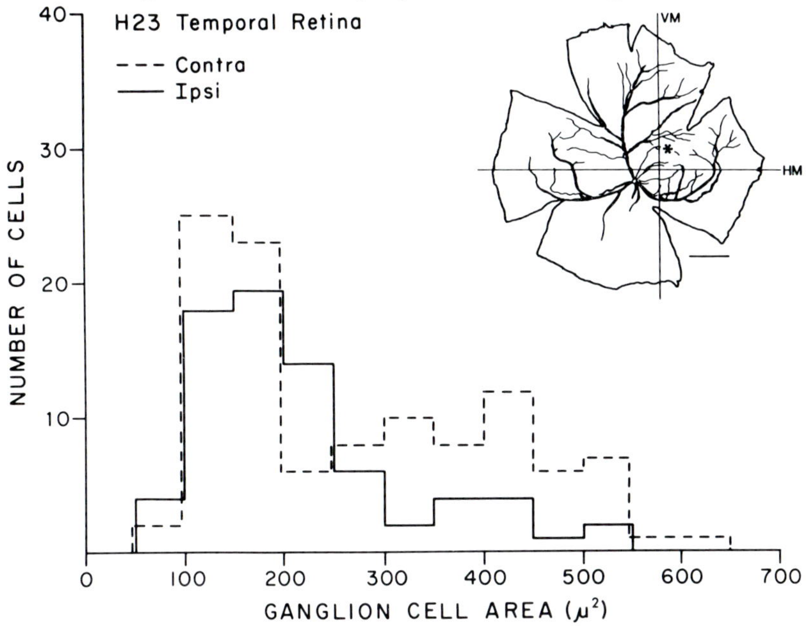

FIG. 1. Distribution of small and medium sized ganglion cells from the temporal hemi-retinae of an 8 month old cat that had received a unilateral visual cortex lesion on the day of birth. The histograms show a 64% loss of medium sized ganglion cells in the retina ipsilateral to the cortical lesion, but no change in the number of small ganglion cells. Large ganglion cells, also unchanged, are not represented. The asterisk in the inset diagram indicates the retinal location where these measurements were made. See text for details. Scale bar is 0.5 cm.

the affected hemi-retina of each eye is compared with the corresponding normal hemi-retina of the other eye, cell size measurements show a loss of medium sized ganglion cells that averages about 60% (Kalil, '80; Pearson et al., '81). There appears to be no degeneration of small ganglion cells, nor is there any measurable change in the size or density of large cells. This selective loss of medium sized ganglion cells predicts a functional deficit in X-type ganglion cells, but not in W- or Y- cells. Recent physiological experiments have confirmed this prediction (Tong et al., '82b).

LATERAL GENICULATE NUCLEUS

When visual cortex is damaged in newborn kittens the postnatal development of the dorsal lateral geniculate nucleus (LGN) is affected dramatically. One very obvious consequence of early cortical removal is the failure of the LGN to complete its normal postnatal rotation in the para-sagittal plane. In the newborn kitten the rostrocaudal axis of the LGN is oriented almost vertically such that the rostral pole of the nucleus lies ventral to the caudal tail. During the first two postnatal months, however, the LGN normally rotates approximately ninety degrees to assume the near horizontal orientation seen in adult animals (Kalil, '78). If visual cortex is removed at birth, the rotation does not take place. Instead, the LGN remains permanently in the vertical orientation characteristic of newborn animals.

Neurons in the LGN undergo rapid retrograde degeneration following removal of visual cortex at birth. As early as 3.5 days postoperatively the layers of the affected LGN have shrunk to about one-half normal thickness, and most of the neuronal cell bodies stain very palely with cresyl violet. Although clear examples of dying neurons are not numerous at 3.5 days survival, a massive invasion of the LGN at this time by glial cells suggests that significant cell death is occuring. However, not all neurons in the LGN are undergoing retrograde change. Thus in the 3.5 day old kitten there are cells sprinkled throughout the LGN that appear to be unaffected by the cortical damage. In contrast to their neighbors these cells are not atrophic and Nissl staining is normal.

When cats with visual cortex removed at birth are examined as adults, it is clear that the LGN is severely shrunken, total volume being only 14% of normal. Despite this, the basic laminar organization of the nucleus is preserved. Moreover, throughout the LGN large neurons can be seen that appear to have survived the effects of early cortical damage. These large, spared neurons are scattered through laminae A and A1, and are especially prominent ventral to lamina A1 in the region of the C layers (see Fig. 1, Murphy and Kalil, '79).

Large, surviving neurons are rarely observed in the LGN following visual cortex damage in adult cats. Instead, after postoperative survivals of six months to one year, the cross-sectional areas of LGN neurons are remarkably uniform, but only about one-half normal size.

The differences in LGN cell size between normal adult cats and animals with visual cortex damage, either at birth or in adulthood, are illustrated in the frequency histograms of Figure 2. Each histogram plots the distribution of the cross-sectional areas of 200 neurons from lamina A of the LGN after embedding in low viscosity nitrocellulose. The top panel shows that in normal adult cats the cell size distribution in the LGN is unimodal, but with a broad range. In the particular animal shown here, mean cell size is $201\mu^2 \pm 5.0$ (S.E.), and the range in size is about fivefold. The center panel illustrates the relative uniformity of LGN cell size 12 months after removal of visual cortex in an adult cat. Mean cell size is $107\mu^2 \pm 1.8$ (S.E.), about 50% smaller than normal, and almost all of the cells are between $50\mu^2$ and $150\mu^2$. At first glance, the distribution of cell sizes in the LGN of an adult cat with visual cortex damage at birth (bottom panel) does not appear to differ significantly from that seen after an adult lesion. Thus after a newborn lesion, mean cell size is also about 50% smaller than normal, $113\mu^2 \pm 5.4$ (S.E.), and most of the cells are also between $50\mu^2$ and $150\mu^2$. Nevertheless, the cell size distribution histogram confirms the qualitative observation mentioned above, namely that after a lesion of visual cortex on the day of birth the presence of large, spared neurons in the LGN is a characteristic and distinguishing feature.

RETINOGENICULATE PROJECTION

Despite the severe retrograde degeneration seen in the LGN following removal of visual cortex at birth,

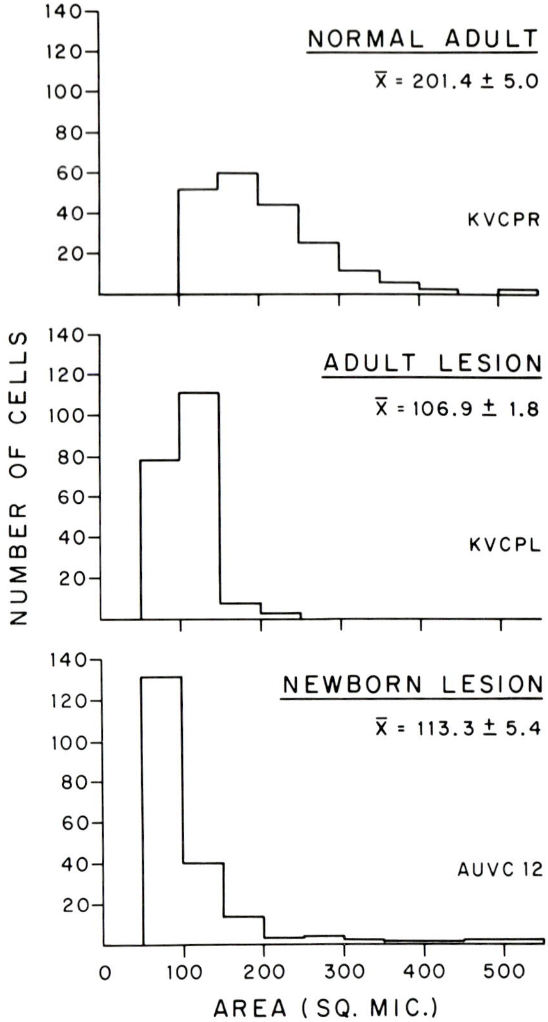

FIG. 2. Cell size distributions in lamina A of the LGN in a normal cat (top panel) and in cats with lesions of visual cortex. The middle panel shows LGN cell sizes 12 months after a cortical lesion in an adult cat and the bottom panel illustrates results from a cat that survived 14 months following damage to visual cortex on the day of birth. Note that in cats with cortical lesions the mean cross-sectional area of lamina A cells is about one-half that of normal size, whether the lesion occurs at birth or in adulthood. However, large neurons are found in lamina A only after early lesions.

electrophysiological recordings show that many cells in the degenerated nucleus continue to receive input from the retina (Murphy and Kalil, '79). The organization of this input is adequate to maintain a normal retinotopic map at a gross level, although the fine grain topography appears to be disrupted. Thus, recordings from single microelectrode penetrations show that closely spaced cells may have disparate receptive field locations, in contrast to the precise registration seen in normal cats. In addition, the average receptive field diameter of neurons recorded in the degenerated LGN is about three times larger than normal.

Light Microscopy

In order to clarify the anatomical basis for these physiological results, several studies were undertaken to map the retinogeniculate pathway at light and electron microscopic levels. With the light microscope, the retinogeniculate projection in adult cats with a lesion of visual cortex at birth was investigated using the Fink-Heimer degeneration technique (Fink and Heimer, '67), following removal of one eye, or the autoradiographic method for tracing axonal connections (Edwards and Hendrickson, '81), after intraocular injection of ^{3}H proline.

In the Fink-Heimer material, degenerating optic axons in the LGN ipsilateral to the early visual cortex lesion are very dense, but they do not show the clear organization of the normal retinogeniculate projection. The normal organization can be seen in the LGN contralateral to the early visual cortex lesion. Here, degenerating retinogeniculate axons tend to run in an orderly fashion along the "lines of projection" (Sanderson, '71), which define the retinotopic map. However, in the LGN ipsilateral to the cortical lesion axons from the retina form a disorganized feltwork that contains many fibers passing across the lines of projection. In addition, preterminal axonal degeneration is especially heavy in the vicinity of large surviving geniculate neurons.

The autoradiographic method yields similar results. With this technique, label transported from the retina to the normal LGN is distributed evenly throughout the appropriate target lamina. By contrast, the retinal projection to the degenerated LGN is marked by patches of label, indicating a nonuniform pattern of innervation. Under high power it can be seen that these patches are frequently composed of discrete clusters of silver grains arrayed close to the proximal dendrites of large surviving geniculate cell bodies.

Thus these two independent light microscopic methods show that retinal ganglion cells that survive an early visual cortex lesion maintain an axonal projection to the degenerated LGN. Although this pathway is dense, it lacks the orderly arrangement of the normal intrageniculate retinal projection, and instead it appears that axons from the retina may be converging onto surviving lateral geniculate neurons. Such convergence could explain the abnormal scatter and large size of LGN cell receptive fields in cats with early lesions of visual cortex.

The retinogeniculate projection in cats with visual cortex damage as adults has been studied only with autoradiographic techniques at the light microscopic level. The projection appears denser than normal, but this can be attributed undoubtedly to the overall shrinkage in the volume of the LGN with consequent compacting of retinal axon terminals. Aside from this, there is no obvious light microscopic evidence to suggest that the retinogeniculate pathway is altered by damage to visual cortex in adult cats. With the electron microscope, however, it can be seen that retinogeniculate synaptic organization is modified significantly after removal of visual cortex at birth or in adulthood (Kalil and Behan, '82).

Electron Microscopy

In the normal adult cat, axon terminals from the retina can be recognized unequivocally because they contain round synaptic vesicles and pale mitochondria, and are relatively large in size. These axon endings, generally called RLP terminals (Guillery, '69), make synaptic contacts in the LGN with only two classes of postsynaptic profiles. One class, which can be identified by the presence of numerous ribosomes, are dendrites of LGN neurons. However, the origins of the second class of postsynaptic profiles, named F because they contain flattened or pleomorphic synaptic vesicles, are not fully understood, although both intra- and extra-geniculate sources have been proposed

(Famiglietti and Peters, '72; Sterling and Davis, '80; Montero and Scott, '81). An important fact to keep in mind is that during development, RLP terminals do not begin to make significant numbers of synaptic contacts with F profiles until near the beginning of the second postnatal month (Kalil and Scott, '79). Thus, at birth, retinogeniculate synaptic patterns are very immature, consisting chiefly of simple contacts between RLP terminals and dendrites. Therefore, an early visual cortex lesion may interfere with the development of retinogeniculate synaptic patterns. By contrast, visual cortex removal in adult cats can only affect synaptic patterns that have already matured.

To determine whether visual cortex removal in infants or adults does indeed alter retinogeniculate connectivity, normal cats were analyzed first to establish quantitative baselines for specific aspects of retinogeniculate synaptic patterns. In brief, large area photographic montages from layer A or the C layers of the LGN were used to measure the percentage of RLP terminals that can be seen to make synapses when single thin sections are studied. In addition, the number of synapses made by each RLP terminal, as seen in a single thin section, with dendrites or F profiles was also counted.

In lamina A of normal cats 60% of the RLP terminals observed in the plane of a single thin section make synaptic contacts. On average, each RLP makes about 1.5 contacts, with a strong preference for dendrites since RLP-dendritic contacts are observed approximately 3.2 times as often as RLP synapses with F profiles. In montages from the C layers of the normal cat, about 65% of the RLPs make synaptic contacts. The mean number of contacts per RLP is 1.2, and in the C layers the ratio of RLP contacts with dendrites to RLP contacts with F profiles (D:F) is more than twice that of lamina A. This strong bias in favor of contacts with dendrites reflects a marked reduction in the relative number of RLP synapses with F profiles in the C layers.

When visual cortex is removed at birth, the percentage of RLP terminals that make synapses in the LGN remains essentially unchanged from normal. In montages from lamina A of the degenerated LGN 64% of the RLPs make synaptic contacts, while in the C layers the average is 63%. Also unchanged from normal is the mean number of synapses made by each RLP in lamina A (1.4/RLP), while in the C layers there appears to be a slight increase (1.5/RLP). In sharp contrast to these results, however, is the change that occurs in the patterning of retinogeniculate synapses. Following visual cortex damage in the newborn, retinal terminals develop fewer contacts with dendrites, but appear to compensate for this reduction by developing an unusually large number of synapses with F profiles. In lamina A, the number of RLP synapses with dendrites is 30% less than normal whereas the number of contacts with F profiles is 60% greater, giving a D:F ratio of about 1.3, instead of 3.2 as found in the normal cat. In the C laminae a similar net change in synaptic patterning occurs, but it results from a different emphasis. Here, the number of RLP synapses with dendrites is reduced modestly, 17%, but the number of synapses with F profiles is increased more than threefold. Thus in infant operated cats the D:F ratio of RLP synapses in the C layers is approximately 1.7, while it is typically greater than 8.0 in normal cats.

Plasticity of retinogeniculate circuitry is found also in cats that received visual cortex damage as adults. As in infant operated animals, removal of visual cortex in adults does not influence the percentage of RLP terminals that make synapses nor the mean number of synapses made by each terminal. In lamina A, however, the average number of synapses made by RLP terminals with dendrites is reduced by one-half, but remarkably this loss is offset by a near doubling of the number of RLP contacts with F terminals. This change in RLP synaptic affiliation results in a D:F ratio of 0.6 in lamina A of adult operated cats, and demonstrates a clear rearrangement of synaptic contacts in the adult visual system (for other examples of plasticity in the LGN after removal of visual cortex in the adult cat, see Hámori and Silakov, '80 and Somogyi et al., '82). Reorganization that is similar in kind, but less pronounced, also occurs in the C layers. The number of RLP synapses with dendrites is reduced by only about 10%, but as in lamina A the loss of dendritic contacts is compensated for precisely by an increase in the number of RLP synapses with F profiles.

THALAMOCORTICAL PROJECTION

In view of the cellular and synaptic changes that occur in the retina and LGN following removal of visual cortex, it is natural to question whether thalamocortical relations are also altered. To study this problem, anterograde and retrograde transport methods were used to trace the retino-thalamo-cortical projection, and to identify cells in the thalamus that participate as relay neurons in this pathway.

The anterograde studies were accomplished by making large, bilateral, intraocular injections of ^{3}H proline in a series of adult cats. These cats had had visual cortex removed unilaterally at different postnatal ages that ranged from newborn to adult. After the injections, about 14 to 16 days were allowed for the transneuronal transport of labeled materials from the retina to the cortex, via the thalamus. The brains were then prepared to study axonal connections with standard autoradiographic methods. In each animal the projection from thalamus to cortex ipsilateral to the visual cortex lesion was compared with that seen on the intact side of the brain. This analysis was limited to the posteromedial division of the lateral suprasylvian cortex (PMLS, Palmer et al., '78), which is an important secondary visual area in the cat (Sherman and Spear, '82) and receives projections from several thalamic nuclei related to vision (Raczkowski and Rosenquist, '80; Symonds et al., '81; Tong et al., '82a).

An attempt was made to compare the densities of the thalamo-PMLS projections on each side of the brain quantitatively by scoring them according to a five part ordinal scale. The values on this scale ranged from zero, indicating that no transported label was observed, to four, indicating that labeling of PMLS cortex matched that seen in areas 17 and 18 of the same brain.

In each brain studied the thalamo-PMLS projection in the intact hemisphere is either very weak or nonexistent. Mean scores for individual animals ranged from 0.0 to 0.8, with an overall mean for all cats of 0.4. By contrast, on the operated side of the brain a strong thalamo-PMLS projection (mean score of 3 cats is 2.8) is seen in cats with lesions of visual cortex made on the day of birth (Fig. 3). The enhanced thalamo-PMLS projection is heaviest along the lower bank of fundus of PMLS cortex where the central and paracentral visual fields are represented (Palmer et al., '78). It is interesting that the corresponding regions in areas 17 and 18 were almost always destroyed completely by the early visual cortex lesions.

No abnormal thalamo-PMLS projection is seen in cats

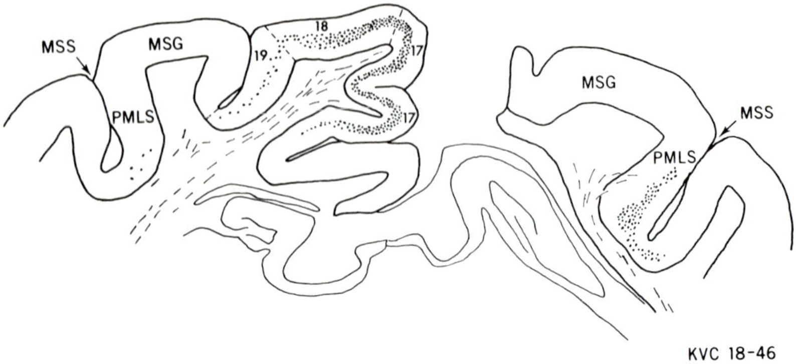

FIG. 3. Distribution of thalamocortical axon terminals as revealed by anterograde transneuronal transport following bilateral intraocular injections of ^{3}H proline in an adult cat that had received a lesion of visual cortex on the day of birth. On the intact side of the brain (left) a strong projection is seen from thalamus to visual cortex, especially to areas 17 and 18, and a very weak projection can sometimes be traced to the PMLS visual area. In the operated hemisphere (right) a marked thalamo-PMLS projection is evident. MSG, middle suprasylvian gyrus; MSS, middle suprasylvian sulcus; PMLS, posteromedial lateral suprasylvian visual area.

that received visual cortex lesions as adults. In these animals the projection on the operated side of the brain is as weak as that on the intact side. Although the adult brain lacks the plasticity seen in the infant, measurable enhancement of the retino-thalamo-PMLS projection can be demonstrated throughout a relatively long period in development. Thus an enhanced thalamo-PMLS projection is evident in cats that received visual cortex damage as late as 18 weeks postnatal. The enhancement that develops following lesions at this age is weaker (mean score 1.9) than that seen after operations in younger animals (Tong et al., '84), but nevertheless it demonstrates an apparent lability in thalamocortical connectivity at an age when one might expect these connections to be immutable.

Although anterograde transneuronal autoradiography is a useful method to establish that retino-thalamo-PMLS connections can be enhanced by the removal of visual cortex, this technique cannot distinguish which of the thalamic cell groups that receive input from the retina contributes to the enhanced projection. Fortunately, this problem can be tackled by using the retrograde transport of horseradish peroxidase (HRP) to label thalamic projection neurons after injections of PMLS cortex. Studies in normal cats show that several cell groups in the thalamus that receive significant retinal input in turn send axons to PMLS cortex: chiefly, these cells are located in the C layers of the LGN, the medial interlaminar nucleus, and the geniculate wing or retinal recipient zone of the pulvinar (see Tong et al., '82a for a review).

To identify the cells of origin of the enhanced thalamo-PMLS projection in adult cats with visual cortex lesions at birth, injections of HRP were made in PMLS cortex bilaterally to compare the retrograde labeling of thalamic neurons on the intact and operated sides of the brain (Kalil et al., '79). The most interesting changes were found in the LGN. On the intact side of the brain the projection from the LGN to PMLS cortex is modest and arises principally from small cells in the ventral C layers (C_1-C_3), although an occasional large neuron in lamina C is also labeled. Cells in the A laminae are never labeled, in agreement with results from normal cats (Tong et al., '82a).

On the operated side of the brain the distribution of labeled neurons in the degenerated LGN is markedly different from normal. First, there is a large increase in the number of labeled small cells in the C layers. Second, all of the large, spared cells in the C layers, which are particularly numerous in this region of the LGN, as mentioned earlier, are heavily labeled. Third, many of the large, spared cells that are scattered throughout the A laminae are also labeled with HRP. Collectively, these retrograde labeling results suggest that damage to visual cortex at birth produces a twofold modification of the geniculocortical pathway. First, the normally sparse input from the C layers of the LGN to PMLS cortex is increased, and secondly a new projection to PMLS cortex develops from the large, spared cells in the A laminae. However, it should be pointed out that the accuracy of this interpretation rests upon the assumption that the normal projection from the LGN to PMLS cortex is the same in the infant as in the adult. Until this identity is demonstrated, the possibility remains that the increased projection from the C layers of the LGN and the new projection from the A layers may in fact represent an immature pattern of connections that is not eliminated when visual cortex is removed at birth.

In cats with lesions of visual cortex as adults, retrograde labeling results present no interpretative problems. The distribution of labeled neurons in the intact and operated sides of the brain are essentially identical.

CONCLUSIONS AND SUMMARY

Several general conclusions can be drawn from the work that has been described above.

1. When visual cortex is removed in newborn or adult cats, massive retrograde degeneration occurs in the LGN. However, when visual cortex is removed at birth, some neurons in the LGN do not degenerate. At maturity, these spared neurons are found chiefly in the vicinity of the C layers in the degenerated LGN and also scattered throughout the A layers. Since large surviving neurons are not found routinely in the LGN following removal of visual cortex in adult cats, their presence in infant operated animals is a salient feature.

2. When visual cortex is removed at birth, retrograde cell degeneration is not confined to the LGN, but also occurs transsynaptically in the retina. Cell degeneration is very specific, affecting only medium sized ganglion cells, and causing a loss of about 60%. Correlative neurophysiological studies show that functionally this loss is restricted to X-type retinal ganglion cells. In cats with lesions of visual cortex made during adulthood, there is no loss of retinal ganglion cells.

3. The retinogeniculate projection in cats with early, but not late, visual cortex lesions lacks the precise organization that can be seen with the light microscope in normal animals. For example, axons from surviving retinal ganglion cells frequently cross the lines of projection in the LGN as they travel to their target cells. In addition, the terminals of these axons often appear to converge near large, spared LGN cells, which may serve to "focus" the retinogeniculate projection on these neurons.

With the electron microscope, it is clear that the patterning of retinogeniculate synapses is altered after early <u>or</u> late damage to visual cortex. In both instances there appears to be no loss in the total number of synapses made by individual retinal terminals, but a large change occurs in the relative numbers of synapses that occur with dendrites or F profiles. Under normal conditions, retinogeniculate terminals make synaptic contacts much more frequently with dendrites than with F profiles, but after cortical removal this balance is reversed. Contacts with dendrites are lost, and replaced, essentially one for one, by contacts with F profiles.

4. Perhaps the most striking change that takes place in the central visual pathway after early removal of visual cortex is the development of an enhanced retino-thalamic projection to the PMLS visual area of the lateral suprasylvian cortex. A retino-thalamo-PMLS pathway exists in the normal cat, but it is very weak. When visual cortex is removed at birth, the strength of the thalamo-PMLS projection, as revealed by anterograde transneuronal autoradiography, is enhanced dramatically. Retrograde tracing studies suggest that this enhancement may result in part from a marked increase in the number of cells in the C layers of the degenerated LGN that project to PMLS, as well as from a new projection by the large, spared cells in the A layers. Although enhancement of the thalamo-PMLS pathway is greatest in cats with visual cortex lesions at birth, an increase in the density of the projection is seen when lesions are made as late as 18 weeks postnatal. However, neither anterograde nor retrograde tracing methods provide any evidence for an enhancement of the thalamo-PMLS projection following removal of visual cortex in adult cats.

5. In summary, the structure of the central visual pathway changes, in many different ways, in response to visual cortex damage in either infants or adults. Certain changes are common regardless of age at the time of cortical damage. For example, retrograde degeneration and atrophy of LGN neurons ensues after early or late cortical lesions. It is arguable, however, that after an early lesion the retrograde reaction is more severe, because it also occurs transneuronally and affects retinal ganglion cells. In other instances, the changes in young animals and adults are similar, but the mechanisms by which they are produced are undoubtedly different. Thus the normal pattern of

retinogeniculate synapses is modified along the same lines in both infant and adult operated cats. However, in the adult operated animal the new synaptic pattern is formed by reorganizing synapses existing at the time of the lesion, whereas in the infant operated cat the modified retinogeniculate pattern is created by novel synaptic development. Lastly, some structural alterations occur following lesions in young animals, but not in adults. These changes, which are seen in the thalamocortical pathway, may enhance the opportunity for communication between the retina and the PMLS visual area of cortex, and therefore may provide a compensatory link that offsets, at least partially, the loss of visual cortex.

ACKNOWLEDGEMENTS

I would like to thank D. Owen, M. Ellis-Metz and M. Rhodes for technical assistance, A. Weber for helpful comments on the manuscript, and R. Olson for typing it. This work was supported by research grant EYO 1331 from the National Eye Institute.

REFERENCES

Boycott BB, Wässle H (1974). The morphological types of ganglion cells of the domestic cat's retina. J Physiol 240:397.

Edwards SB, Hendrickson A (1981). The autoradiographic tracing of axonal connections in the central nervous system. In Heimer L, Robards MJ (eds): "Neuroanatomical Tract-Tracing Methods," New York: Plenum Press, p 171.

Famiglietti EV, Peters A (1972). The synaptic glomerulus and the intrinsic neuron in the dorsal lateral geniculate nucleus of the cat. J Comp Neurol 144:285.

Fink RP, Heimer L (1967). Two methods for selective silver impregnation of degenerating axons and their synaptic endings in the central nervous system. Brain Res 4:369.

Guillery RW (1969). The organization of synaptic interconnections in the laminae of the dorsal lateral geniculate nucleus of the cat. Z Zellforsch 96:1.

Hámori J, Silakov VL (1980). Plasticity of relay neurons in dorsal lateral geniculate nucleus of the adult cat: morphological evidence. Neuroscience 5:2073

Hughes A (1981). Population magnitudes and distribution of the major modal classes of cat retinal ganglion cells as estimated from HRP filling and a systematic survey of the soma diameter spectra for classical neurones. J Comp Neurol 197:303.

Kalil RE (1978). Development of the dorsal lateral geniculate nucleus in the cat. J Comp Neurol 182:265.

Kalil RE (1980). Retrograde degeneration of retinal ganglion cells following removal of visual cortex in the newborn kitten. Soc Neurosci Abstr 6:790.

Kalil RE, Scott G (1979). Development of retinogeniculate synapses in the dorsal lateral geniculate nucleus of the cat. Soc Neurosci Abstr 5:791.

Kalil RE, Tong L, Spear PD (1979). Reorganization of the geniculocortical pathway in the cat following neonatal damage to visual cortex. Invest Ophthal Vis Sci Suppl 16:157.

Kalil RE, Behan M (1982). Synaptic organization of the lateral geniculate nucleus of the cat after removal of visual cortex. Invest Ophthal Vis Sci Suppl 24:228.

Montero VM, Scott GL (1981). Synaptic terminals in the dorsal lateral geniculate nucleus from neurons of the thalamic reticular nucleus: a light and electron microscope autoradiographic study. Neuroscience 6:2561.

Murphy EH, Kalil RE (1979). Functional organization of lateral geniculate cells following removal of visual cortex in the newborn kitten. Science 206:713.

Palmer LA, Rosenquist AC, Tusa RJ (1978). The retinotopic organization of lateral suprasylvian visual areas in the cat. J Comp Neurol 177:237.

Pearson HE, Labar DR, Payne BR, Cornwell P, Aggarwal N (1981). Transneuronal retrograde degeneration in the cat retina following neonatal ablation of visual cortex. Brain Res 212:470.

Raczkowski D, Rosenquist AC (1980). Connections of the parvocellular C laminae of the dorsal lateral geniculate nucleus with the visual cortex in the cat. Brain Res 199:447.

Sanderson KJ (1971). The projection of the visual field to the lateral geniculate and medial interlaminar nuclei in the cat. J Comp Neurol 143:101.

Sherman SM, Spear PD (1982). Organization of visual pathways in normal and visually deprived cats. Physiol Rev 62:740.

Somogyi J, Hámori J, Silakov VL (1982). Free postsynaptic sites in the lateral geniculate nucleus of adult cats following chronic decortication. Cell Tissue Res 225:437.

Spear PD (1979). Behavioral and neurophysiological consequences of visual cortex damage. In Sprague JM, Epstein AN (eds): "Progress in Psychobiology and Physiological Psychology, Vol. 8" New York: Academic Press, p 45.

Spear PD, Kalil RE, Tong L (1980). Functional compensation in lateral suprasylvian visual area following neonatal visual cortex removal in cats. J Neurophysiol 43:851.

Sterling P, Davis T (1980). Neurons in cat lateral geniculate nucleus that concentrate exogenous $[^3H]$-γ-aminobutyric acid (GABA). J Comp Neurol 192:749.

Stone J, Keens J (1980). Distribution of small and medium-sized ganglion cells in the cat's retina. J Comp Neurol 192:235.

Symonds LL, Rosenquist AC, Edwards SB, Palmer LA (1981). Projections of the pulvinar-lateral posterior complex to visual cortical areas in the cat. Neuroscience 6:1995.

Tong L, Kalil RE, Spear PD (1982a). Thalamic projections to visual areas of the middle suprasylvian sulcus in the cat. J Comp Neurol 212:103.

Tong L, Kalil RE, Spear PD (1984). Critical periods for functional and anatomical compensation in the lateral suprasylvian visual area following removal of visual cortex in cats. J. Neurophysiol, in press.

Tong L, Spear PD, Kalil RE, Callahan E (1982b). Loss of retinal X-cells in cats with neonatal or adult visual cortex damage. Science 217:72.

Development of Visual Pathways in Mammals, pages 275–287

FUNCTIONAL CONSEQUENCES AND COMPENSATION FOLLOWING NEONATAL DAMAGE TO VISUAL CORTEX IN CATS

Peter D. Spear

Department of Psychology
University of Wisconsin
Madison, WI 53706 USA

INTRODUCTION

When adult cats receive damage to visual cortical areas 17, 18, and 19, they show severe deficits in their ability to perform many visual tasks. By comparison, many of the deficits are less severe, or even fail to occur, in cats that received damage to areas 17, 18, and 19 neonatally (see Spear, 1979, 1984, for reviews). Thus, there is greater sparing or recovery of visual function following neonatal visual cortex damage than following adult damage. These behavioral observations imply that the remaining visual pathways possess a greater ability to compensate for damage to areas 17, 18, and 19 in the neonatal brain than in the adult brain.

How does this compensation occur? Are there differences in the neural consequences of visual cortex damage in neonates and adults that might underly the differences in behavioral sparing or recovery? In this presentation, I will describe some physiological studies from my laboratory that have been directed at answering this question. These studies were carried out in collaboration with Thomas Baumann, Lillian Tong, Ronald Kalil, and Eileen Callahan.

Where would one begin in looking for functional compensation following neonatal visual cortex damage? Electrophysiological mapping studies indicate that, outside of cortical areas 17, 18, and 19, there are at least 10 additional topographically organized visual areas of cortex in the cat (Palmer, Rosenquist, Tusa, 1978; Tusa, Palmer,

1980). A number of observations suggest that one of these areas, the posteromedial lateral suprasylvian (PMLS) visual cortex, is the best place to look for physiological compensation following neonatal removal of areas 17, 18, and 19. For example, behavioral studies indicate that PMLS cortex is involved in recovery from visual cortex damage in adult cats (Spear, 1979), and there is indirect evidence that PMLS cortex is involved in recovery from neonatal visual cortex damage as well (Spear, 1984). In addition, anatomical studies described by Ronald Kalil in this symposium indicate that there is an enhanced, and possibly new, projection from the lateral geniculate nucleus to PMLS cortex following neonatal, but not adult, visual cortex damage (Kalil, Tong, Spear, 1979). Therefore, we began our neurophysiological studies in the PMLS area of cortex.

RESPONSES OF PMLS NEURONS IN NORMAL ADULT CATS

Before considering the effects of visual cortex lesions, it is necessary to briefly describe the responses of PMLS neurons in normal adult cats (Smith, Spear, 1979; Spear, Baumann, 1975). Approximately 90% of the cells in PMLS cortex respond to light, and these cells can be divided into four classes on the basis of their receptive field properties. The percentage of cells in each class is summarized in Figure 1A, left. Approximately 80% of the cells are direction selective. They respond best to moving stimuli and the response is dependent upon the direction of stimulus movement. The remaining 20% of PMLS cells are about evenly divided among the other three receptive field classes. Those in the movement sensitive class respond maximally to moving stimuli, but they are not sensitive to the direction of movement. Those in the stationary class respond as well or better to stationary flashing stimuli as to moving stimuli (and also lack direction selectivity). Cells in the indefinite class give very weak responses to light or have diffuse receptive fields that cannot be defined precisely. Most PMLS cells have large receptive fields (mean of 15-20 deg. diameter). In addition, most of the cells are sensitive to stimulus size as a result of spatial summation or receptive field surround inhibition. However, unlike cells in visual cortical areas 17, 18, and 19, few if any PMLS cortex cells are orientation selective using standard tests.

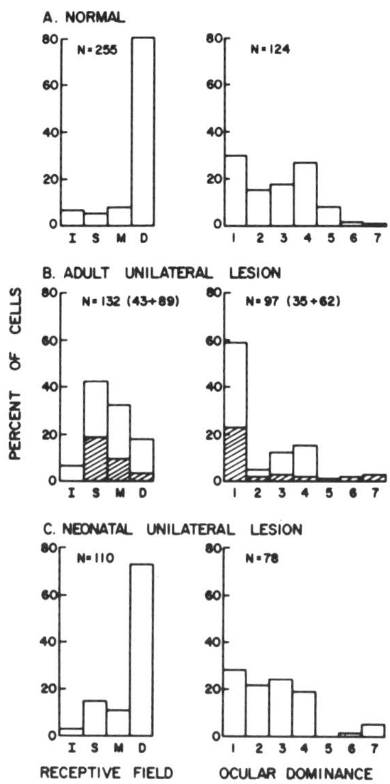

Fig. 1. Summary of response properties of posteromedial lateral suprasylvian cortex cells in three groups of cats. A: Normal adult cats. B: Cats with damage to areas 17, 18, and 19 received as adults. Crosshatched portions of each bar are results from cats with long-term lesions (6.5-8 mo. survival, N=43), and open portions are results from cats with shorter survival times (0.5-1 mo., N=89). The height of each bar represents the sum of the crosshatched and open portions. C: Cats with damage to areas 17, 18, and 19 received on the day of birth. Left shows the percentage of responsive cells (N, number of responsive cells in the sample) in each of four receptive field classes described in the text: I, indefinite; S, stationary; M, movement sensitive; D, direction selective. Right shows the percentage of cells in each of seven ocular dominance groups: 1 and 7, driven exclusively by the contralateral or ipsilateral eye, respectively; 4, driven about equally by either eye or driven only by binocular stimulation; 2, 3, 5, and 6, binocularly driven with intermediate degrees of ocular dominance.

Figure 1A, right, summarizes the ocular dominance characteristics of PMLS cells that have their receptive fields in the binocular overlap region of the visual field. About 70% of the cells can be driven by both eyes, and there is an overall dominance by the contralateral eye.

EFFECTS OF VISUAL CORTEX LESIONS IN ADULTS

We next investigated the effects of removing areas 17, 18, and 19 in adult cats (Spear, Baumann, 1979a; Spear, Kalil, Tong, 1980). Lesions of various kinds were made and recordings were carried out in PMLS cortex some time later. For the present discussion, only the effects of large unilateral lesions of areas 17, 18, and 19 will be considered. Following such lesions, three main changes occur in ipsilateral PMLS cortex. These are summarized in Figure 1B. First, there is a marked reduction in the percentage of direction selective cells; only about 20% now display this property. Second, there is an increase in the percentage of cells that respond best to stationary flashing stimuli. Thus, the loss of direction selective cells is not simply accompanied by an increase in the percentage of cells that are movement sensitive but not direction selective. Rather, a large proportion (about 35%) of the cells now give their optimal response to stationary flashing lights and are in the stationary receptive field class. Third, there is a decrease in the proportion of cells that respond to the ipsilateral eye, from about 70% in normal adult cats to about 40% in cats with adult visual cortex lesions. Correspondingly, the percentage of cells that respond exclusively to the contralateral eye increases. On the basis of these changes, it can be concluded that visual cortex provides most PMLS cortex neurons with direction selectivity, inhibition of responses to stationary flashing stimuli, and the ability to respond to the ipsilateral eye.

As can be seen in Figure 1B, similar results are observed in PMLS cortex when the visual cortex lesions are made a few weeks prior to recording as when they are made many months earlier. Indeed, the effects of lesions made only a few hours prior to recording are similar to those of bilateral lesions followed by visual discrimination training to establish that behavioral recovery has occurred (Spear, Baumann, 1979a,b). Thus, there is no evidence that functional compensation for the loss of visual cortex inputs can occur in PMLS cortex of adult cats.

EFFECTS OF NEONATAL VISUAL CORTEX LESIONS

We next recorded from PMLS neurons in cats that had received neonatal visual cortex lesions to determine if the physiological effects differ from those following adult lesions (Spear, Kalil, Tong, 1980). Unilateral lesions of areas 17, 18, and 19 were made on the day of birth and recordings were carried out in PMLS cortex when the animals were 7-9 mo. old. Histological analysis of the cortical lesions and the thalamic retrograde degeneration indicates that the neonatal lesions were as large or larger than the lesions made in adult cats.

Figure 1C shows the results from cats with neonatal visual cortex lesions. The percentage of direction selective cells is normal in these animals and significantly higher than that in cats with adult visual cortex lesions. Likewise, the percentages of cells in the other receptive field classes are normal. The percentage of cells that respond to the ipsilateral eye also is normal and significantly higher than that in cats with adult lesions.

We specifically investigated the possibility that PMLS neurons might develop certain properties of cells in areas 17, 18, 19 following neonatal removal of these areas. However, no massive appearance of anomalous properties is seen. Receptive field size remains large and the visuotopic organization is characteristic of normal PMLS cortex. In addition, only a small percentage (about 5%) of the neurons are orientation selective.

These results indicate that none of the changes that occur in PMLS cortex after adult visual cortex lesions are present after neonatal lesions made on the day of birth. The response properties of PMLS neurons are normal, despite the absence of visual cortex. This represents a physiological compensation in PMLS cortex that occurs after neonatal, but not adult, visual cortex damage. Additional experiments indicate that this compensation is not due to callosal inputs from the opposite hemisphere (Tong, Spear, Kalil, 1983). Therefore, it appears to be formed on the basis of inputs from ipsilateral thalamocortical pathways. Presumably, the functional compensation is related to the enhanced thalamocortical inputs to PMLS cortex that were described by Ronald Kalil in this symposium.

CRITICAL PERIOD FOR THE FUNCTIONAL COMPENSATION

What is the critical period for this functional compensation? That is, how late in development can one make a visual cortex lesion and still get functional compensation in PMLS cortex? To address this question, we prepared additional animals with unilateral lesions made at 1 day, 2, 4, 8, 12, 18, or 26 weeks of age or as adults (Tong, Spear, Kalil, 1981). Recordings were made from neurons in PMLS cortex at least 6 mo. later.

Several unexpected findings were observed. First, the critical period for development of normal binocularity is very long. The percentage of cells that respond to the ipsilateral eye is the same (i.e., normal) following visual cortex lesions made at 18 weeks of age as that following lesions on the day of birth. The loss of response to the ipsilateral eye that is produced by lesions in adult animals does not appear until lesions are made at 26 weeks of age. Thus, the critical period for the development of responses to the ipsilateral eye in PMLS cortex following damage to visual cortex is over between 18 and 26 weeks of age.

Second, the critical period for the development of direction selectivity among PMLS cells following removal of visual cortex is shorter than the critical period for development of normal binocularity. A normal percentage of direction selective cells is present in PMLS cortex of animals that received visual cortex lesions at 12 weeks of age but not of animals that received lesions at 18 weeks of age. In the latter animals the percentage of direction selective cells is not significantly different from that following lesions in adult animals. Thus, the critical period for development of direction selectivity in absence of visual cortex ends between 12 and 18 weeks of age. Results for cells in the movement sensitive and stationary receptive field classes are similar.

Third, following lesions made between two and eight weeks of age, about 33% of the PMLS cells display orientation selectivity. Fewer orientation selective cells are seen following lesions made at 12 weeks of age, and only about 5% of the cells are orientation selective following lesions made at 18 and 26 weeks of age. Thus, the results following lesions at 18 and 26 weeks of age are similar to those following lesions made at one day of age or in adult cats. However, lesions incurred between 2-8 weeks of age result in the development of an abnormally large number of cells with orientation selectivity. These cells are unlike those in areas 17, 18, and 19 in that they have the large receptive fields that are characteristic of normal PMLS neurons. In addition, they seem to be somewhat more broadly tuned. Nevertheless, they are clearly selective to the orientation of a flashing stimulus and they are found in PMLS cortex of every animal that received a visual cortex lesion between 2-8 weeks of age.

Several conclusions can be drawn from this experiment. The critical period for the development of functional compensation in PMLS cortex following a visual cortex lesion is very long; it extends to between 18 and 26 weeks of age. However, the critical period ends at different ages for different response properties. It ends first for orientation selectivity, then for direction selectivity, and last for binocularity. The presence of orientation selectivity represents the development of an anomalous property for many PMLS neurons. Curiously, this property develops for many cells following lesions made at 2-8 weeks of age but not following lesions made on the day of birth. A possible explanation for this finding will be discussed below.

EFFECTS OF VISUAL CORTEX LESIONS UPON RETINAL GANGLION CELLS

In order to fully understand the nature and mechanisms of the changes that occur in the visual pathways following visual cortex damage, one must know the nature of the inputs to the system. Earlier in this symposium, Ronald Kalil presented anatomical data showing that there is transneuronal retrograde degeneration of retinal ganglion cells after neonatal visual cortex damage (see also Pearson, Labar, Payne, Cornwell, Aggarwal, 1981). These anatomical data suggest that the retinal degeneration is selective to one physiological class of ganglion cells. In the normal cat retina, physiological studies indicate that there are three classes of ganglion cells (see Stone, Dreher, Leventhal, 1979, for a review). X-cells have the smallest receptive fields, the highest spatial resolution, linear spatial summation, and axons with intermediate conduction velocities. Y-cells have large receptive fields, poorer spatial resolution, nonlinear spatial summation, and axons with the fastest conduction velocities. W-cells are a heterogeneous group that generally have large receptive fields, the poorest spatial resolution, sluggish responses to visual stimulation, and axons with the slowest conduction velocities. Anatomical studies suggest that the transneuronal retrograde degeneration following neonatal visual cortex damage is specific to the X-cells, but the evidence is indirect.

We wished to obtain direct information about the physiological properties of the retinal inputs to the visual system after visual cortex damage. To do this, we made intraocular recordings from retinal ganglion cells in cats that had received visual cortex lesions on the day of birth or as adults (Tong, Spear, Kalil, Callahan, 1982). Recordings were carried out 10-14 mo. after the lesions were made, and the cells were classified using a battery of tests. We found that the absolute retinal sampling density of Y- and W-cells is normal following both neonatal and adult damage to visual cortex. The basic receptive field properties of these cells also appear to be normal. In contrast, there is a 78% loss of X-cells in the retina following neonatal visual cortex lesions. A loss of X-cells also occurs following adult visual cortex lesions, but in this case there is only a 22% loss.

We next studied the critical period for the marked X-cell loss to determine how it relates to changes in PMLS cortex (Callahan, Tong, Spear, 1982). Following visual cortex lesions made at two weeks of age, there is a 37% loss of retinal X-cells. This is significantly less than the loss that occurs following lesions made on the day of birth but not significantly different from the small loss that occurs after adult lesions. Similarly, following lesions made at four weeks of age, there is only a 16% loss of retinal X-cells.

Together, these results indicate that visual cortex damage incurred on the day of birth produces a marked transneuronal retrograde degeneration that is restricted to the X-cell functional class of retinal ganglion cells. This marked loss of retinal X-cells no longer occurs following lesions made at 2-4 weeks of age or in adults; only a small loss of retinal X-cells occurs in these animals.

SUMMARY

Figure 2 provides very simplified schematic diagrams that summarize the results that have been presented. These diagrams show functional information that is transmitted along different portions of the visual pathways in normal cats and cats with neonatal or adult visual cortex damage. In normal cats, retinal X-, Y-, and W-cells from both eyes project to the thalamus (specifically, the lateral geniculate nucleus). The lateral geniculate nucleus in turn projects X-, Y-, and W-cell information for both eyes to visual cortex (areas 17, 18, and 19). Our lesion studies indicate that visual cortex provides PMLS cortex with information necessary for the formation of direction selectivity and inhibition of responses to stationary flashing stimuli for both eyes. Thalamic inputs to PMLS cortex (independent of visual cortex) provide responsiveness primarily for the contralateral eye and lack information necessary for the formation of direction selectivity and inhibition of responses to stationary flashing stimuli for most cells.

Following visual cortex lesions in adult cats, the retinal ganglion cells are nearly normal. Some loss of retinal X-cells does occur, but this is relatively small. As already noted, the thalamic inputs to PMLS cortex that remain after visual cortex has been removed provide PMLS cells with responses primarily to the contralateral eye and lack information necessary for direction selectivity and inhibition of responses to stationary flashing stimuli. There is no functional compensation in PMLS cortex following longstanding visual cortex damage in adult cats.

When a cat grows up from birth without visual cortex, several changes occur. X-cells in the retina are virtually lost. Despite this, enhanced projections develop from the thalamus to ipsilateral PMLS cortex. Correspondingly, physiological compensation occurs in PMLS cortex: the cells there develop normal binocularity, direction selectivity, and inhibition of responses to stationary flashing stimuli. The critical period for the functional compensation extends to between 4-6 months of age, but the end of the critical period differs for different response properties. Moreover, with lesions made at 2-12 weeks of age, but not at one day

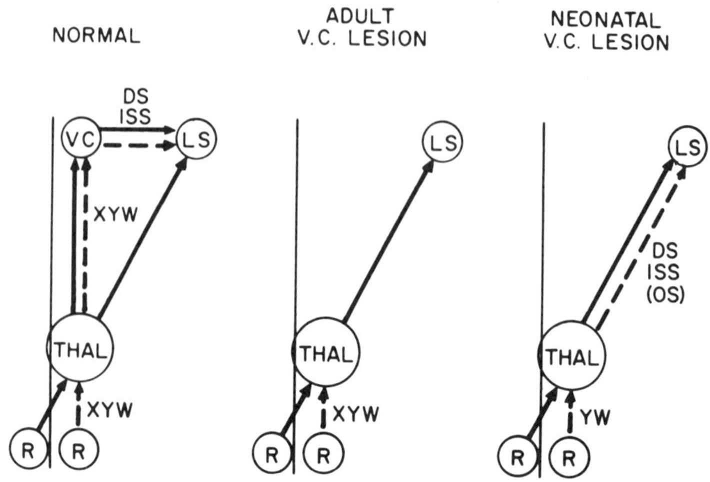

Fig. 2. Schematic diagrams to summarize the results that have been presented. The left-hand drawing represents the inputs from the two eyes (R) to the thalamus (THAL) and from there to ipsilateral visual cortical areas 17, 18, and 19 (VC) and posteromedial lateral suprasylvian cortex (LS) in normal cats. The thin vertical line represents the midline. The middle and right-hand drawings represent the changes in these inputs following damage to cortical areas 17, 18, and 19 in adult and neonatal cats, respectively. These diagrams are intended to show what functional information is provided to LS cortex neurons by the primary geniculocortical pathways and by the thalamic-LS pathways independent of visual cortex. Details of the pathways and the thalamic nuclei involved have been omitted for the sake of clarity. Continuous lines with arrowheads represent contralateral eye inputs and dashed lines with arrowheads represent ipsilateral eye inputs. DS, information required for the formation of direction selectivity. ISS, information required for inhibition of responses to stationary flashing stimuli. OS, information for the formation of orientation selectivity. X, Y, and W represent the X-, Y-, and W-cell pathways.

of age, a significant proportion of cells develop orientation selectivity. Thus, the critical period for the development of this anomalous property also begins later than others. A possible reason for this is that development of orientation selectivity in PMLS cortex requires retinal X-cell inputs, which are markedly reduced after lesions made at one day of age but not later.

GENERAL CONCLUSIONS

It is clear that visual cortex damage has widespread effects upon the visual system that go beyond a simple loss of the damaged tissue. Some of these effects are orthograde, due to a loss of efferents from visual cortex (e.g., from visual cortex to PMLS cortex). In addition, some of the effects are retrograde, and these can occur as far peripherally as the retina.

Nevertheless, the visual system can compensate to some extent for the effects of early visual cortex damage. In addition, the physiological compensation that occurs need not be haphazard or potentially maladaptive; it can be very orderly, as in the case of PMLS cortex.

This compensation can provide information about the rules of neural development. For example, the inputs to PMLS cortex in normal cats and cats that had visual cortex damage on the day of birth are very different. The most obvious difference is that visual cortical inputs to PMLS cortex are present in one case and absent in the other (see Figure 2). Nevertheless, the major receptive field properties of the PMLS neurons are essentially the same in the two conditions. This suggests that receptive field properties of neurons are not determined simply by the nature of their inputs. Rather, information contained within the recipient cells in some way influences what their response properties will be.

The physiological compensation that occurs in PMLS cortex may be part of the neural basis of the superior behavioral recovery or sparing in animals with neonatal visual cortex damage. If so, the results presented here have two implications for behavioral studies. First, the amount and kind of behavioral recovery that is possible may depend upon the age at which the early damage has occurred, since different neural effects have different critical periods. Second, behavioral abilities following early visual cortex damage may never be completely normal, since even at their best most PMLS cells never attain the precise selectivity of the visual cortex cells that were lost.

Callahan EC, Tong L, Spear PD (1982). The critical period for effects of visual cortex lesions on retinal X-cells in cats. Invest Ophthalmol Vis Sci Suppl 22:236.

Kalil RE, Tong L, Spear PD (1979). Reorganization of the geniculocortical pathway in the cat following neonatal damage to visual cortex. Invest Ophthalmol Vis Sci Suppl 18:157.

Palmer LA, Rosenquist AC, Tusa RJ (1978). The retinotopic organization of lateral suprasylvian visual areas in the cat. J Comp Neurol 177:237.

Pearson HE, Labar DR, Payne BR, Cornwell P, Aggarwal N (1981). Transneuronal retrograde degeneration in the cat retina following neonatal ablation of visual cortex. Brain Res 212:470.

Smith DC, Spear PD (1979). Effects of superior colliculus removal on receptive-field properties of neurons in lateral suprasylvian visual area of the cat. J Neurophysiol 42:57.

Spear PD (1979). Behavioral and neurophysiological consequences of visual cortex damage: Mechanisms of recovery. In Sprague JM, Epstein AN (eds): "Progress in Psychobiology and Physiological Psychology," Vol. 8, New York: Academic Press, p 45.

Spear PD (1984). Neural mechanisms of compensation following neonatal visual cortex damage. In Cotman CW (ed): "Synaptic plasticity and remodeling," New York: Guilford Press, in press.

Spear PD, Baumann TP (1975). Receptive-field characteristics of single neurons in lateral suprasylvian visual area of the cat. J Neurophysiol 38:1403.

Spear PD, Baumann TP (1979a). Effects of visual cortex removal on receptive-field properties of neurons in lateral suprasylvian visual area of the cat. J Neurophysiol 42:31.

Spear PD, Baumann TP (1979b). Neurophysiological mechanisms of recovery from visual cortex damage in cats: Properties of lateral suprasylvian visual area neurons following behavioral recovery. Exp Brain Res 35:161.

Spear PD, Kalil RE, Tong L (1980). Functional compensation in lateral suprasylvian visual area following neonatal visual cortex removal in cats. J Neurophysiol 43:851.

Stone J, Dreher B, Leventhal A (1979). Hierarchical and parallel mechanisms in the organization of visual cortex. Brain Res Rev 1:345.

Tong L, Spear PD, Kalil RE (1981). Critical period for functional compensation in lateral suprasylvian visual area following neonatal visual cortex removal in cats. Soc Neurosci Abs 7:675.

Tong L, Spear PD, Kalil RE (1983). Role of the corpus callosum in functional compensation in the lateral suprasylvian visual area after early visual cortex damage in cats. Invest Ophthalmol Vis Sci Suppl 24:228.

Tong L, Spear PD, Kalil RE, Callahan EC (1982). Loss of retinal X-cells in cats with neonatal or adult visual cortex damage. Science 217:72.

Tusa RJ, Palmer LA (1980). Retinotopic organization of areas 20 and 21 in the cat. J comp Neurol 193:147.

Development of Visual Pathways in Mammals, pages 289–302

DEVELOPMENTAL CHANGES IN SUSCEPTIBILITY TO RETINAL GANGLION CELL LOSS AFTER LESIONS OF VISUAL CORTEX IN PRIMATES AND OTHER MAMMALS

Rosalyn E. Weller and Jon H. Kaas

Department of Psychology
Vanderbilt University
Nashville, Tennessee 37240

After lesions of primary visual cortex in mammals, neurons in the lateral geniculate nucleus projecting to this cortex rapidly degenerate, and ganglion cells of the retina projecting to the degenerated lateral geniculate nucleus may also die. The loss of retinal ganglion cells as a result of such transneuronal retrograde degeneration was first suggested by reports of reduced thickness of the optic nerves several months after lesions of visual cortex in kittens and puppies (Ganser 1882; von Monakow 1889; cited in Cowan 1970). Much later, instances of retinal ganglion cell loss were sometimes reported for humans with longstanding cortical damage (Walsh 1947; Haddock, Berlin 1950; Van Buren 1963b; also see Gills, Wadsworth 1967). Modern investigations of the phenomenon began with quantitative descriptions made from serial sections of retinas of macaque monkeys with lesions of striate cortex. Van Buren (1963a) found a marked reduction in the thickness of the ganglion cell layer in the affected halves of the retinas of a monkey with a long-term unilateral lesion of striate cortex, and Cowey (1974) reported a loss of up to 80% of the ganglion cells from a monkey examined 8 years after a bilateral lesion.

Our preliminary investigations of ganglion cell loss began with two macaque monkeys with large but incomplete bilateral removals of striate cortex performed within the first postnatal week. When we studied retinal ganglion cell loss and retinal projections in these monkeys as 4-5 year old adults, we found up to a 75% loss of ganglion cells in the affected parts of the retinas (Figure 1A; Figure 2A), and systematically altered retinal projections

(Weller, Kaas 1977; Weller et al. 1979). In the intact portions of the lateral geniculate nuclei of these monkeys, the retinal projections appeared to be completely normal. However, in the degenerated portions of the lateral geniculate nuclei, the retinal terminations were markedly reduced in the parvocellular layers, while being of normal density in the magnocellular layers, the S or superficial layers, and the interlaminar zones (Figure 3; see Kaas et al. 1978 for normal patterns of retinal projections and nomenclature for geniculate layers). We concluded that only ganglion cells projecting to the parvocellular geniculate layers had degenerated, and that neurons projecting to the magnocellular layers, the S layers, and the interlaminar zones had not. The preservation of some retinal ganglion cells was not due to the selective presence of neurons in some of the degenerated layers since extremely few neurons remained in all degenerated portions of the lateral geniculate. We hypothesized that neurons projecting to the parvocellular layers were lost because they projected to only the lateral geniculate nucleus, and had no other "sustaining" target, while neurons projecting to the magnocellular layers, interlaminar zones, and S layers, had other sustaining targets, probably in the superior colliculus, and therefore they were preserved. Because neurons in the magnocellular layers of the lateral geniculate nucleus of monkeys had been found to have Y-like physiological properties, and the parvocellular layers, X-like properties (Sherman et al. 1976; Dreher et al. 1976), similar to the X and Y ganglion cell classes of cats, we further proposed that it was the X-like cell class that was lost from the retina. While the branching of Y-like and W-like ganglion cells to both the lateral geniculate nucleus and the superior colliculus in primates is presently inferred from indirect evidence (see Weller et al. 1981; Leventhal et al. 1981), the concept that cells without a collateral branch to a sustaining target are susceptible to loss while those cells with such a branch are not susceptible is consistent with traditional views of the nature of transneuronal retrograde degeneration (see Cowan 1970).

Experiments on other species of mammals soon suggested that factors other than the presence or absence of an extrageniculate axon branch may be important in ganglion cell loss. When tree shrews and squirrels were examined at various periods after lesions of visual cortex as adults, ganglion cell loss and alterations of retinal termination patterns in the lateral geniculate nucleus were not obvious (Weller, Kaas, unpublished experiments). Similarly, Perry and Cowey (1979) detected no ganglion cell loss after visual cortex lesions in rats. However, most or all ganglion cells in these species may project to both the lateral geniculate nucleus and the superior colliculus, and thus little or no ganglion cell loss may result for that reason. Such an explanation was not so easily applied when prosimian primates, galagos, were examined 3 months to 3 years after striate cortex lesions as adults. They also had no obvious ganglion cell loss, and both degenerated parvocellular and magnocellular geniculate layers had apparently normal densities of retinal terminations (Weller et al. 1981). Prosimian primates such as galagos have parvocellular geniculate layers with X-like neurons (Norton, Casagrande 1982), and there is no reason to suspect that the X-like inputs differ in collaterals to other targets from X-like retinal cells in monkeys. Thus, there was evidence for species differences in susceptibility to ganglion cell loss, even in primates known to have large numbers of X-like ganglion cells.

The above results on adult mammals also suggested that age at the time of the lesion might also be an important variable. Investigators recently have found that cats lesioned as infants lose ganglion cells, while cats lesioned as adults do not (Kalil 1978; Pearson et al. 1981; Calahan et al. 1982). Electrophysiological studies indicate that the loss is of X-cells (Tong et al. 1982), so the type of cell lost is similar in cats and monkeys.

Finally, survival time after the cortical lesion is likely to be a significant variable. Questions about the relative importance of species, age at the time of cortical lesion, and also postlesion survival time led us to further experiments on monkeys. Some of the results have been briefly summarized elsewhere (Weller, Kaas 1980; 1982).

<u>Macaque monkeys with lesions as infants</u>. Lesions were placed in striate cortex of four macaque monkeys at about 3 weeks of age, and each monkey was allowed to survive for a different length of time (Figure 2A). The results were compared with those of our previous study of monkeys surviving for 5 years after lesions as infants (Weller et al. 1979) and of a similar monkey with a 1 year survival studied by Dineen and Hendrickson (1981). The results

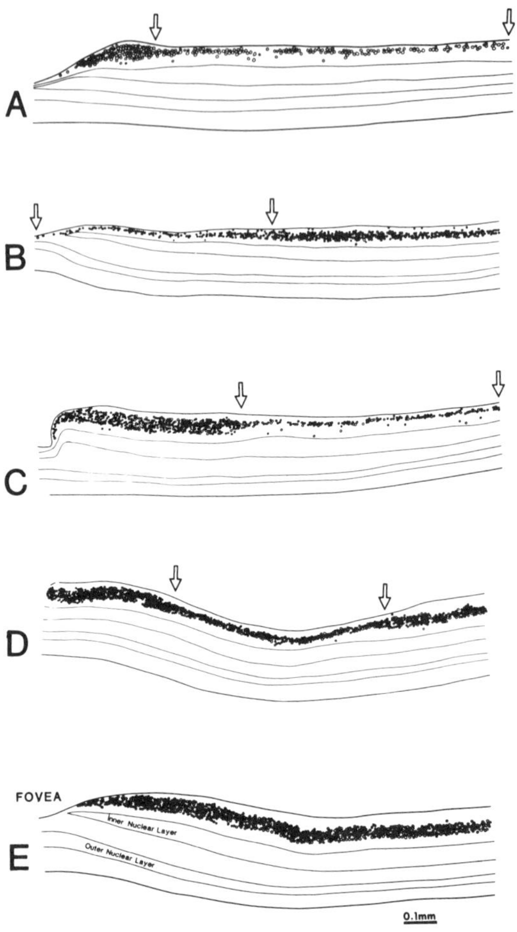

Fig. 1. Parafoveal distributions of ganglion cells of the retina of macaque monkeys with (A-D) and without (E) lesions of striate cortex. Drawings are from horizontal 8μm sections. The ganglion cell loss was found in the region between the arrows. The location of loss varied because cortical lesions varied. Monkeys ablated as infants were examined (A) after 5 years, (B) after one year, and (C) after two months. An animal with a lesion as an adolescent (D) was examined 14 months later. E, a normal adolescent.

indicate that transneuronal ganglion cell loss after striate cortex lesions in infant monkeys is rapid and extensive. Considerable cell loss occurs in as little as 3 weeks, and as much as 65% of the total ganglion cell population is lost within 2 months.

Examples of regions of ganglion cell loss after survivals of 5 years (A), 1 year (B), and 2 months (C) are shown in Figure 1. Estimates of percentage of loss (Figure 2A) were made by comparisons with matched locations in normal parts of retinas from the same animal. All our cases also demonstrated a loss of projections to the degenerated zones of the parvocellular layers of the lateral geniculate nucleus (Figure 3). For unknown reasons, the loss of ganglion cell inputs to the dorsal or outer parvocellular layers appeared to be more severe than that to the inner parvocellular layers. This result suggests some difference in the inputs to inner and outer parvocellular layers.

<u>Macaque monkeys with lesions as juveniles</u>. The initial descriptions of ganglion cell loss in macaque monkeys involved animals lesioned as juveniles or adolescents, where Van Buren (1963a; 1963b) found a 70% loss after a survival of 4 years, and Cowey (1974) reported an 80% loss after a survival of 8 years. We placed lesions in striate cortex of 3 macaque monkeys of unknown ages. Two were obviously immature juveniles, and one appeared to be somewhat older, but still immature. After 6 months survival the retinas of one of the juveniles showed a slight (17%) loss of ganglion cells, with counts still within what could be a normal range (Figure 2B). After a 1 year survival, the apparently older adolescent animal was similar, with a possible loss of 19%. The other juvenile was allowed to survive for 14 months, and it showed a ganglion cell loss of 57% (Figure 1D), and clearly reduced projections to the degenerated parvocellular layers of the lateral geniculate nucleus. This latter result is comparable to that of Dineen and Hendrickson (1981), who ablated striate cortex of one hemisphere in a 1 year old macaque monkey, and ablated the other side 3 months later. After 10 more months, they found an 80% ganglion cell loss corresponding to the lesion at 1 year of age, and a 65% loss for the lesion performed at 15 months of age. In sum, the results strongly suggest that through the juvenile period, macaque monkeys are susceptible to marked ganglion cell loss,

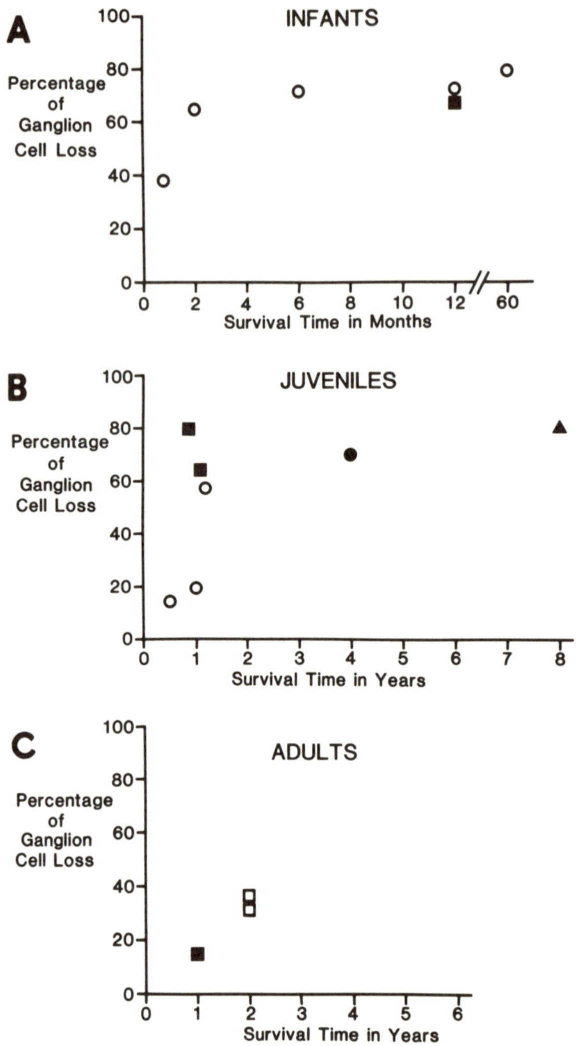

Fig. 2. Ganglion cell loss in macaque monkeys lesioned as infants (A), juveniles (B) and adults (C). Postlesion survival times are indicated. Open circles, present results; closed circle, from Van Buren 1963a; open squares, from Dineen et al. 1982; closed squares, from Dineen, Hendrickson 1981; closed triangle, from Cowey 1974.

specifically of those cells projecting to the parvocellular geniculate layers, and this loss can be as extensive or nearly as extensive as that produced by lesions in infants. However, the ganglion cell loss that does occur may not be so rapid. After 6 months survival, little loss was evident in a monkey with cortex removed as a juvenile, while a near maximal loss was found only two months after a lesion was made in an infant.

Macaque monkeys with lesions as adults. We have no results from fully mature animals, although the preservation of ganglion cells in our monkey with cortex ablated as an apparent adolescent suggests that even young adults would have little ganglion cell loss over moderate survival periods. Dineen and Hendrickson (1981) found only a 15% loss in a monkey lesioned as an adult and then surviving 1 year (Figure 2C), and Dineen et al. (1982) reported a 30-40% loss in two monkeys lesioned as adults and surviving 2 years postoperatively. Thus, fully adult macaque monkeys show some loss of ganglion cells, but only after extensive postoperative survival times.

Results from other primates. For comparisons with results obtained from adult macaque monkeys and humans, where some ganglion cell loss occurs, we examined adult New World owl and squirrel monkeys at various times after receiving lesions as adults. No ganglion cell loss or alteration in retinogeniculate projections was observed in 2 owl monkeys surviving 4 and 14 months postoperatively or in 3 squirrel monkeys with survival times ranging from 7 to 17 months. These results are comparable to those reported for prosimian galagos where postlesion survivals of up to 3 years resulted in no obvious ganglion cell loss (Weller et al. 1981). Although the number of cases is limited, we conclude that the X-like ganglion cells in adult New World monkeys, and in adult prosimian primates, like X-cells in adult cats, are not lost in appreciable numbers after cortical lesions, even after long survival times.

A critical question, of course, is whether or not ganglion cells are lost in New World monkeys and prosimians after lesions as infants or juveniles. As of yet, this information is limited. We did examine a squirrel monkey lesioned at 3 weeks of age and allowed to survive 6 months longer. In this case, there was no obvious ganglion cell

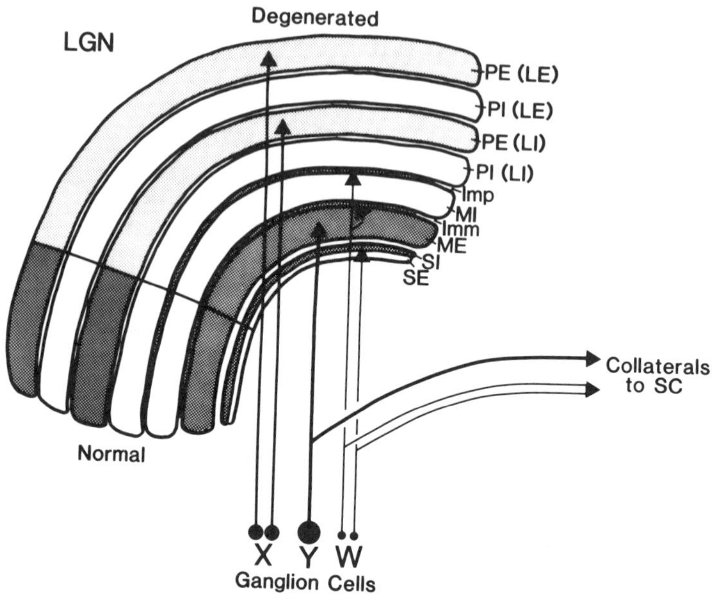

Fig. 3. We hypothesize that sustaining collaterals to the superior colliculus, SC, prevent the loss of Y-like and W-like ganglion cells of the retina of macaque monkeys while X-like cells, without collaterals, are susceptible to loss. Retinal projections from the contralateral eye (shaded) terminate in leaflets of the external parvocellular layers (PE-LE and LI), the external magnocellular layer, ME, the internal superficial layer, SI, and the interlaminar zones, Imp and Imm (see Kaas et al. 1978 for terminology) in the normal portion of the nucleus (left), but terminations are greatly reduced in the degenerated parts of the parvocellular layers (right).

loss. Since this result contrasts to those from macaque monkeys, where a 6 month survival after a lesion as an infant resulted in a 65% ganglion cell loss, it seems likely that New World monkeys lose any susceptibility they might have for ganglion cell loss soon after birth, or earlier.

The significance of the results. We are left with clear evidence that at least three variables are important in the occurrence and extent of retinal ganglion cell loss after lesions of visual cortex in mammals. First, species vary: only higher primates, humans and Old World monkeys, have been shown to lose ganglion cells after lesions as juveniles and adults. Second, for macaque monkeys and cats at least, age at the time of lesion is a critical variable. Cats are susceptible to ganglion cell loss for only a short period after birth, and macaque monkeys appear to be highly susceptible as infants, but much less so after the juvenile period. Third, survival time is important and it interacts with the age of the animal at the time of lesion. Cell loss for macaque monkeys is rapid in infants, much slower for juveniles, and quite slow for adults. Longer times are required for equivalent losses in older animals. The time required for a marked cell loss changes from weeks or less to years.

We are not certain why any of the three factors above are important. For the present, we favor the following explanation.

1. After striate cortex lesions, retinal ganglion cells without collaterals to the superior colliculus are susceptible to loss. Susceptibility, of course, means that ganglion cells have some way of determining the presence or absence of target cells. But, the ganglion cells apparently do not distinguish between the lateral geniculate nucleus and superior colliculus targets at this time, and feedback from either target is sufficient.

2. Some extrageniculate collaterals may not be able to maintain a ganglion cell. At least some, and perhaps most, X-cells of the retina in cats have an extrageniculate branch that apparently goes to the pretectum (see Sur, Sherman 1982; Fukuda, Stone 1974). Since it is the X-cells that are lost from the retina after cortical lesions (Tong et al. 1982), X-cells with extrageniculate collaterals may degenerate. Perhaps some collaterals form

an inadequate number of synapses, are too insubstantial to transport a sustaining signal, or the extrageniculate target is not adequate. It is not yet known if X-like cells in primates have collaterals to the pretectum. However, retinal inputs to the pretectum appear to be reduced after ganglion cell loss (Dineen et al. 1982).

3. Some species variation in susceptibility to ganglion cell loss can be expected from species differences in proportions of ganglion cells projecting to the lateral geniculate nucleus without collaterals to the superior colliculus. Some mammals, perhaps rats, may have so few cells without collaterals that a detectable loss may not occur at any age. Within cats and primates, where susceptible X-cells or X-like cells occur in substantial numbers, the proportion of such cells varies from about 40% in cats to around 80% or more in macaque monkeys. Fewer would be expected in nocturnal prosimians and the nocturnal owl monkey, and more might be expected in humans, with a visual system specialized for detailed vision and object recognition abilities apparently related to the X-like cell relay.

4. The sustaining influence of the feedback from target cells is very critical in early stages of maturation of ganglion cells, and it becomes progressively less critical. We see no functional reason for this postulated factor at postnatal ages, but it seems reasonable to suggest that feedback is critically related to the process that occurs earlier in the developing visual system when ganglion cells that fail to find targets die (see Cunningham 1982). If the normal developmental mechanism extends past the normal time of cell death and decreases with time, then the importance of age for the ganglion cell loss after striate lesions is explainable without postulating other age dependent mechanisms.

Of course, lesions in very young, typically fetal, mammals introduce other age related factors. At early stages of development, neurons are capable of sprouting and finding new sustaining targets, and such plasticity can prevent cell death (see Perry, Cowey 1982). They are also capable of retracting or losing branches that fail to establish connections. This latter factor may account for the complete loss of retinal input to the degenerated lateral geniculate nucleus found after cortical ablations in fetal monkeys (Ogren et al. 1983).

5. Mammalian species vary in maturation rates, and the "juvenile" requirement for the sustaining feedback from target cells is gradually lost at species dependent rates from the time of normal ganglion cell loss. It is commonly argued that the evolution of higher primates was dependent on a certain degree of neoteny, or the retention of "juvenile" traits in the adult. Thus, higher primates in particular should have extended the period of susceptibility of ganglion cell loss, not for any functionally important reasons per se, but as part of a general retention of juvenile features. We expect that prosimians, New World monkeys, Old World monkeys, and humans have step-wise extensions of the period of susceptibility to ganglion cell loss. Indeed, only humans have been shown to have extensive losses of ganglion cells after lesions in adults (Haddock, Berlin 1950).

This research was supported by NEI grant EY02686.
Dr. Weller's present address is Department of Neurobiology and Behavior, State University of New York at Stony Brook, Stony Brook, NY 11794.

Callahan EC, Tong L, Spear PD (1982). The critical period for effects of visual cortex lesions on retinal X-cells in cats. Supplement to Invest Ophthal Vis Sci 22:236.

Cowan W (1970). Anterograde and retrograde transneuronal degeneration in the central and peripheral nervous system. In Nauta W, Ebbesson S (eds): "Contemporary Research Methods in Neuroanatomy", New York: Springer-Verlag.

Cowey A (1974). Atrophy of retinal ganglion cells after removal of striate cortex in rhesus monkey. Perception 3: 257.

Cunningham TJ (1982). Naturally occuring neuron death and its regulation by developing neural pathways. In Bourne GH, Danielli JF (eds): "International Review of Cytology", New York: Academic Press, pp. 163.

Dineen J, Hendrickson, AE (1981). Age-correlated differences in the amount of retinal degeneration after striate cortex lesions in monkeys. Invest Ophthal Vis Sci 21: 749.

Dineen J, Hendrickson AE, Keating EG (1982). Alterations of retinal inputs following striate cortex removal in adult monkey. Exper Brain Res 47:446.

Dreher B, Fukada Y, Rodieck W (1976). Identification, classification, and anatomical segregation of cells with X-like and Y-like properties in the lateral geniculate nucleus of Old World primates. J Physiol 258:433

Fukuda Y, Stone J (1974). Retinal and central projections of Y-, X-, and W-cells of the cat's retina. J. Neurophysiol. 37:749.

Gills, Jr. JP, Wadsworth JAC (1967). Retrograde transsynaptic degeneration of the inner nuclear layer of the retina. Invest Ophthalmol 6:437.

Haddock JN, Berlin L (1950). Transsynaptic degeneration in the visual system. Report of a case. Arch Neurol 64:66.

Kaas J, Huerta M, Weber J, Harting J (1978). Patterns of retinal terminations and laminar organization of the lateral geniculate nucleus of primates. J Comp Neurol 182:517.

Kalil R (1978). Projection of the retina to the lateral geniculate nucleus in the cat following neonatal ablation of visual cortex. Neurosci Abstr 5:633.

Leventhal AG, Rodieck RW, Dreher B (1981). Retinal ganglion cell classes in the Old World monkey: Morphology and central projections. Sci 213:1139.

Norton TT, Casagrande VA (1982). Laminar organization of receptive-field properties in lateral geniculate nucleus of bushbaby (Galago crassicaudatus). J Neurophysiol 47:715.

Ogren MP, Rakic P, Goldman-Rakic P (1983). Consequences of prenatal striate cortex lesions on retinogeniculate projections in the monkey. Supplement to Invest Ophthalmol Vis Sci 24:64.

Pearson HE, Labar DR, Payne BR, Cornwell P, Aggarwal N (1981). Transneuronal retrograde degeneration in the cat retina following neonatal ablation of visual cortex. Brain Res 212:470.

Perry VH, Cowey A (1979). The effects of unilateral cortical and tectal lesions on retinal ganglion cells in rats. Exper Brain Res 35:85.

Perry VH, Cowey A (1982). A sensitive period for ganglion cell degeneration and the formation of aberrant retino-fugal connections following tectal lesions in rats. Neurosci 7:583.

Sherman M, Wilson J, Kaas J, Webb S (1976). X- and Y-cells in the dorsal lateral geniculate nucleus of the owl monkey (Aotus trivirgatus). Sci 192:475.

Sur M, Sherman SM (1982). Retinogeniculate terminations in cats: Morphological differences between X and Y cell axons. Sci 218:389.

Tong L, Spear PD, Kalil RE, Callahan, EC (1982). Loss of retinal X-cells in cats with neonatal or adult visual cortex damage. Sci 217:72.

Van Buren JM (1963a). Trans-synaptic retrograde degeneration in the visual system of primates. J Neurol Neurosurg Psychiat 27:402.

Van Buren JM (1963b). The retinal ganglion cell layer. Springfield, IL: Charles C Thomas, Publisher.

Walsh FB (1947). Clinical neuro-opthalmology. Baltimore: The Williams and Wilkins Company.

Weller R, Kaas J (1977). Evidence for selective loss of X-cell input to the dorsal lateral geniculate nucleus of monkeys with long-term ablations of striate cortex. Neurosci Abstr 3:581.

Weller R, Kaas J (1980). Loss of retinal ganglion cells and altered retinogeniculate projections in monkeys with striate cortex lesions. Supplement to Invest Ophthal Vis Sci 19:2.

Weller RE, Kaas JH (1982). Retinal degeneration and loss of retinogeniculate projections in infant and adult monkeys surviving variable times after lesions of striate

cortex. Supplement to Invest Ophthal Vis Sci 22:46.
Weller RE, Kaas JH, Ward J (1981). Preservation of retinal ganglion cells and normal patterns of retinogeniculate projections in prosimian primates with long-term ablations of striate cortex. Invest Ophthal Vis Sci 20:139.

Weller RE, Kaas JH, Wetzel AB (1979). Evidence for the loss of X-cells of the retina after long-term ablation of the visual cortex in monkeys. Brain Res 160:135.

Development of Visual Pathways in Mammals, pages 303–314

THE HOST RETINAL PROJECTION TO TECTAL TRANSPLANTS IN RATS

Alan R. Harvey* and Raymond D. Lund†

* Unit of Human Physiology, Centre for Neuroscience, Flinders University, Bedford Park, S.A. 5042, Australia.
† Department of Cell Biology and Anatomy, School of Medicine, University of Pittsburgh, Pittsburgh, PA 15261, U.S.A.

INTRODUCTION

It is now well established that in a number of mammalian species, tissue from various parts of the embryonic central nervous system (CNS) can be successfully transplanted into the brains of neonatal or adult hosts (e.g. Björklund, Stenevi 1979; Das, Hallas, Das 1980; Gash, Sladek 1980; Perlow 1980; Lund, Harvey, Jaeger, McLoon 1982). The grafts invariably survive the transplantation procedure, grow and differentiate within the host and in many cases establish neural connections with the host CNS.

Our work in this area has concentrated upon transplantation of parts of the visual system in rats (Lund *et al.* 1982). In particular, we have been grafting embryonic tectal tissue onto the midbrain of neonatal rats (Lund, Hauschka 1976; Harvey, Lund 1981; Lund, Harvey 1981; Harvey, Golden, Lund 1982; Harvey, Lund 1983). Our aims have been to determine (1) how the grafts grow and mature in the host, (2) what sort of neural connectivity is developed between the grafted tissue and the host CNS, and (3) what factors are important in influencing the formation of these connections.

To illustrate these points, the present paper concentrates upon a description of how optic axons from the host retinae grow into and innervate tectal transplants. The factors which influence the pattern of host retinal innervation of grafts will be discussed and compared with the factors which apparently guide optic fibre growth during normal development.

METHODS

The experiments described in this paper involve the transplantation of foetal tectal tissue from albino or pigmented rat embryos, usually of 15 days gestational age (the day after mating is E0). The tectae (containing presumptive superior colliculus and perhaps some developing inferior colliculus) are dissected out in cold tissue culture medium. They are then injected onto the midbrain of, in most cases, neonatal host rats through a glass pipette inserted beneath the skull. The host animals are then returned to their mothers and allowed to survive for various periods of time before the interrelationship between the host and graft is examined anatomically, histochemically or in some instances, physiologically.

Usually, tectal grafts can be identified in 80 to 90% of host brains, although in some litters all hosts contain recognizable transplants. Figure 1 shows a tectal transplant lying over the left superior colliculus of a host rat. This graft was transplanted into a 7 day old host.

It is important to stress at this point that:

a) Thymidine labelling studies have shown that there is little or no migration of neurons from the host to the transplant or vice versa (Lund, Harvey 1981). Grafts can, in most instances, be reliably identified in cellular and fibre stained material and these preparations are sufficient for delineating the boundaries of the transplant.

b) Most of the donor embryonic tissue is derived from E15 embryos. At this time, approximately 20% of tectal cells have not completed their final division and have yet to migrate away from the ventricular proliferative zone (e.g. Mustari, Lund, Graubard 1979; Altman, Bayer 1981).

c) At E15, the tectal tissue is naive in that it has not yet been influenced by ingrowing axons from areas extrinsic to the tectum. Optic axons, for example, do not begin to grow into the superior colliculus until E16 (Lund, Bunt 1976).

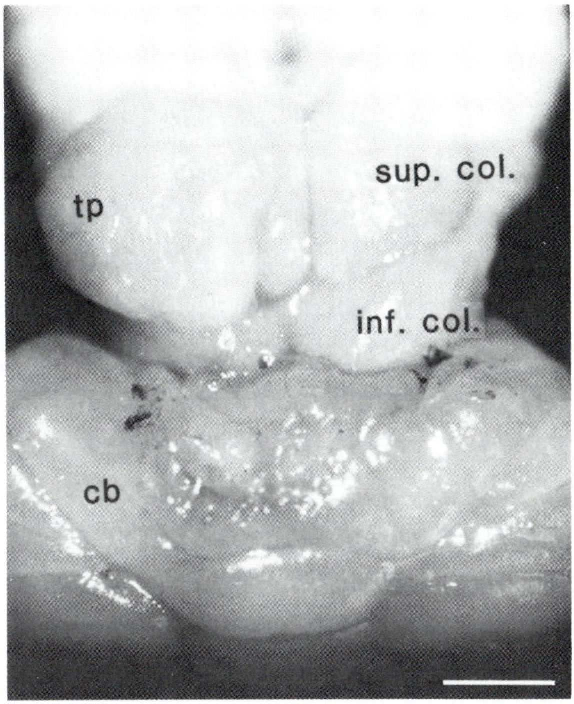

Fig. 1. Photomicrograph of the dorsal surface of an adult rat brain showing a transplant (tp) lying over the left superior colliculus (sup.col). Rostral is to the top. Inf.col., host inferior colliculus; cb, host cerebellum. Scale = 2 mm.

RESULTS

The Development of the Host Retinal Projection to Tectal Transplants

The development of host retinal input to tectal tissue transplanted into newborn rats has been studied by injecting the host eyes with horseradish peroxidase (HRP) at various times shortly after transplantation (Harvey, Lund 1983). The graft is then examined for the presence of anterogradely labelled retinal fibres using the tetramethylbenzidine (TMB) chromogen. The host retina projects only to those grafts which are contiguous with the host superior colliculus. By 4 days after transplantation, HRP-labelled optic axons can be seen extending into the grafts as much as 350 μm from the host superior colliculus. The fibres are only found adjacent to the surface of the transplanted tissue. Over the next few days host optic axons are found at successively greater distances from the host/transplant connection. Patches of what appear to be terminal arborizations of host retinal axons can be discerned as early as 5 to 6 days after transplantation and the pattern of retinal innervation of tectal grafts appears to be mature by 12 to 14 days. At no time during the development of innervation do retinal fibres grow over the whole surface or invade the whole graft. Rather, the axons tend to fasciculate into a number of discrete bundles and grow into circumscribed areas of the graft which are always located adjacent to the transplant surface. The development of the host retinal projection precedes morphological maturation of the transplants.

The Mature Pattern of Retinal Input and its Relation to Transplant Morphology

An example of host retinal input to a mature graft is shown in Fig. 2. Note the localized area (arrowed) containing a high density of HRP-labelled retinal axons. This is typical of the host optic projection to mature tectal transplants and can be shown using both ^{3}H-proline autoradiography and degeneration methods after eye removal (Lund, Harvey 1981). In a given transplant, there may be as many as 8 to 10 of these localized areas of dense retinal projection. Such patches can be located in regions far distant from the host/transplant connection. It is clear therefore that ingrowing retinal fibres do not necessarily innervate the first cells they encounter in the grafted tissue.

Of particular interest is the observation that areas in the grafts which receive a high density of retinal innervation have characteristic cytoarchitectural features (Lund, Harvey 1981). In neurofibrillar preparations

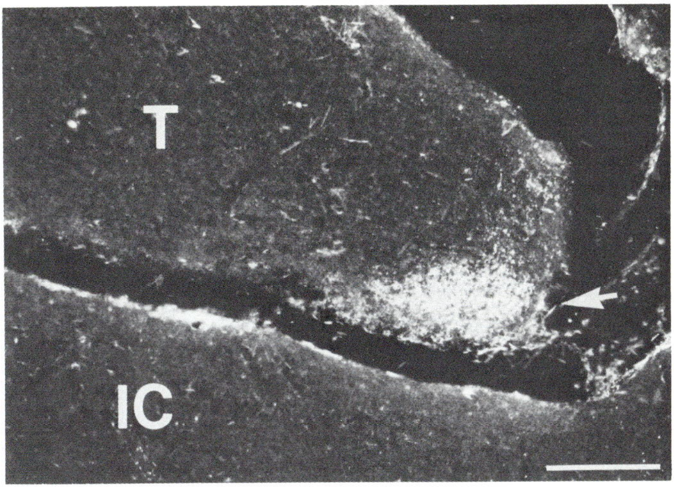

Fig. 2. Dark field photomicrograph of a TMB-stained coronal section showing a patch of HRP-labelled host retinal fibres (arrow) in a mature tectal transplant (T). IC, host inferior colliculus. Scale = 100 μm.

(e.g. Fig. 3B), much of the transplant contains a high density of stained fibres which run in various directions throughout the graft. However, local relatively fibre-free areas are also seen both within the transplants and at the transplant surface. In Nissl stained material, these fibre-free regions often contain a preponderance of cells with small somata and lack the large multipolar neurons found elsewhere in the graft. Patches of host retinal input are always confined to one or more of the superficially located fibre-free areas (see Figs. 1 to 3 in Lund, Harvey 1981). It is significant that the morphological characteristics of retino-recipient areas in tectal grafts are shared by the stratum griseum superficiale (SGS) of the rat superior colliculus *in situ*, a layer which normally receives optic input.

Acetylcholinesterase (AChE) Studies

A feature of the normal SGS is that it contains a relatively high concentration of AChE (e.g. Palkovitz, Jacobowitz 1974; Silver 1974). One of us (ARH) has examined the distribution of AChE in tectal grafts using the method of Karnovsky and Roots (1964) to determine whether AChE is concentrated in the fibre-free regions and to examine how its presence is related to host retinal innervation (Harvey, MacDonald, 1983).

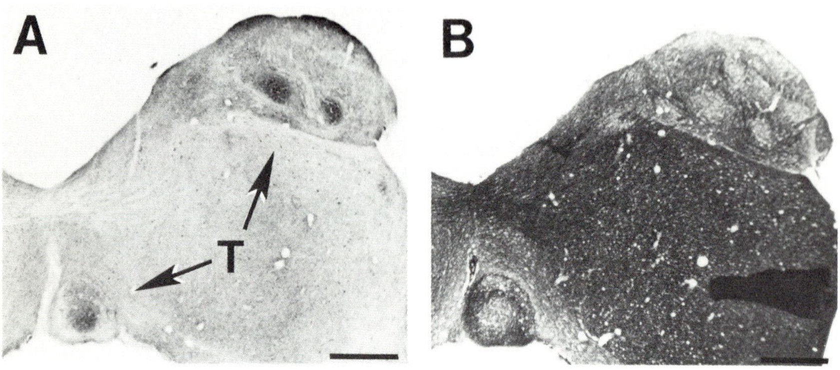

Fig. 3. Photomicrographs of tectal transplants (T) in the midbrain of a host rat. The main graft lies on the dorsal surface of the host inferior colliculus. A smaller fragment is located in the ventricle. A, coronal section stained for AChE; B, adjacent coronal section stained for neurofibrils. Scale for A and B = 500 μm.

An example of an AChE stained section of a tectal transplant is shown in Fig. 3A along with an adjacent section stained for neurofibrils (Fig. 3B). Note that the areas of dense AChE staining in both the graft over the host inferior colliculus and the graft fragment in the ventricle correspond to relatively fibre-free regions in the adjacent silver stained section. This particular graft did not connect with the host superior colliculus and received no retinal input. It is clear therefore that in rats the development of AChE staining in tectum is not dependent upon optic input. However, where retinal input has been traced to grafts, patches of innervation correspond to areas which stain densely for AChE. This latter observation, taken together with the cytoarchitectural features described earlier, suggests that host retinal fibres preferentially innervate specific areas of the graft which contain presumptive SGS cells.

Not all of the superficial fibre-free, AChE-rich areas receive retinal input. This may be explained by the observation that at no time do retinal fibres grow over the entire transplant surface. Similar areas located deep within the graft neuropil also never receive host optic input, perhaps because ingrowth of retinal axons is restricted to the transplant surface (Harvey, Lund 1983).

Ingrowth of Fibres from the Two Eyes

In some cases, tectal grafts receive an input from each of the host eyes (Lund, Harvey 1981). This can be tested by injecting one eye with ^{3}H-proline, removing the other eye and then processing alternate sections for autoradiography and degeneration. Fibres from each eye project into fibre-free regions in the graft, but areas of high density innervation are non-overlapping, with the input from the right and left eyes lying adjacent to each other (Lund, Harvey 1981). This segregation of retinal fibres occurs even within the fibre tracts running from the host to the transplant and is similar to that found in binocular innervation of central visual structures during normal development.

Comparison with Other Host Inputs

Studies utilising the retrograde transport of HRP have shown that, in addition to retina, a large number of host regions project into tectal tissue transplanted into newborn hosts (Harvey, Lund 1981). To gain a better understanding of the factors which influence the mode of ingrowth of host optic fibres into grafts, it is important to compare this ingrowth with that of fibres from other host areas. If there are differences in the way different host areas grow into the transplants it might suggest that axons are following specific cues rather than being directed by purely mechanical factors.

The host cortical projection to tectal grafts has been examined using degeneration methods (Lund, Harvey 1981; Harvey *et al.* 1982). Cortical degeneration is found throughout the grafts, but is sparse or absent in fibre-free areas, apparently independent of whether these areas are innervated by host retinal fibres. The projection from serotonergic nuclei in the host brainstem has also been studied by one of us (ARH) using a monoclonal antibody to 5-dihydroxytryptamine (5-HT) (kindly donated by Dr. A.C. Cuello). The density of serotonergic innervation varies from one graft to another, however 5-HT containing fibres can be found within the fibre-free and fibre-dense regions of the transplants. Unlike retinal axons, ingrowing cortical and serotonergic fibres are not restricted to the transplant surface.

Ultrastructural and Physiological Studies

It is clearly important to determine whether ingrowing host fibres form synapses with transplanted tectal neurons and whether the input is functional. We have shown that host retina and cortex make asymmetric synaptic contact primarily with dendritic spines or small dendrites of cells within the grafts (Lund, Harvey 1981; Harvey *et al*. 1982). For retina, the maximum density of degenerating terminals after eye removal is similar to that found in the SGS of normal rats.

In a series of physiological experiments, attempts have been made to activate neurons in transplants by electrically stimulating the host occipital cortex and optic nerves (Harvey *et al*. 1982). About 12% of the cells in tectal grafts could be activated after cortical stimulation, however no neurons were excited after optic nerve stimulation. The lack of physiological evidence for retinal innervation of tectal transplants is almost certainly due to the fact that very few of the recording microelectrode penetrations passed through retino-recipient areas in the grafts (see Fig. 7, Harvey *et al*. 1982). Only two of the 227 units tested by optic nerve stimulation were located in superficial fibre-free areas in the grafts. More experiments are required to define the nature of the visual influence on transplant activity.

Retinal Innervation of Tectal Tissue Grafted into Older Hosts

In order to determine how the age of the host affects its ability to innervate tectal grafts, tissue has been transplanted to the midbrain of older host rats. McLoon and Lund (1983) have been unable to trace any host retinal input to embryonic tectal tissue grafted onto the adult superior colliculus. One of us (ARH) is currently examining the host innervation of tectal tissue transplanted to the midbrain of 3, 7, 14 and 21 day old hosts (Harvey 1984). Retinal fibres

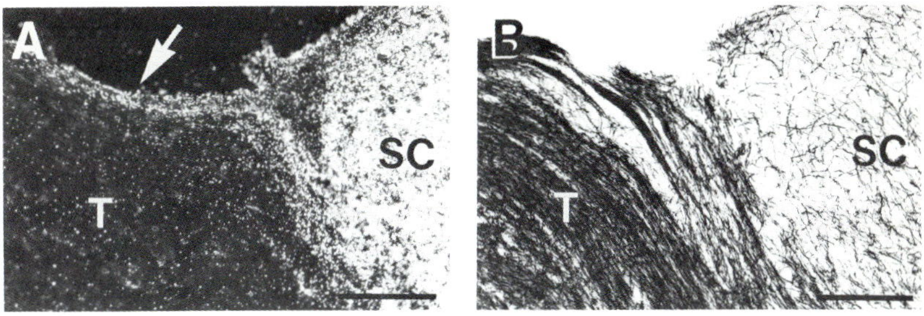

Fig. 4. Photomicrographs of a tectal transplant (T) connecting to the caudal pole of the host superior colliculus (SC). The graft was transplanted into the host three days after birth. ^{3}H-proline was injected into the host eyes four months later. A, dark-field view of a saggital section prepared for autoradiography showing host optic axons (arrow) growing into the graft. B, adjacent saggital section stained for neurofibrils. Scale for A and B = 100 μm.

have been traced into tectum grafted into 3 day old hosts (Fig. 4), but only rarely in 7 day and 14 day host rats. The decrease in optic innervation of transplants with age seems to parallel the maturation of retinal projections to their normal target site, the host superior colliculus (e.g. Lund, Lund 1972).

SUMMARY AND CONCLUSIONS

A number of techniques have been used to examine the way host retinal fibres grow into and innervate tectal grafts. The data obtained from these studies can be summarized as follows:

(1) Host retinal fibres grow into embryonic tectal grafts relatively soon after transplantation.

(2) Ingrowing retinal fibres are found only adjacent to the surface of the transplants. This restriction to the graft surface does not appear to be due to mechanical barriers *per se*, since other host projections (from cerebral cortex and brainstem) grow into the whole depth of the transplanted tissue. By comparison, in normal development, the first formed fascicles of optic axons

growing over the developing diencephalon and midbrain are also initially limited to the surface of these tissues. (Lund, Bunt 1976; Frost, So, Schneider 1979; Edwards, Schneider, Caviness 1981).

(3) Retinal fibres growing into tectal grafts tend to fasciculate into a small number of bundles. Again this is similar to the behaviour of optic axons during normal development (Lund, Bunt 1976; Edwards *et al.* 1981).

(4) Ingrowing retinal axons do not initially form widespread connections which later retract.

(5) Retinal fibres do not necessarily innervate the first cells they encounter in the transplants. Projections are confined to discrete, superficially located areas in the graft which morphologically and histochemically resemble the retino-recipient layers of normal superior colliculus.

(6) Within the retino-recipient areas in the grafts, optic axons form synapses with transplanted neurons.

(7) In binocularly innervated grafts, fibres from the two host eyes tend to segregate out from one another.

In conclusion, the behaviour of host optic axons as they grow into and innervate tectal grafts resembles the way retinal fibres innervate central visual targets during normal development. It seems likely therefore, that within tectal transplants, host optic axons follow cues similar to those which influence their normal growth patterns in intact brains. The transplant technique thus provides an opportunity to manipulate and analyse the factors which guide optic fibre growth during normal development of the mammalian visual system.

ACKNOWLEDGEMENTS

We gratefully acknowledge the secretarial assistance of J. Gotch.

In the U.S.A., this work was supported by grants (EY 01950, 03326 and 03414) to R.D. Lund from NIH. A.R. Harvey was supported in Australia by grants from the

NH& MRC, FMC Research Foundation, the Neurosurgical Research Foundation of S.A. Inc. and the Channel 10 Children's Medical Research Foundation of S.A. Inc.

REFERENCES

Altman J, Bayer SM (1981). Time of origin of neurons of the rat superior colliculus in relation to other components of the visual and visuomotor pathways. Exp Brain Res 42: 424.

Björklund A, Stenevi U (1979). Reconstruction of brain circuitries by neural transplants. TINS 2:301.

Das GD, Hallas BH, Das KG (1980). Transplantation of brain tissue in the brain of rat I. Growth characteristics of neocortical transplants from embryos of different ages. Am J Anat 158:135.

Edwards MA, Schneider GE, Caviness VS Jr (1981). Relationship between developing optic axons and their target cells in the superficial superior colliculus of the mouse. Soc Neurosci Abstr 7:733.

Frost DO, So K-F, Schneider GE (1979). Postnatal development of retinal projections in Syrian hamsters: a study using autoradiographic and anterograde degeneration techniques. Neuroscience 4:1649.

Gash D, Sladek JR Jr (1980). Vasopressin neurons grafted into Brattleboro rats: Viability and activity. Peptides 1:11.

Harvey AR (1984). Transplantation of foetal tectum into young host rats. Neurosci Lett Suppl (in press).

Harvey AR, Golden GT, Lund RD (1982). Transplantation of tectal tissue in rats III. Functional innervation of transplants by host afferents. Exp Brain Res 47:437.

Harvey AR, Lund RD (1981). Transplantation of tectal tissue in rats II. Distribution of host neurons which project to transplants. J comp Neurol 202:505.

Harvey AR, Lund RD (1983). Transplantation of tectal tissue in rats IV. Maturation of transplants and development of host retinal projection. Dev Brain Res (in press).

Harvey AR, MacDonald AM (1983). Acetylcholinesterase in tectal tissue transplanted into newborn rats. Neurosci Lett Suppl 9:S49.

Karnovsky MJ, Roots L (1964). A 'direct-colouring' thiocholine method for cholinesterases. J Histochem Cytochem 12:219.

Lund RD, Bunt AH (1976). Prenatal development of central optic pathways in albino rats. J comp Neurol 165:247-264.

Lund RD, Harvey AR (1981). Transplantation of tectal tissue in rats I. Organization of transplants and pattern of distribution of host afferents within them. J comp Neurol 201:191.

Lund RD, Harvey AR, Jaeger CB, McLoon SC (1982). Transplantation of embryonic neural tissue to the tectal region of newborn rats. In Morrison AR, Strick PL (eds): "Changing Concepts of the Nervous System," New York: Academic Press, p 361.

Lund RD, Hauschka SD (1976). Transplanted neural tissue develops connections with host rat brain. Science 193: 582.

Lund RD, Lund JS (1972). Development of synaptic patterns in the superior colliculus of the rat. Brain Res 42:1.

McLoon SC, Lund RD (1983). Development of fetal retina, tectum and cortex transplanted to the superior colliculus of adult rats. J comp Neurol 217:376.

Mustari MJ, Lund RD, Graubard K (1979). Histogenesis of the superior colliculus of the albino rat: a tritiated thymidine study. Brain Res 165:39.

Palkovits M, Jacobowitz DM (1974). Topographic atlas of catecholamine- and acetylcholinesterase-containing neurons in the rat brain II. Hindbrain (mesencephalon, rhombencephalon). J comp Neurol 157:29.

Perlow MJ (1980) Functional brain transplants. Peptides 1:101.

Silver A (1974). "The Biology of Cholinesterases." Amsterdam and Oxford: North Holland.

Development of Visual Pathways in Mammals, pages 315–327

TROPHIC RELATIONSHIPS DURING VISUAL SYSTEM DEVELOPMENT

Timothy J. Cunningham and Forrest A. Haun

Department of Anatomy
The Medical College of Pennsylvania/EPPI Division
3200 Henry Avenue, Philadelphia, PA 19129, U.S.A.

Neurons in the developing central nervous system are particularly vulnerable to lesions that damage their target structures or their afferent input. Several instances of the sensitivity of young neurons to lesions have been documented in the visual system of mammals, where damage to one visual structure often produces a profound and rapid loss of neurons in connecting cell populations -- populations that are one or more synapses away. For example, lesions to the visual cortex in neonatal rodents and cats result in rapid degeneration of neurons in the dorsal lateral geniculate nucleus (Cunningham, et al., 1979; Labar et al., 1981). In cats and primates, visual cortex ablation also results in retrograde transneuronal death of some retinal ganglion cells (Pearson, et al., 1981, Kalil, this volume, Weller and Kaas, this volume). Similarly, deafferentation of the rodent subcortical visual centers by enucleation at birth produces substantial neuron loss in the dorsal lateral geniculate and superior colliculus, with this loss evident as early as one week after the eye is removed (Delong and Sidman, 1962; Heumann and Rabinowicz, 1979; Cunningham and Giordano, 1982).

The reason young visual system neurons are so sensitive to axotomy, target deprivation, or removal of developing afferents is not clear. One idea, derived mainly from studies on neurons with peripherally directed axons, is that neuron death induced by lesions may actually represent an exaggeration of naturally occurring neuron death. It is presumed that neurons die naturally because they fail to achieve and maintain sufficient contact with afferents or targets during the period of synaptogenesis (Cunningham, 1982; Oppenheim, 1981).

In the visual centers of many mammals, naturally occurring neuron death has now been documented. Importantly, it is found during the same early periods when neurons in these regions are particularly susceptible to the effects of lesions (see, for example, Pearson, et al., 1983; Cunningham, et al., 1981; Potts et al., 1982; Lam et al., 1982; Perry and Cowey, 1979; Giordano et al., 1979; Sengelaub and Finlay, 1982; Finlay , et al., 1982; Heumann and Rabinowicz, 1979; Giordano and Cunningham, 1982). These data support the suggestion that in the developing mammalian visual system, as in other developing central and peripheral systems, one cause of both normal and lesion-induced neuron death is simply a failure by the neuron, during critical stages of its development, to compete successfully for some limited survival-promoting resource in its innervating or target structures[1].

More particular evidence that such a scheme applies to developing visual neurons comes from studies in which some visual structures are destroyed but the input or target contact of others is increased. In these cases neurons which would die normally are rescued. For example, the expansion of the target contacts of one eye in rodents (which results from removal of the opposite eye at birth), reduces the amount of normal cell death in the surviving eye (Jeffery and Perry, 1981; Sengelaub and Finlay , 1981). Counts of optic axons from the surviving eye of cats and monkeys, enucleated prenatally, suggest the same effect occurs in these species (Williams, et al., 1983; Rakic and Riley, 1983). Experiments on infant rats indicate that retinal afferents can exert similar controls over the survival of neurons in the subcortical visual centers (Cunningham, et al., 1979). After rapid destruction of the dorsal lateral geniculate nucleus by visual cortex lesions in newborn rats, visual centers caudal to the geniculate receive an excessive innervation from the retina. In the nucleus of the optic tract and in the superior colliculus, this excessive innervation sustains neurons that would usually die during normal development.

[1]We have suggested elsewhere (Cunningham, 1982) that "simple competition" can not explain all instances of natural neuron death or neuron death after lesions. Competition does appear to explain best the death of some cells in early stages of development. As the cell matures, other factors, such as the quantitative relationship of afferent and target connectivity may be involved.

It is reasonable to suppose that the basis of a neuron's dependence on its afferent and target structures is the availability of trophic factors supplied by these connecting neuron populations[2]. Support for this argument is derived from two lines of research. First, there are a number of studies showing relatively specific trophic support of both central and peripheral neurons _in vitro_ (see Perez-Polo, et. al., 1983 for review), including a recent report on trophic interactions of neonatal rat retinal ganglion cells in culture (McCaffery, et. al., 1982). Second, and more directly applicable to _in situ_ development, are the _in vivo_ studies on axotomized or target deprived peripheral neurons. It has been recognized for more than a century that young peripheral neurons are especially susceptible to isolation from their targets by axotomy or surgical removal of the periphery (see Lieberman, 1974 for review). It is frequently suggested that the removal of trophic or maintenance factors, derived from the periphery and acting on the cell at critical stages of development, is the basis of the greatly exaggerated neuron death found after these manipulations (e.g. Prestige, 1967). Direct evidence for the importance of such factors has come from experiments with Nerve Growth Factor (NGF), a specific trophic agent for sensory and sympathetic ganglion cells that promotes the survival and differentiation of these cells _in vivo_ (Cohen, et al., 1954; Greene and Shooter, 1980; Bradshaw, 1978). For example, if superior cervical ganglion cells are axotomized during the developmental period when the cells are particularly susceptible to the lesion, exogenous NGF is able to delay the degeneration of some cells for at least a week (Banks and Walter, 1977; Hendry and Campbell, 1976).

Are such trophic and maintenance factors operative in the central nervous system, in particular the mammalian visual system? Furthermore, given the complexity of interconnections in the visual system (compared to the more well-defined connections of peripheral neurons) will it be possible to define specific trophic relationships between particular populations of visual system neurons? We have begun to study these questions in the geniculocortical system of infant rats, and describe here a relatively specific neuronotrophic influence on neurons of the dorsal lateral geniculate nucleus (dLGN) developing _in situ_.

---

[2]Here we use "trophic" in its most general sense, to include agents that support neuron survival and maturation.

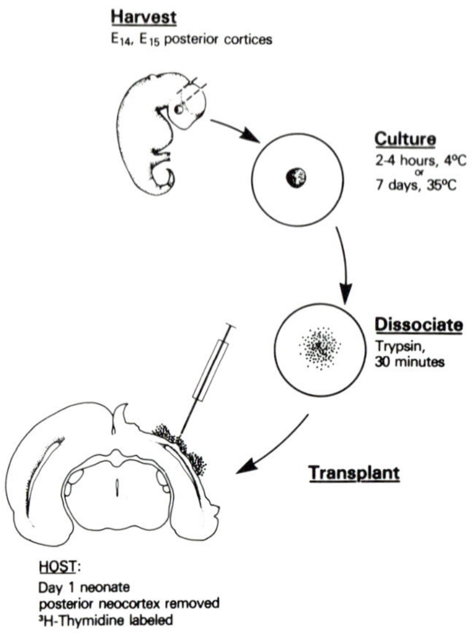

Fig. 1: Diagram summarizing procedures followed in the transplantation experiments. (See text)

The experiments were designed to determine if transplants of fetal posterior cortex tissue placed into the lesion cavity of a newborn rat with a posterior cortical lesion, would act as a source of trophic support, and therefore affect the rapid degeneration of the dorsal lateral geniculate that usually occurs after such a lesion (Cunningham et. al., 1979). The experimental procedures we followed are illustrated in Figure 1. Transplant tissue was obtained from fetal rats (Embryonic days 14 or 15 [E14, E15]) and either stored in culture media at $4^{\circ}C$ for a few hours or cultured at $35^{\circ}C$ for a week. The preculturing step was added to our original design in an attempt to make up for the disparity in age between the tissue that was removed by the lesion (newborn), and the tissue that was transplanted (E14, E15). We hoped that preculturing the tissue would allow transplantation of more differentiated cortical cells without paying the price of reduced viability found in direct transplants of older cortical tissue (Das et. al., 1980). Cortical explants from 3 or 4 donors were cultured together in one Falcon "Primaria" plastic petri dish (coated to inhibit fibroblast outgrowth). The culture media consisted of Ham's F-10 nutrient medium with 5% porcine serum and 450mg% glucose, and was changed every other day. Culturing occurred in a 90% air/10% CO_2 atmosphere. All direct and precultured transplants were first trypsin-dissociated, then transplanted into the lesion cavity of a newborn rat in which the lesion had removed at least the posterior half of the neocortex. In some host rats, neurons of the dLGN had been labeled by pulse injections of ^{3}H-thymidine into pregnant mothers. The injections were made on E14 or E16, ages which span the middle-to-late period of dLGN neurogenesis (Lund and Mustari, 1977). We measured the surviving dLGN volume and counted labeled cells, comparing rats with transplants to rats with lesions only. In the lesion only control animals the lesion cavity was filled with gelfoam soaked in fresh culture media. We also compared rats with cortical transplants to rats with cerebellar transplants in order to test for specificity of the effects of the cortical tissue. Regardless of the origin of tissue, the volume of the transplanted cell suspension was 3.5-4.5mm^3.

The host rats were perfused with buffered formalin on postnatal days 6 or 10 and the brains sectioned for Nissl staining and radioautography. Two rats with lesions plus cortical transplants received intraocular injections of ^{3}H-proline 24 hours before sacrifice in order to confirm the boundaries of the dLGN (first defined on cytoarchitectonic grounds), by its distinct pattern of retinal innervation.

The cortical lesions were examined in sectioned material and found to have removed all dLGN projection areas. The lesions usually extended ventrally to the rhinal fissure and rostrally into somatosensory cortex. Planimetric measurements of the average volume of surviving non-visual cortex were similar in all groups of rats (39% of contralateral intact neocortical volume in cortical transplant animals, 37% in both cerebellar transplant and control animals). Examination of the transplant cells in the lesion cavity showed them to be viable and usually attached to the host subiculum or hippocampus.

Cortical Transplants vs. No Transplants

The appearance of the host dLGN in 6 day old rats with lesions and transplants is strikingly different than its appearance in lesion only controls (Fig. 2A, B). Although the nucleus is severely affected in both groups of rats, it is always larger and more distinct in rats with cortical transplants. We quantitated this difference between the two groups of rats by measuring the total volume of surviving dLGN, and expressing these measurements as a percent of the dLGN volume on the unoperated side of the thalamus. These

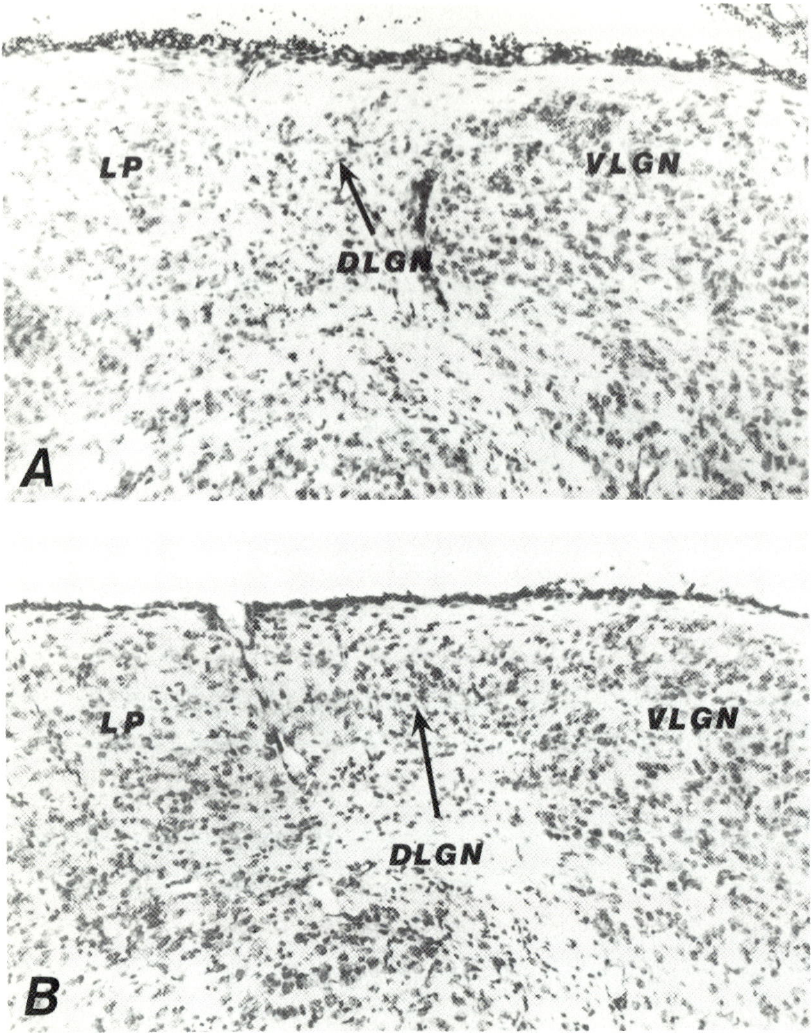

Fig. 2: Coronal sections through corresponding levels of right posterior thalamus in P6 rats with large right posterior cortex lesions at birth. A. Lesion-only rat with small, ill-defined dorsal lateral geniculate (dLGN). The lateral posterior nucleus (LP) is also affected. B. Lesion-plus-cortical transplant rat. The dLGN is more distinct and contains a larger area of normal appearing cells. LP appears less severely affected. VLGN- ventral lateral geniculate. Cresyl violet stain. Lateral is at top of the micrographs.

volume measurments show more than a two-fold increase in total dLGN volume in rats with lesions and cortical transplants, compared to lesion-only controls (Fig. 3). The increased survival of the nucleus was temporary, however, because few if any surviving neurons could be identified in the region of the dLGN in 10-day-old rats.

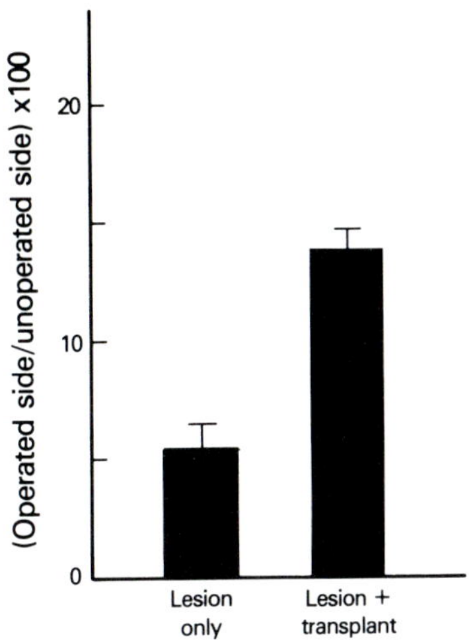

Fig. 3: Graph of the remaining volume of dLGN in P6 rats with lesions only (n=8) and rats with posterior cortex transplants (n=10). In rats with transplants there is more than a 2.5 fold increase in surviving geniculate volume (expressed as percent of the unoperated side). $p < .001$, Kruskal-Wallis ANOVA.

Since at postnatal day 6 there was no obvious difference in dLGN neuron packing density between the two groups of rats, we assumed that the increased volume we observed in the animals with transplants was due to increased neuron survival. In order to confirm increased neuron numbers in these rats, and to determine if there were particular geniculate populations rescued by the transplant, we counted geniculate neurons heavily labeled with ^{3}H-thymidine on E14 or E16 (Fig. 4). The counts show that E14 generated neurons are almost completely depleted in rats with and without cortical transplants. However, counts of the later generated E16 labeled neurons show a more than two-fold increase in cortical transplant animals compared with lesion-only controls. These counts of late generated cells therefore give results comparable to the simple dLGN volume measurement, which indicates that volume measurement provides a rapid and reliable assay of the transplants' effectiveness. We used volume measurements to investigate the effects of allowing the cortical neurons to differentiate in culture prior to transplantation, and to test for specificity by comparing the effectiveness of cortical transplants with cerebellar transplants.

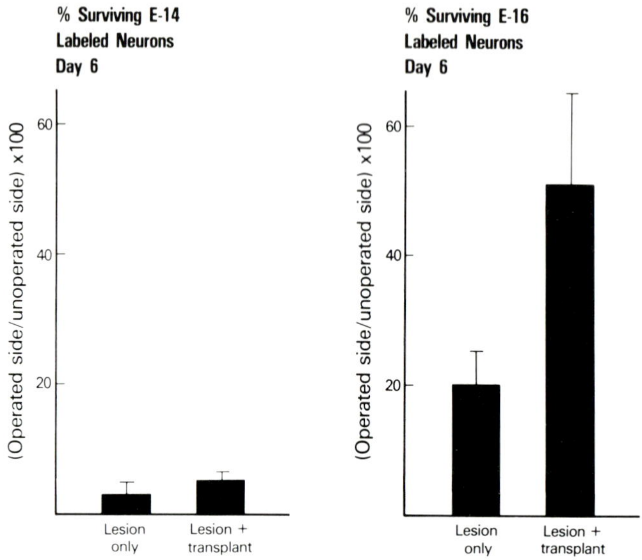

Fig. 4: Graphs of the number of heavily labeled dLGN neurons ($\geq$15 grains/nucleus) on the ipsilateral (operated) side in P6 rats with posterior cortex lesion at birth, expressed as percent labeled neurons on the unoperated side. Counts were corrected by the method of Abercrombie (1946) after measuring diameters of labeled nuclei. Labeling was with ^{3}H-thymidine on embryonic days 14 or 16 (E14, E16). In rats with cortical transplants there is 2.5 fold mean increase in the number of surviving E16 generated neurons compared to lesion-only controls ($p < .025$, Mann-Whitney U test). A similar comparison of E14 generated neurons shows no significant difference. (n=3 in all groups).

Precultured Transplants vs. "Same Day" Transplants

The embryonic posterior cortical tissue appeared to grow and differentiate in culture as evidenced by the increase in size of single explants and the development of "fibrous cables" between explants (Hatten and Sidman, 1978). Measurements of the volume of surviving dLGN in 6 day old rats show that the precultured transplants significantly increase the amount of surviving geniculate when compared to the direct (or "same day") transplants (Fig. 5).

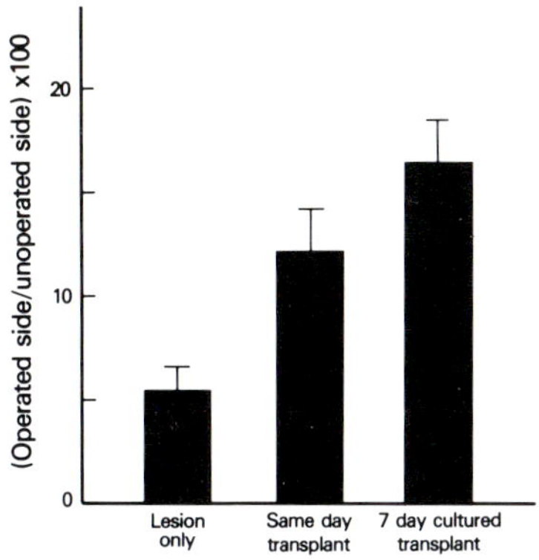

Fig. 5: Graph of remaining volume of dLGN in rats where cortical tissue was harvested and transplanted into lesion cavity on the same day (n=6), or cultured for 7 days prior to transplantation (n=4). Lesion-only control volumes are presented for comparison. Significantly more dLGN volume remains with precultured transplants compared to same day transplants ($p < .05$, Mann-Whitney *U* test).

Cortical vs. Cerebellar Transplants

In the 6 day old rats, the cerebellar transplants appeared as viable as the cortical transplants. However, volume of surviving dLGN in rats with cerebellar transplants was not significantly different than the volume obtained in the lesion only control rats (Fig. 6). In three of the five rats with cerebellar transplants the cerebellar tissue was cultured for 7 days prior to transplantation. The culturing conditions were the same as those used for the cortical tissue. Nevertheless the volume of remaining dLGN is comparable in all cerebellar transplant animals whether the preculturing step was added or not.

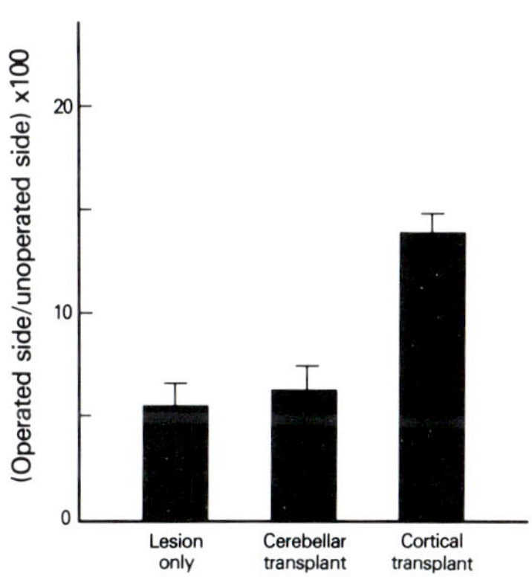

Fig. 6: Graph of remaining volume of dLGN in rats with cortical transplants (n=10), cerebellar transplants (n=5), and lesion-only control rats (n=8). Rats with cerebellar transplants are not significantly different from control rats.

Summary and Conclusions

The results of these experiments show a relatively specific influence of fetal CNS transplants on the survival of neonatal dLGN neurons. This neuronotrophic specificity is reflected in both the type of tissue transplanted and in the population of host neurons affected by the transplant.

The fact that late-generated neurons survive preferentially in 6 day old rats with transplants suggests several possible mechanisms underlying their temporary rescue by the transplanted cortical tissue. It may be that the younger donor tissue (E14, 15) produces trophic factors which are most effective on the younger (late-generated) host tissue. If this is the case then the larger effect we find after allowing the tissue to differentiate in culture could be due to survival of more earlier-generated host neurons because more differentiated transplant tissue may produce trophic factors which support these older neurons. We are currently testing this possibility. It is also possible that the late generated dLGN cells have not invaded the damaged cortical area at the time of the lesion, and therefore are simply target deprived without being axotomized. Consistent with this notion is the fact that the late generated neurons also survive preferentially in lesion only control animals (although to a much smaller degree than in transplant animals). The transplants' effect, then, may be limited to undamaged developing neurons whose axons are approaching the target zone. The precultured tissue may provide additional trophic support that also sustains earlier generated axotomized geniculate neurons. Furthermore, some of the surviving cells could be geniculate interneurons which survive because of a delayed degeneration of their targets, that is, the cortically projecting neurons.

On the basis of these results we suggest that the susceptibility of young visual system neurons to lesions is due to removal of trophic factors derived from the damaged area. These factors are likely to operate only during restricted periods of the neuron's maturation, as suggested by our finding that the transplant specifically supports late generated dLGN neurons. Since the support is temporary, it is also likely that the cells must eventually make and receive connections to ensure further maturation and continued survival. In the present experiments the dLGN neurons appear unable to achieve these permanent connections and therefore their degeneration is delayed rather than reversed.

The similarity of this delay to that described for sympathetic neurons after application of NGF (see above) suggests that similar mechanisms underlie the action of trophic factors in the central and peripheral nervous systems.

Finally, these results represent an application of recent advances in the technique of CNS neural transplantation to the fundamental and long-standing question of the nature of neuronotropism in the developing mammalian CNS. CNS transplantation seems an especially powerful tool for investigating this question, because (1) the neuronotrophic influence of different types of CNS cells at different maturational stages can be studied in situ, and (2) a CNS system with complex and well-studied interconnections, like the mammalian visual system, is open to study of the trophic relationships among its component structures.

Literature Cited

Abercrombie M (1946). Estimation of nuclear population from microtome sections. Anat Rec 94:238-248.

Banks BEC, Walter SJ (1977). The effects of postganglionic axotomy and nerve growth factor on superior cervical ganglia of developing mice. J Neurocytol 6:287.

Bradshaw RA (1978). Nerve growth factor. Ann Rev Biochem 47:191.

Cohen S, Levi-Montalcini R, Hamburger V (1954). A nerve growth stimulating factor isolated from sarcomas 37 and 180. Proc Nat Acad Sci 40:1014.

Cunningham TJ (1982). Naturally occurring neuron death and its regulation by developing neural pathways. In Bourne GH, Danielli JF (eds): "International Review of Cytology", Vol 72, New York: Academic Press, p 163.

Cunningham TJ, Mohler IM, Giordano DL (1981). Naturally occurring neuron death in the ganglion cell layer of the neonatal rat: Morphology and evidence for regional correspondence with neuron death in the superior colliculus. Dev Brain Res 3:203.

Cunningham TJ, Huddelston C, Murray M (1979). Modification of neuron numbers in the visual system of the rat. J Comp Neurol 184:423.

Das GD, Hallas BH, Das KG (1980). Transplantation of brain tissue in the brain of the rat. I. Growth characteristics of neocortical transplants from embryos of different ages. Am J Anat 158:135.

Delong GR, Sidman, RL (1962). Effects of eye removal at birth on histogenesis of mouse superior colliculus: An autoradiographic analysis with tritiated thymidine. J Comp Neurol 118:205.

Finlay BL , Berg AT, Sengelaub DR (1982). Cell death in the mammalian visual system during normal development: II. Superior Colliculus. J Comp Neurol 204:318.

Giordano DL, Murray M, Cunningham TJ (1980). Naturally occurring neuron death in the optic layers of the superior colliculus of the postnatal rat. J Neurocytol 9:603.

Giordano DL, Cunningham TJ (1982). Optic afferents, neuron maturation and neuron survival in the rat superior colliculus. Dev Brain Res 4:465.

Greene LA, Shooter EM (1980). The nerve growth factor: biochemistry, synthesis and mechanism of action. Ann Rev Neurosci 3:353.

Hatten ME, Sidman RL (1978). Cell reassociation behavior and lectin-induced agglutination of embryonic mouse cells from different brain regions. Exp Cell Res 113:111.

Hendry IA, Campbell J (1976). Morphometric analysis of rat superior cervical ganglion after axotomy and NGF treatment. J Neurocytol 5:351.

Heumann D, Rabinowicz T (1979). Postnatal development of the dorsal lateral geniculate nucleus in the normal and enucleated albino mouse. Exp Brain Res 38:75.

Jeffrey G, Perry VH (1981). Evidence for ganglion cell death during development of the ipsilateral projection in the rat. Dev Brain Res 2:176.

Labar DR, Berman NE, Murphy EH (1981). Short- and long-term effects of neonatal and adult visual cortex lesions on the retinal projection to the pulvinar in cats. J Comp Neurol 197:639.

Lam K, Sefton AJ, Bennett MR (1982). Loss of axons from the optic nerve of the rat during early postnatal development. Dev Brain Res 3:487.

Lieberman AR (1974). Some factors affecting retrograde neuronal responses to axonal lesions. In Bellairs R, Gray EG (eds): "Essays on the Nervous System", Oxford: Clarendon, p 71.

Lund RD, Mustari MJ (1977). Development of the geniculocortical pathway in rats. J Comp Neurol 173:289.

McCaffery CA, Bennett MR, Dreher R (1982). The survival of neonatal rat retinal ganglion cells in vitro is enhanced in the presence of appropriate parts of the brain. Exp Brain Res 48:377.

Oppenheim RW (1981). Neuronal cell death and some related regressive phenomena during neurogenesis: A selective historical review and progress report. In Cowan WM (ed): "Studies In Developmental Neurobiology", New York: Oxford, p. 74.

Pearson HE, Labar DR, Payne BR, Cornwell P, Aggarwal N (1981). Transneuronal retrograde degeneration in the cat retina following neonatal ablation of the visual cortex. Brain Res 212:470.

Pearson HE, Payne BR, Cunningham TJ (1982). Naturally occurring neuron death in the retina of the postnatal cat. Soc Neurosci Abstr 9:322.

Perez-Polo JR, de Vellis J, Harber B (1983). Growth and trophic factors. In "Progress in Clinical and Biological Research", Vol 118, New York: Alan R Liss.

Perry VH, Cowey A (1979). The effects of unilateral cortical and tectal lesions on retinal ganglion cells in rats. Exp Brain Res 35:85.

Potts RA, Dreher B, Bennett MR (1982). The loss of ganglion cells in the developing retina of the rat. Dev Brain Res 3:481.

Prestige MC (1967). The control of cell numbers in the lumbar ventral horns during the development of Xenopus laevis tadpoles. J Embryol Exp Morphol 18:359.

Rakic P, Riley KP (1983). Regulation of axon number in primate optic nerve by prenatal binocular competition. Nature 305:135.

Sengelaub DR, Finlay BL (1982). Cell death in the mammalian visual system during normal development. I. Retinal ganglion cells. J Comp Neurol 204:311.

Sengelaub DR, Finlay BL (1981). Early removal of one eye reduces normally occurring cell death in the remaining eye. Science 213:573.

Williams RW, Bastiani MJ, Chalupa LM (1983). Loss of optic axons in cat optic nerve following fetal unilateral enulceation: An electron microscopic analysis. J Neurosci 3:133.

Environmental Modification of the Visual System

Development of Visual Pathways in Mammals, pages 331–345

CELL GROWTH IN THE LATERAL GENICULATE C-LAMINAE OF VISUALLY NORMAL AND MONOCULARLY DEPRIVED CATS.

Paul D. Wilson, Dean M. Murakami, and John D. Parsons

Department of Psychology
University of California
Riverside, California 92521

Lamina A and A1 of the dorsal lateral geniculate nucleus (dLGN) of the cat can be distinguished in unstained brain sections almost as easily as in Nissl stained sections. In contrast, the several C-laminae ventral to lamina A1 in the dLGN are concealed in Nissl material and can only be identified on the basis of the ocular origin of their retinal afferents. The laminae C, C1, C2 and C3 were first demonstrated with autoradiographic tracing of retinal afferents by Guillery (1970) and by Hickey and Guillery (1974); they can also be clearly revealed with horseradish peroxidase (HRP) as an anterograde tracer, as shown in Figure 1. The physiological evidence reported by Wilson et al., (1976) indicates that laminae A, A1, and the magnocellular portion of lamina C contain a mixture of X- and Y-cells, while the parvocellular part of lamina C and laminae C1 and C2 contain W-cells. There is considerable anatomical evidence consistent with W-cells being confined to the parvocellular C - laminae of the dLGN. Further, the distribution of the relay cells physiologically identified and morphologically characterized by Friedlander et al. (1981) and by Stanford et al. (1981, 1983) is consistent with the segregation of the W-cells in the parvocellular laminae of the dLGN.

Numerous studies over the past twenty years have confirmed the observation by Wiesel and Hubel (1963) that early monocular deprivation reduces soma size of A-laminae cells (reviewed by Sherman and Spear, 1982). Also, several investigators (Garey et al., 1973; Kalil, 1978; Hickey, 1980) have examined the normal postnatal growth of soma sizes in

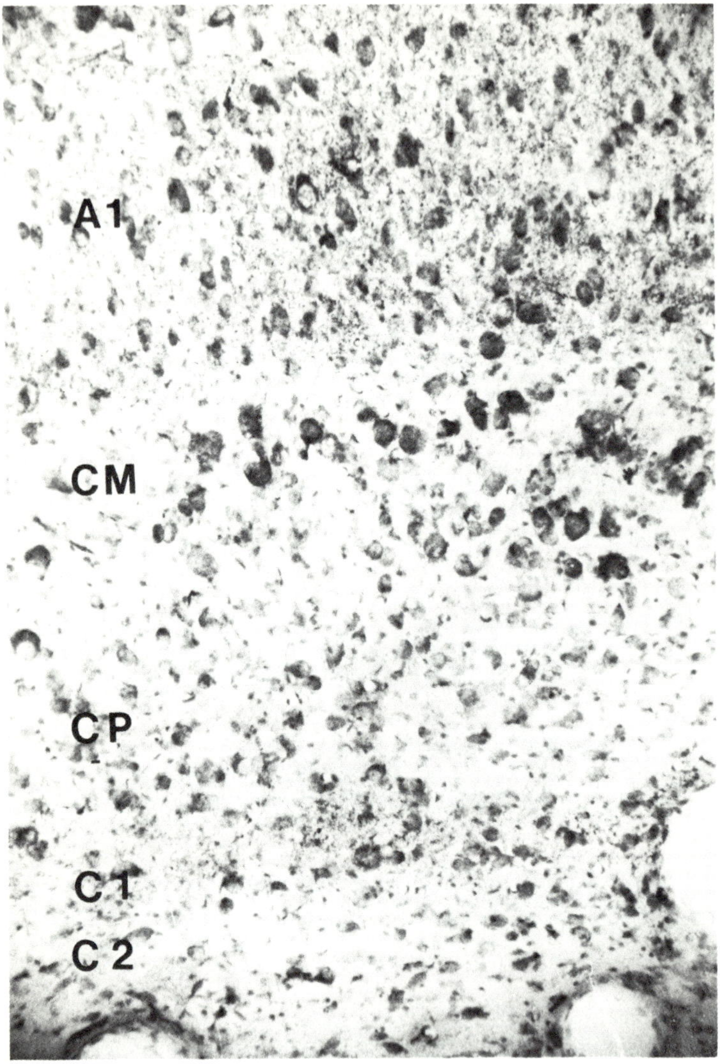

Fig. 1. Photomicrograph of cat dLGN with HRP-labeled projections from the ipsilateral eye to laminae A1 and C1. The tier of large diameter cells constituting lamina CM contrasts with the parvocellular lamina CP.

the A-laminae of nondeprived cats. In contrast, only a few recent studies have examined the effect of visual deprivation on C-laminae cells, and no studies of normal soma growth in the C-laminae have been reported. Perhaps this relative neglect of the C-laminae has resulted from the greater difficulty of distinguishing the C-laminae than the A-laminae. It may also reflect an assumption that the W-cells in the parvocellular C-laminae are immune to the effects of visual deprivation.

CELL GROWTH IN DEPRIVED C-LAMINAE

The effect of monocular deprivation on soma size of C-laminae cells has only recently been examined (Hickey, 1980; Leventhal and Hirsch, 1983; Murakami and Wilson, 1983). Hickey (1980) examined cats monocularly deprived to 12 and 16 weeks of age and measured cells in the C-laminae, with the individual laminae delineated by autoradiographic methods. He found soma sizes markedly affected in lamina C, only slightly affected in lamina C1, and unaffected in lamina C2. Murakami and Wilson (1983) have also examined the effects of monocular deprivation on C-laminae cells. Five cats were monocularly lid sutured at 3 weeks of age and sacrificed for histological study at 16, 20, 32, and 52 weeks of age. Three days before sacrifice, the cat was anesthetized and 15μl of a 30% HRP solution was injected into one eye. Subsequently, 40 μm sections through the dLGN were processed with the TMB method of Mesulam (1978) to reveal the HRP anterogradely transported to retinal terminals, and sections were counterstained with neutral red or thionin. Soma sizes in the dLGN were measured from cell profiles having a clear nucleolus and Nissl substance by taking the mean of the orthogonal long and short diameters through cell profiles. The measured cells were obtained from the middle third, medial to lateral, of coronal sections at the level of the posterior commissure, corresponding to anterior 6.0-6.5 in the atlas of the adult cat dLGN by Sanderson (1971). At least 100 cells were measured for each lamina A, A1, and C, and at least 50 cells were measured for each lamina C1 and C2. These minima were almost always exceeded with the sampling procedures used, which were designed to avoid bias toward dorsal, ventral, medial, or lateral borders of the laminar area examined (see Murakami and Wilson, 1983).

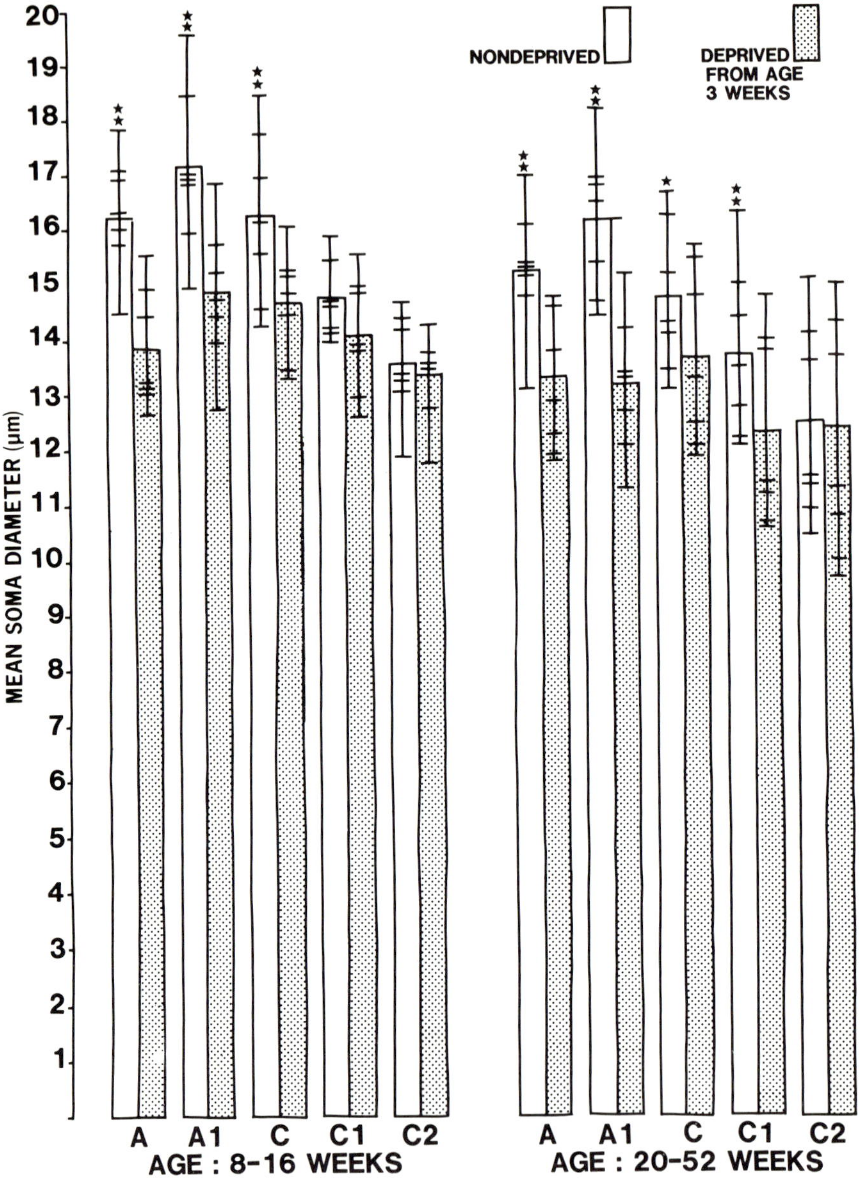

Fig. 2. Mean soma diameters in cats at 8-16 weeks of age and at 20-52 weeks of age. Deprivation began at 3 weeks of age. Individual animals shown by horizontal marks. Laminae significantly different by one-tail T test at $p<.001$ (two stars) and at $p<.01$ (one star).

In these animals, the soma diameters in the deprived layers of the dLGN differed significantly from those in the nondeprived layers for laminae A, A1, C, and C1, but not for lamina C2. However, it was noted that among the five cats, the one cat examined at 16 weeks of age appeared to have no difference between deprived and nondeprived lamina C1, which was consistent with the report of Hickey (1980) that lamina C1 was only slightly affected by monocular deprivation lasting to 12 or 16 weeks of age.

Further investigation (Murakami et al., 1982) has confirmed that marked soma size changes in lamina C1 are only apparent in cats monocularly deprived beyond 16 weeks of age. Figure 2 compares the effect of monocular deprivation on soma sizes in the dLGN of two groups of cats, those deprived until 8, 12 and 16 weeks of age and those deprived until 20, 32, and 52 weeks of age. Monocular deprivation began at 3 weeks of age for all cats. Soma diameters were significantly smaller in the deprived layers than in nondeprived layers for laminae A, A1, and C in both the 8 to 16 week and the 20 to 52 week age groups, but the difference between deprived and nondeprived lamina C1 was significant only in the cats 20 to 52 weeks of age.

Soma sizes differed between deprived and nondeprived lamina C in cats in both age groups, 8 to 16 weeks and 20 to 52 weeks of age. However, as noted above, lamina C can be subdivided into a dorsal magnocellular portion and a more ventral parvocellular portion (Figure 1), and physiological evidence (Wilson et al., 1976; Stanford et al., 1983) suggests that the magnocellular subdivision has predominantly Y-cells and some X-cells, while the parvocellular subdivision of lamina C has mostly or entirely W-cells. Soma diameters were measured separately for these two subdivisions of lamina C in a subset of animals. For each age examined (8, 12, 16, 20, and 52 weeks) cells were smaller in the deprived than in the nondeprived lamina C-magnocellular (CM). Soma sizes also tended to be smaller in the deprived than in the nondeprived lamina C-parvocellular (CP) ($p<.01$), but the difference was not significant for the four animals in the older age group (Fig. 3).

The question might be asked: how consistent are the effects of monocular deprivation on cells in lamina C1? The answer is shown in Figure 3. In each animal monocularly deprived until 20 weeks of age or older, soma sizes were

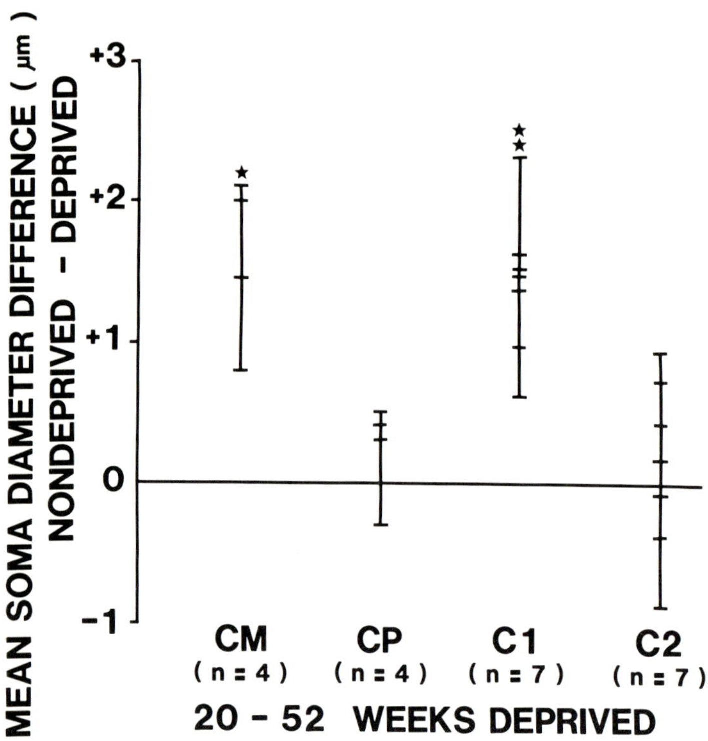

Fig. 3. Differences in mean soma diameter between nondeprived and deprived laminae for individual animals indicated by horizontal marks. Negative values indicate mean soma size greater in deprived than in nondeprived lamina. Cats monocularly deprived at 3 weeks and examined at 20-52 weeks of age. Differences significant for lamina Cl ($p<.001$; two stars) and lamina CM ($p<.01$; one star).

greater for the deprived than the nondeprived lamina Cl. This was the case for soma sizes in lamina C magnocellular (CM) in both age groups, 8 to 16 weeks and 20 to 52 weeks of age, but not for lamina CP or lamina C2 in either age group.

Monocular deprivation markedly affects the size of cells in the binocular segment of the deprived A-laminae, but has much less effect on the monocular segment of lamina A (Guillery and Stelzner, 1970; Hickey, Spear, and Kratz, 1977; Kalil, 1980), indicating that binocular competition contributes substantially to the effects of monocular deprivation. Soma diameters have also been measured in the monocular segment of the parvocellular C-laminae (Murakami and Wilson, 1983) and no difference was found between cell sizes contralateral (deprived) and ipsilateral (nondeprived) to the lid sutured eye. However, these measurements did not include cells from lamina C1, which, like lamina A1, receives ipsilateral retinal input and has no monocular segment. The monocular segment of the parvocellular C-lamina comprises lamina C-parvocellular and lamina C2, neither of which shows effects from monocular deprivation.

In summary, the effects of monocular deprivation might be considered slight if the effects over all parvocellular laminae are summed; but the effects on lamina C1 are clear and consistent in the cats examined at 20 to 52 weeks of age. Leventhal and Hirsch (1983) found no effect from monocular deprivation in parvocellular lamina cells labeled from an injection of HRP into cortical area 19, a major cortical recipient area for parvocellular geniculate projections. However, their methodology may be insufficiently accurate to distinguish lamina C1 from the neighboring lamina C-parvocellular and lamina C2. They did not distinguish these laminae by anterograde tracing methods applied directly in the monocularly deprived cats, but relied on measurements of the dorsoventral extent of the laminae as delineated with autoradiographic methods in normal cats and transposition of these measurements to the brain sections of the deprived cats. Any inconsistency of laminar thickness between the normal and deprived laminae of the cat dLGN would obviate this method. Slight errors in estimating laminar thickness would of course lump neighboring deprived and nondeprived cells. Another explanation of the difference between the findings of Murakami and Wilson (1983, and this chapter) and those of Leventhal and Hirsch (1983) is that the latter were investigating only parvocellular neurons projecting to area 19, and the parvocellular cells affected by monocular deprivation do not project to area 19.

The period during which A-laminae cells are most sensitive to monocular deprivation is between 3 weeks and 12 to 16 weeks of age, as first reported by Wiesel and Hubel (1963) and confirmed by numerous studies. (There is some evidence that the critical period may extend well beyond 16 weeks; Cynader and Mitchell, 1980; Jones et al., 1981). The findings reported here are consistent with a critical period for cells in lamina C-magnocellular similar to that for A-laminae cells. In contrast, lamina C1 cells are affected by monocular deprivation only in cats deprived until 20 weeks of age or older. Three alternative explanations might be considered for the relatively late appearance of the lamina C1 effects. One is that these cells have a later critical period than the cells in the A-laminae and lamina C-magnocellular. An investigation of this by Murakami et al. (1982) found that cats monocularly deprived at 12 weeks of age and examined 21 to 40 weeks later had markedly reduced cell sizes in deprived lamina C1 ($p<.005$), and some reduction in the cell sizes in laminae A ($p<.01$) and A1 ($p<.01$). This finding is consistent with a later critical period for soma size changes in lamina C1 than in the A-laminae. Another possibility is that the cells in lamina C1 have the same critical period as A-laminae cells, but, for reasons not understood, the effects of deprivation are not apparent until the cat is 20 weeks old. A third possibility is that soma size reduction in lamina C1 is only apparent as a cumulative effect from a longer period of deprivation than is needed for soma size changes to appear in the A-laminae.

The results of several experiments (Wiesel and Hubel, 1963, 1965; Guillery and Stelzner, 1970; Guillery, 1972; Sherman et al., 1974) suggest that the effects of monocular deprivation on A-laminae cells is mostly secondary to binocular competition in visual cortex. The effects of monocular deprivation on lamina C1 cells may also be the result of binocular competition, for example in cortical area 19 which is profoundly affected by monocular deprivation (Rapaport et al., 1982; Leventhal and Hirsch, 1983), and which is the principal recipient of the projections of the W-cells in the parvocellular C-laminae (Maciewicz, 1974; Gilbert and Kelly, 1975; Hollander and Vanegas, 1977; LeVay and Ferster, 1977; Ferster and LeVay, 1978; Leventhal et al., 1980). However, at least two considerations argue against this. Area 19 receives considerable input from other cortical areas, including areas 17 and 18 (Gilbert and Kelly, 1975; Lee et al., 1982), as well as thalamic inputs. The effects of monocular deprivation on area 19 might be secondary to effects in areas 17 and 18, as pointed out previously (Rapaport et al., 1982; Leventhal and Hirsch, 1983). Furthermore, it is difficult to explain why cells in lamina C1, but not cells in laminae C-parvocellular or C2, would show effects of binocular competition in cortex. The difference in the effect in ipsilateral recipient lamina C1 and contralateral recipient laminae C-parvocellular and C2 seems to be much greater than the difference between the effect of monocular deprivation in lamina A1 and lamina A.

Another mechanism which might underlie the effects of monocular deprivation on lamina C1 cells is translaminar suppression mediated directly or indirectly by translaminar dendrites. Binocular suppression within the dLGN has been demonstrated physiologically (Singer, 1970; Rodieck and Dreher, 1979), and binocular inhibition may be mediated predominantly by Y-cells (Rodieck, 1973; Singer and Bedworth, 1973). Y - cell dendrites cross laminar boundaries (Friedlander et al., 1981) and perhaps mediate translaminar suppression resulting in soma size reduction in deprived lamina C1. W-cell dendrites also cross laminar boundaries (Stanford et al., 1983) and suppression of lamina C1 cells by W-cells in adjacent laminae is a possibility, but it would not explain the effect being limited morphologically to lamina C1.

NORMAL C-LAMINAE CELL DEVELOPMENT

Soma diameters have been measured for cells in the dLGN of cats raised with normal visual experience and sacrificed at 1 to 20 weeks of age (Parsons et al., 1982). The histological procedures were the same as in Murakami and Wilson (1983). Figure 4A shows the mean soma diameter for cells in laminae A, A1, and CM in kittens 1 to 20 weeks old and in adult cats. The growth in soma size is closely corresponding for these three laminae, A, A1, and CM, and is generally consistent with the values previously reported for A-laminae development (Garey et al., 1973; Kalil, 1978; Hickey, 1980). The mean soma diameter in these laminae sharply increased between 1 and 4 weeks postnatally, and continued to increase much more gradually in subsequent weeks. Cells in lamina A and A1 attained close to the adult size cell (shown on right of Fig. 4A) by 12 to 20 weeks, while the mean soma size in lamina CM at 20 weeks appeared to

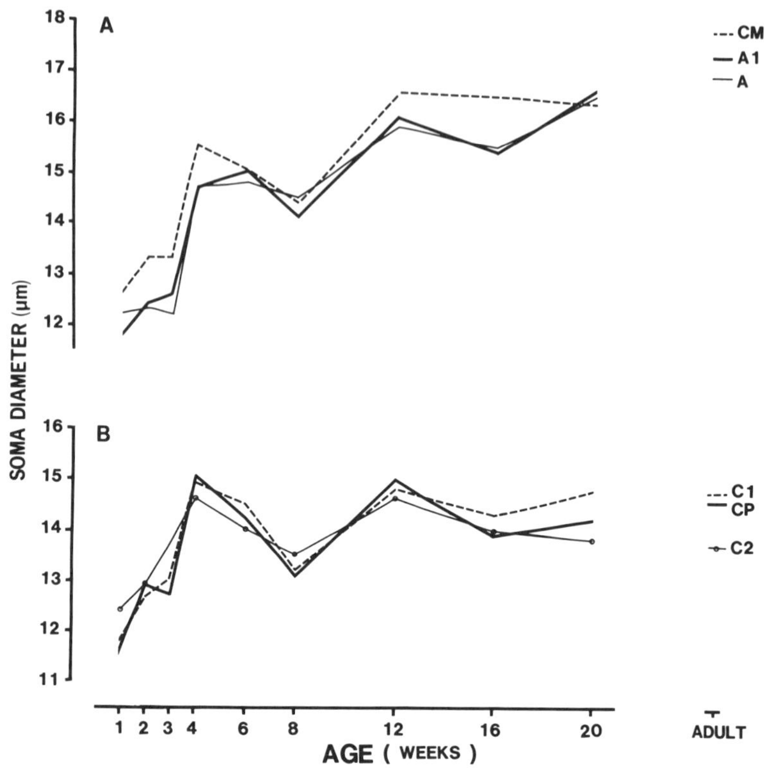

Fig. 4. (A) Mean soma diameter for cells in laminae A, A1 and CM (C-magnocellular) and (B) for cells in laminae CP (C-parvocellular), C1 and C2 in visually normal kittens 1 to 20 weeks of age. Mean soma diameters in adult cats at the right.

be less than that found in the adult cat. There is evidence that lamina CM contains predominantly Y-cells. Wilson et al. (1976) identified Y-cells and X-cells in the dorsal tier of lamina C, and Stanford et al. (1983) found only Y-cells in the upper tier. The growth curve for soma diameters in lamina CM suggests that at least some of these cells may mature relatively late.

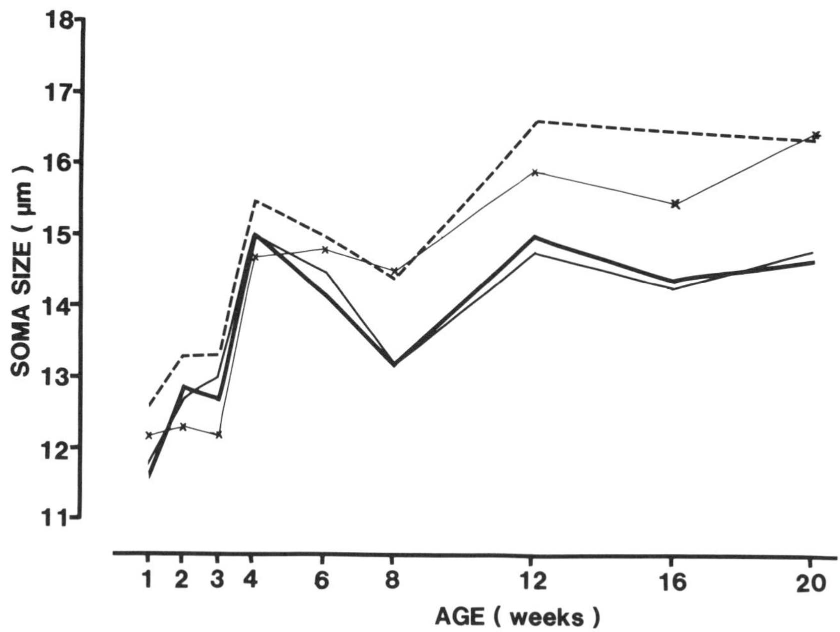

Fig. 5. Comparison of mean soma diameter growth in laminae A, CM, CP and Cl, as in Fig. 4.

Figure 4B illustrates the closely corresponding growth curves for mean soma size of cells in laminae CP, Cl, and C2, although it should be noted that lamina C2 cells average smaller than cells in laminae CP and Cl. The cells in these parvocellular laminae have attained fully the adult soma size by 4 weeks of age, in contrast to the cells in laminae A, Al, and CM. Figure 5 shows this contrast. The graphs indicate a reduction in soma size within all laminae at 8 weeks of age. This may merely reflect chance sampling of animals with considerable between-animal variability, and the data for all laminae at age 8 weeks coming from the same three animals.

Winkler and Potter (1914) divided the cat's dLGN into pars A and B on the basis of differences in cell sizes, noting pars B had predominantly small cells. A layer of large cells was observed at the dorsal border of pars B and called central interlaminar nucleus by Thuma (1928) and lamina magnocellularis by Rioch (1929), who contrasted it with lamina parvocellularis more ventrally. Hickey and Guillery (1974) noted the presence of large diameter cells in the dorsal part of lamina C. We have combined the terminology of Rioch (1929) with that of Guillery (1970; Hickey and Guillery, 1974) in subdividing lamina C into C-magnocellular (CM) and C-parvocellular (CP). The findings summarized in this chapter and data from previous reports indicate that lamina C comprises in fact two laminae, a dorsal magnocellular lamina (CM) and ventral parvocellular lamina (CP). Lamina CM contains predominantly or entirely Y- and X-cells, whereas lamina CP contains W-cells. Lamina CM is markedly affected by early monocular deprivation, while lamina CP is little affected, and the cells in laminae CM and CP have different growth rates. These findings seem to require that the laminated dLGN be considered to have six distinguishable retinorecipient laminae: A, A1, CM, CP, C1, and C2, in addition to a lamina C3, which Hickey and Guillery (1974) showed as a non-retinorecipient lamina.

References Cited:

Cynader M, Mitchell D E (1980). Prolonged sensitivity to monocular deprivation in dark-reared cats. J Neurophysiol 43:1026.

Ferster D, LeVay S (1978). The axonal arborizations of lateral geniculate neurons in the striate cortex of the cat. J Comp Neurol 182:923.

Friedlander M J, Lin C-S, Stanford L R, Sherman S M (1981). Morphology of physiologically identified neurons in the lateral geniculate nucleus of the cat. J Neurophysiol 46:80.

Garey L J, Fisken R A, Powell T P S (1973). Observations on the growth of cells in the lateral geniculate nucleus of the cat. Brain Res 52:359.

Gilbert C D, Kelly J P (1975). The projections of cells in different layers of cat visual cortex. J Comp Neurol 163:81.

Guillery R W (1970). The laminar distribution of retinal fibers in the dorsal lateral geniculate nucleus of the cat: A new interpretation. J Comp Neurol 138:339.

Guillery R W (1972). Binocular competition in control of geniculate cell growth. J Comp Neurol 144:117.

Guillery R W, Stelzner D J (1970). The differential effects of unilateral lid closure upon the monocular and binocular segments of the dorsal lateral geniculate nucleus in the cat. J Comp Neurol 139:413.

Hickey T L (1980). Development of the dorsal lateral geniculate nucleus in normal and visually deprived cats. J Comp Neurol 189:467.

Hickey T L, Guillery R W (1974). An autoradiographic study of retinogeniculate pathways in the cat and fox. J Comp Neurol 156:239.

Hickey T L, Spear P D, Kratz K E (1977). Quantitative studies of cell size in the cat's dorsal lateral geniculate nucleus following visual deprivation. J Comp Neurol 172:265.

Hoffmann K-P, Stone J, Sherman S M (1972). Relay of receptive field properties in dorsal lateral geniculate nucleus of the cat. J Neurophysiol 35:518.

Hollander H, Vanegas J (1977). The projection from lateral geniculate nucleus onto the visual cortex in the cat. A quantitative study with horseradish peroxidase. J Comp Neurol 173:519.

Jones K R, Spear P D, Tong L (1981). Period of susceptibility to effects of monocular deprivation: differences between striate and extrastriate cortex. Soc Neurosci Abstr 7:142.

Kalil R (1980). A quantitative study of the effects of monocular enucleation on cell growth in the dorsal lateral geniculate nucleus of the cat. J Comp Neurol 189:483.

Kalil R (1978). Dark rearing in the cat: effects on visuomotor behavior and cell growth in the dorsal lateral geniculate nucleus. J Comp Neurol 178:451.

Lee C W F, Ho H T, Dreher B (1982). Area 21a and posteromedial lateral suprasylvian area (PMLS) of the cat visual cortex: one or two areas? Proc Austral Physiol Pharm Soc 13:195P.

LeVay S, Ferster D (1977). Relay cell classes in the lateral geniculate nucleus of the cat and the effects of visual deprivation. J Comp Neurol 172:563.

Leventhal A G, Hirsch, H V B (1983). Effects of visual deprivation upon the geniculocortical W-cell pathway in the cat: area 19 and its afferent input. J Comp Neurol 214:59.

Leventhal A G, Keens J, Tork I (1980). The afferent ganglion cells and cortical projections of the retinal recipient zone (RRZ) of the cat's 'pulvinar complex.' J Comp Neurol 194:535.

Maciewicz R J (1974). Afferents to the lateral suprasylvian gyrus of the cat traced with horseradish peroxidase. Brain Res 78:139.

Mesulam M M (1978). Tetramethyl benzidine for horseradish histochemistry: a noncarcinogenic reaction product for visualizing neural afferents and efferents. J Histochem Cytochem 26:106.

Murakami D M, Wilson P D, Yarosh, C A (1982). Effect of early and late monocular deprivation on cell size in the C laminae of the cat dLGN. Soc Neurosci Abstr 8:667.

Murakami D M, Wilson P D (1983). The effect of monocular deprivation on cells in the C laminae of the cat lateral geniculate nucleus. Dev Brain Res 9:353.

Parsons J E, Murakami D M, Wilson P D (1982). Normal development of the cat lateral geniculate nucleus. Soc Neurosci Abstr 8:667.

Rapaport D H, Dreher B, Rowe M H (1982). Lack of binocularity in cells in area 19 of cat visual cortex following monocular deprivation. Brain Res 246:319.

Rioch D M (1929). Studies on the diencephalon of carnivora I. The nuclear configuration of the thalamus, epithalamus and hypothalamus of the dog and cat. J Comp Neurol 49:1.

Rodieck R W, Dreher B (1979). Visual suppression from the nondominant eye in the lateral geniculate nucleus: a comparison of cat and monkey. Exp Brain Res 35:465.

Sanderson K J (1971). The projection of the visual field to the lateral geniculate and medial interlaminar nuclei in the cat. J Comp Neurol 143:101.

Sherman S M, Spear P D (1982). Organization of visual pathways in normal and visually deprived cats. Physiol Rev 62:738.

Sherman S M, Guillery R W, Kaas J H, Sanderson K J (1974). Behavioral, electrophysiological and morphological studies of binocular competition in cats. J Comp Neurol 158:1.

Singer W, Bedworth N (1973). Inhibitory interaction between X and Y units in the cat lateral geniculate nucleus. Brain Res 49:291.

Stanford L R, Friedlander M J, Sherman S M (1981). Morphology of physiologically identified W-cells in the C laminae of the cat's lateral geniculate nucleus. J Neurosci 1:578.

Stanford, L R, Friedlander M J, Sherman S M (1983). Morphological and physiological properties of geniculate W-cells of the cat: a comparison with X- and Y-cells. J Neurophysiol 50:582.

Thuma B D (1928). Studies on the diencephalon of the cat. I. The cytoarchitecture of the corpus geniculatum laterale. J Comp Neurol 46:173.

Wiesel T N, Hubel D H (1963). Effects of visual deprivation on morphology and physiology of cells in the cat's lateral geniculate body. J Physiol (Lond) 26:978.

Wiesel T N, Hubel D H (1965). Comparison of the effects of unilateral and bilateral eye closure on cortical unit responses in kittens. J Neurophysiol 28:1029.

Wilson P D, Rowe M H, Stone J (1976). Properties of relay cells in the cat's lateral geniculate nucleus: a comparison of W-cells with X- and Y-cells. J Neurophysiol 39: 1193.

Winkler C, Potter A (1914). "An Anatomical Guide to Experimental Researches on the Cat's Brain." Amsterdam; Versluys.

Development of Visual Pathways in Mammals, pages 347–361

EFFECTS OF MONOCULAR DEPRIVATION UPON VISUAL CORTICAL AREAS 17, 18, AND 19 IN THE CAT

Audie Leventhal, Ph.D.

Department of Anatomy
University of Utah
Salt Lake City, Utah 84132

An animal's two eyes have different views of the world which are integrated in the visual cortex. The capability to perform this synthesis develops normally only if post-natal visual experience is provided to both eyes simultaneously (Wiesel and Hubel, '63; Wiesel and Hubel, '65). If one eye or the other is favored, many of the resulting effects are hypothesized to be the outcome of a competition among the afferent pathways from the two eyes for control of individual cortical cells (Guillery and Stelzner, '70).

Evidence for this comes from studies in which one eye is deprived of patterned visual stimulation by suturing the eyelids closed (monocular deprivation, MD)(reviewed in Hirsch and Leventhal, '78; Movshon and Van Sluyters, '81). Most of these studies have looked at the effects of MD on cells in area 17 of the visual cortex; monocular deprivation results in a striking decrease in the proportion of neurons in area 17 that can be activated by the deprived eye (Wiesel and Hubel, '63). However, area 17 is only one of many cortical visual areas and it is the only one to receive direct inputs from a particular class of relay cells, the X-type relay cells in the dorsal lateral geniculate nucleus (LGNd) (Stone and Dreher, '73; Dreher et al., '80). For these reasons, the results of studies of area 17 may not be representative of the effects of visual deprivation on cortical development in general.

In this article, I will describe the binocular interactions normally present in cortical areas 17, 18 and 19 as well as the effects of MD upon the binocularity of cells in these three areas of cat visual cortex. The effects of MD upon the LGNd relay cells providing the afferent input to these areas will also be described. Areas 18 and 19 receive the bulk of their LGNd relay cell afferents from Y and W cells, respectively (Dreher et al., '80) and the receptive field properties of cells in these areas are distinctive (Hubel and Wiesel, '65). Comparisons of the effects of MD upon areas 17, 18 and 19 should clarify the mechanisms mediating deprivation induced changes in cerebral cortex.

Overall, 1130 cells were studied in areas 18 and 19 of normal cats and of cats deprived of vision in one eye by monocular lid-suture from birth to 9-14 months of age (Leventhal and Hirsch, '83; Hirsch and Leventhal, '83). Cats were prepared for electrophysiological recordings using standard techniques. Cells were recorded from the cortical representation of the area centralis and from the cortical representations of more peripheral regions of the retina. The electrode was moved least 75 um between units to reduce sampling bias and to record from a large region of cortex in each cat.

The ocular dominance of each cell was determined quantitatively. Ocular dominance ratios were determined by dividing the response evoked from the dominant eye by the response evoked from the non-dominant eye. Evoked responses were determined by subtracting the unit's spontaneous activity from its total response. A ratio of one indicates that the two eyes activated the cell equally, while higher ratios indicate an imbalance in the ability of the two eyes to activate the cell.

Cells studied were placed into one of 5 groups: group 1, only the contralateral eye activated the cell; group 2, both eyes activated the cell but the contralateral eye was more effective (ratios > 1.5); group 3, both eyes activated the cell to a similar degree (ratios < 1.5); group 4, both eyes activated the cell but the ipsilateral eye was more effective (ratios > 1.5); group 5, only the ipsilateral eye activated the cell. Cells in groups 1 and 5 are considered monocular, cells in group 2 and 4 are considered weakly binocular, cells in group 3 are considered strongly binocular.

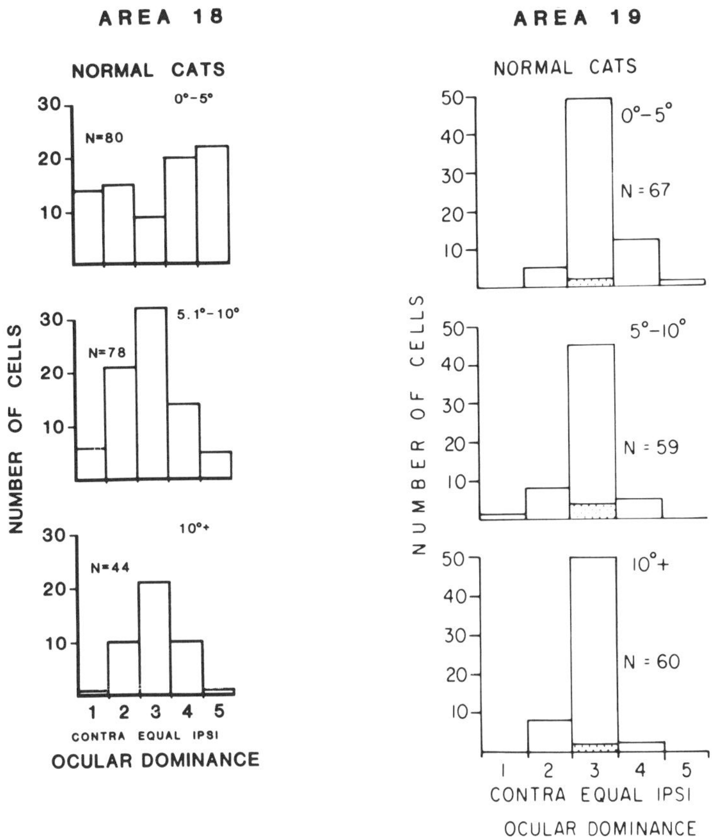

Figure 1. Ocular dominance distributions of cells in areas 18 and 19 of normal cats. Distributions for cells subserving different parts of the visual field (eccentricities) are shown separately. In area 18 of the normal cat, most cells with receptive fields close to the representation of the area centralis, but not in other regions, are monocular or weakly binocular. In area 19 of the normal cat, most cells, regardless of eccentricity, are binocular and activated equally well by the two eyes.

BINOCULAR INTERACTIONS IN AREAS 18 AND 19 OF NORMAL CATS

The distributions of the ocular dominances of cells recorded from areas 18 and 19 of normal cats are shown in

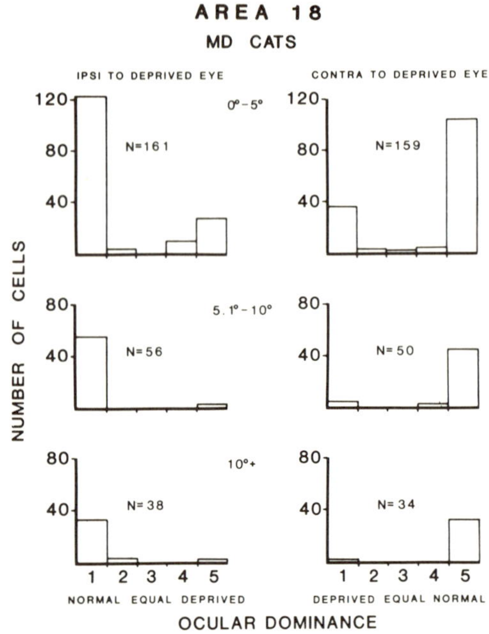

Figure 2. Ocular dominance distributions of cells in area 18 of MD cats. Distributions for cells subserving different eccentricities are shown separately as are the distributions for cells in the hemispheres ipsilateral and contralateral to the deprived eye. In area 18 of MD cats, most cells in both hemispheres are activated only by the normal eye. However, 20% of the cells within the representation of the central 5° of visual angle are activated only by the deprived eye. Outside of this region, far fewer cells are activated by the deprived eye.

Figure 1. Cells with receptive fields located at different distances from the representation of the area centralis are shown separately.

In normal cats, most area 18 cells with receptive fields within 5° of the area centralis are activated monocularly (45%) or are weakly binocular (44%). Very few cells in this region (11%) are strongly binocular. In marked contrast, few area 18 cells with receptive fields between 5 and 10° from the area centralis are monocular (13%) and

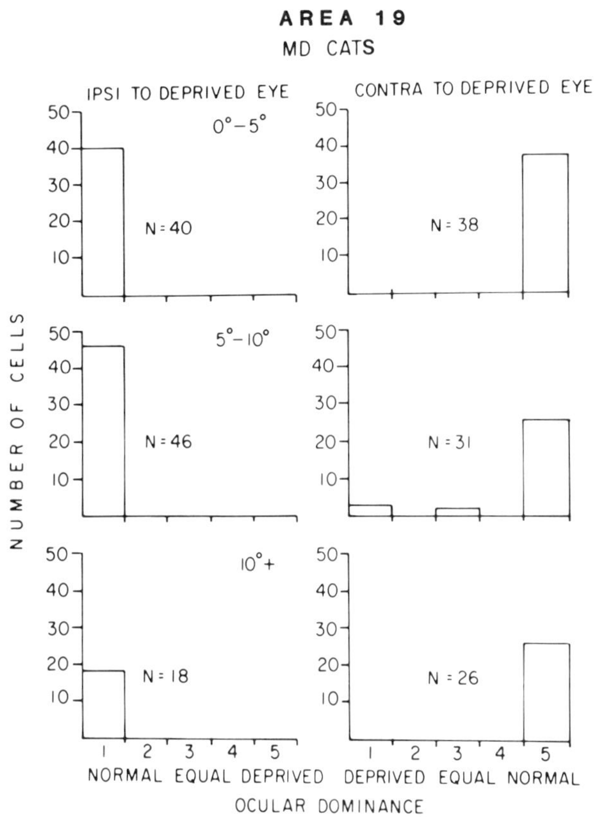

Figure 3. Ocular dominance distributions of cells in area 19 of MD cats. Distributions for cells subserving different eccentricities are shown separately as are the distributions for cells in the hemispheres ipsilateral and contralateral to the deprived eye. In area 19 of MD cats virtually all cells in both hemispheres, regardless of eccentricity, are activated only by the normal eye. However, in MD cats, all cells activated by the deprived eye were recorded from the hemisphere contralateral to the deprived eye.

many are strongly binocular (42%). In regions of area 18 subserving parts of the retina more than 10° from the area centralis, the proportion of monocular cells is still lower

(4%) while the proportion of strongly binocular cells is even higher (58%).

In area 19 of normal cats, most cells regardless of the sizes or the eccentricities of their receptive fields exhibit ocular dominance ratios of less than 1.5. All penetrations in area 19 were reconstructed in our studies (Leventhal and Hirsch, '83) and most cells in all layers of area 19 were found to be strongly binocular. Overall, we found that very few cells in area 19 exhibit ocular dominance ratios over 5.0. A comparison of the results for area 19 and our previous results for area 17 (Leventhal and Hirsch, '78, '80) showed two differences in the binocularity of cells in areas 17 and 19. First, a higher proportion of cells in area 19 than in area 17 is influenced equally by the two eyes (ratios < 1.5). Second, in area 17, but not in area 19, receptive field width and binocularity are correlated; in area 17 cells with narrow receptive fields tend to be the least binocular. For the most part this is because S-type cells in layer 4 of area 17 have narrow receptive fields and tend to be monocular or unequally activated by the two eyes (Gilbert, '77; Leventhal and Hirsch, '78). Finally, in area 19, but not in area 17, some cells can only be activated if both eyes are stimulated simultaneously.

EFFECTS OF MD UPON AREAS 18 AND 19

In regions of area 18 subserving the central 5° of the visual field in both hemispheres of MD cats, most cells (74%) are activated only by the normal eye; 20% are activated only by the deprived eye and only 6% are activated binocularly (Figure 2). Cells activated by the deprived eye are found in all layers of area 18 and are not concentrated in layer 4 as is reported to be the case in area 17 of MD cats (Shatz and Stryker, '78). In area 18 in both hemispheres of MD cats, the dominance of the normal eye is significantly stronger in regions representing parts of the retina between 5 and 10° from the area centralis. In these regions, 94% of the cells are activated only by the normal eye, 5% of the cells are activated only by the deprived eye, and only 1% of the cells are activated binocularly. Cells with receptive fields more than 10° from the area centralis show similar patterns. In these regions 91% of the cells are activated only by the normal eye; 5% are activated only by the deprived eye, and 4% are activated binocularly.

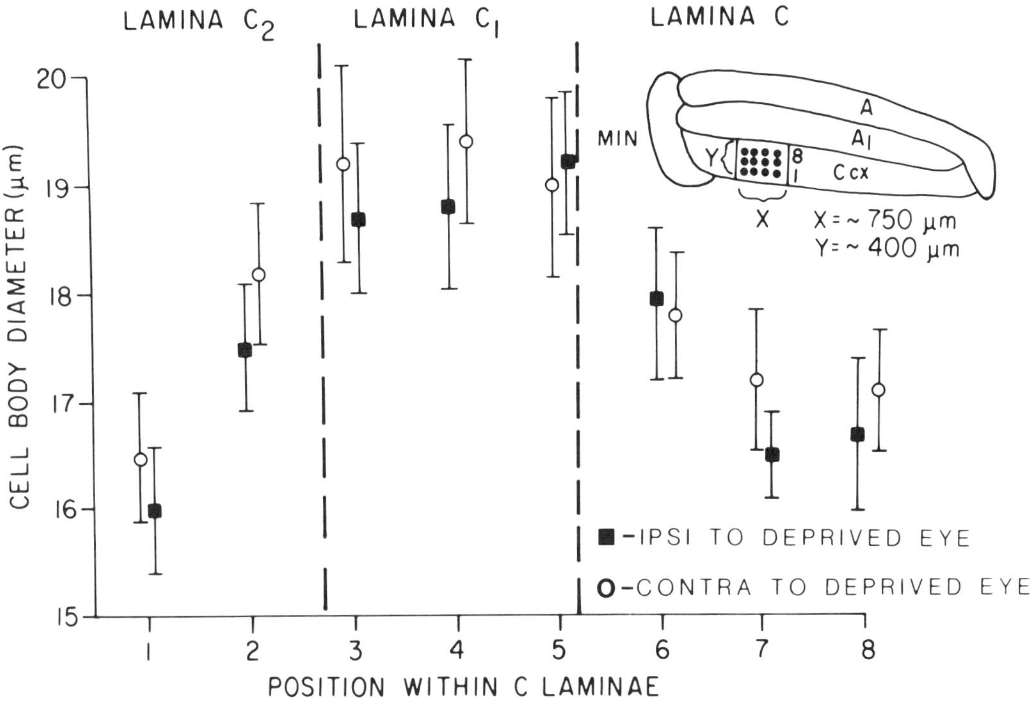

Figure 4. Cell body diameters of relay cells in the parvocellular C laminae of the cat's LGNd labeled by electrophoretic injections of HRP into area 19. Forty-one units were recorded from area 19 in the two hemispheres of this cat. All of these were driven exclusively by the normal eye. Cell body diameter was recorded as the mean of the cell's longest axis and orthogonally shorter axis. The distances from the center of each cell to the bottom of lamina A_1 as well as to the border between the optic tract and bottom of the LGNd were recorded. Cells were then placed into eight 50 μm bins, each successively further from the optic tract (see inset). The approximate positions of the borders of lamina C_1 are indicated by dashed lines. These were determined from a corresponding section through the LGNd of a cat that received an injection of ^{3}H-proline into one eye. All points are based upon samples of 20-40 cells with visible nucleoli. Standard deviations are indicated by lines through the points. Solid squares and open circles indicate cells sampled from the hemispheres ipsilateral and contra-

lateral to the deprived eye, respectively. Notice that in both hemispheres, relay cells in a position to receive ipsilateral afferents (lamina C_1) are larger than those in a position to receive contralateral afferents (laminae C and C_2). Cells in a position to receive afferents from the deprived eye are not smaller than those in a position to receive afferents from the normal eye.

In both normal and MD cats, penetrations into parts of area 18 representing the central 5° of visual angle show clear evidence of ocular dominance columns (Hubel and Wiesel, '62, '65). That is, as the electrode is advanced down the lateral bank of the lateral gyrus, alternating clusters of cells subserving the left and right eyes are encountered. In normal cats these clusters are approximately equal in size while, in MD cats, clusters subserving the normal eye are much larger than those subserving the deprived eye. There is no obvious evidence of ocular dominance columns in regions of area 18 subserving the peripheral retina in either normal or MD cats. This is because in normal cats many cells in these regions are activated equally well by the two eyes while in MD cats virtually all cells subserving peripheral vision are activated only by the normal eye.

In area 19 of MD cats, virtually all cells (97%), regardless of the eccentricity of their receptive fields, their laminar location, or the hemisphere from which they are recorded are activated only by the normal eye (Figure 3). Cells activated by the normal eye in MD cats exhibit normal properties. However, in MD cats 34% of cells are S type while only 12% of area 19 cells in normal cats are S type.

EFFECTS OF MD UPON THE SIZES OF RELAY CELLS PROJECTING TO AREAS 17, 18 AND 19

In the deprived laminae of the LGNd of MD cats relay cells projecting to areas 17 and 18 are severely shrunken. Since MD has such a profound effect upon the binocularity of area 19, it is of interest to know what effect MD has upon the relay cells in the parvocellular C laminae which provide the LGNd afferent input to area 19.

To accomplish this, we made small electrophoretic injection of HRP into area 19 of both hemispheres of the MD cats in which cortical recordings were carried out (Leventhal and Hirsch, '83). The sizes of retrogradely labeled cells in the parvocellular C laminae were then determined (Figure 4). Surprisingly, our results indicate that cells in the deprived parvocellular C laminae projecting to area 19 in MD cats are not smaller than those in the normal parvocellular C laminae (see also Murakami and Wilson, '80). In contrast, in the same sections cells in the deprived A laminae and the large cells in the top of the deprived magnocellular C lamina (which presumably project to areas 17 and 18) are noticably shrunken.

DIFFERENCES IN EFFECTS OF MONOCULAR DEPRIVATION UPON AREAS 17, 18 AND 19

The loss of binocularity in the visual cortex of MD cats appears to be more severe in area 19 than in areas 17 or 18 (see also Rapaport et al, '82). Significant numbers of cells activated by the deprived eye have been recorded from the area centralis representation in area 18 and from layer 4 of area 17 of MD cats (Shatz and Stryker, '78). There are very few cells activated by the deprived eye in area 19 of MD cats and there is no evidence that they are concentrated either in layer 4 or in any other layer. Thus, monocular deprivation produces a more complete loss of responsiveness to the deprived eye in area 19 than it does in areas 17 and 18.

The differences in the effects of MD on cells in areas 17, 18 and 19 must be viewed in light of several differences in the ocular dominance of cells in areas 17, 18 and 19 of normal adult cats. First, in normal cats, there are more monocular cells in areas 17 and 18 than in area 19. In area 17, monocular cells are most common in layer 4 (Gilbert, '77; Leventhal and Hirsch, '78) and in area 18 they are found near the projection of the area centralis. Second, in area 17, but not area 19, receptive field width is correlated with binocularity; in area 17 cells with narrow receptive fields are less binocular than cells with wide receptive fields. Third, there are significant numbers of cells in area 19 that can only be activated if the two eyes are stimulated simultaneously (see also Hale et al., '78). Areas 17 and 18 contain few if any cells of this type.

Overall, in area 18 of normal cats, we find that 45%, 13%, and 4% of cells 0-5°, 5.1-10°, and more than 10° from the area centralis representation, respectively, are activated monocularly; about half of the cells in each group are activated by each eye. In area 18 of MD cats, 20%, 5%, and 5% of the cells are activated only by the deprived eye in regions 0-5°, 5.1-10°, and more than 10° from the area centralis representation, respectively. The proportion of monocular cells activated by each eye in area 18 of normal cats is, therefore, very close to the proportion of cells activated by the deprived eye in MD cats. The same thing is apparently true in areas 17 and 19 since layer 4 of area 17 normally contains monocular cells in normal cats as well as cells activated by the deprived eye in MD cats while area 19 normally contains virtually no monocular cells and in MD cats area 19 contains virtually no cells activated by the deprived eye.

Taken together, the foregoing results for areas 17, 18 and 19 suggest that MD does not affect cells destined to be monocular in the normal adult in the same fashion as other cells; deprivation does not prevent the deprived eye from activating cortical cells it alone would normally activate. However, MD does affect the overall proportion of cells in areas 17, 18 and 19 activated only by the normal eye. This increase can be explained by a shift in the ocular dominance of virtually all cells in areas 17, 18 and 19 destined normally to be activated by both eyes. In these areas the normal eye is able to "take over" effectively all cells it would normally activate but is unable to activate cells it would not normally activate. Since the distribution of monocular cells normally differs in areas 17, 18 and 19, MD has different effects upon these areas. However, the same basic mechanism seems to underlie the changes produced by MD in these three areas; binocular competition may be the only factor underlying the increased proportion of cells activated only by the normal eye in areas 17, 18 and 19 of MD cats.

DIFFERENCES IN THE EFFECTS OF MD UPON THE SIZES OF RELAY CELLS PROJECTING TO AREAS 17, 18 AND 19

The A laminae of the cat's LGNd contain only X- and Y-type relay cells (Hoffmann et al., '72); the sizes of these cells are reduced by MD and these cell types may be affected differently by monocular deprivation (Sherman et al., '72). X- and Y-type cells are not found in the parvocellular C laminae (Wilson et al., '76). Instead, these layers contain a variety of other cell types, referred to collectively as W-cells (Rowe and Stone, '77), and receive inputs from a variety of ganglion cell types (discussed in Leventhal, '82). Our findings that relay cells in the parvocellular C laminae projecting to area 19 do not shrink in MD cats therefore indicates that LGNd W-cells are not affected by monocular deprivation in the same way as are X- and Y-cells. This is probably not because cells in the parvocellular C laminae are too small to shrink significantly. Kalil ('80) has shown that small cells in the monocular segment of lamina A shrink up to 33% in monocularly enucleated cats. Thus, there is now evidence for differences in the effects of monocular deprivation on X-, Y-, and W-type cells in the LGNd.

FACTORS UNDERLYING THE EFFECTS OF MONOCULAR DEPRIVATION ON AREA 19 AND ITS AFFERENTS

We interpret our results as evidence that the binocularity of cells in areas 17, 18 and 19 may be generated in different ways (Leventhal and Hirsch, '83). This suggestion is based on the following observations. In area 17 of normal cats (Gilbert, '77; Leventhal and Hirsch, '78), but not in area 19 of normal cats, layer 4 contains a relatively high proportion of monocular cells. In area 17, these monocular cells have distinctive characteristics. They have narrow receptive fields and most are S type. Furthermore, in MD cats, layer 4 of area 17 (Shatz and Stryker, '78) contains a reduced but significant proportion of cells that can be activated by the deprived eye. There is no concentration of deprived eye cells in layer 4 of area 19. Consistent with this, ocular dominance columns can be demonstrated in areas 17 and 18 of normal cats and the sizes of ocular dominance columns subserving the deprived eye are reduced in MD cats. Ocular dominance columns are not evident during physiological recordings from area 19 of normal or MD cats. Thus, in areas 17 and 18 of normal and MD cats, the afferent pathways from the two eyes are distinct, especially in layer 4. In contrast, the afferent pathways from the two eyes are much less distinct in area 19 of normal or MD cats.

It seems very likely that relay cells in the LGNd A laminae of the normal cats converge onto individual cells in areas 17 and 18 and that these relay cells compete for control of cortical cells (Guillery and Stelzner, '70). A variety of cell types in area 17 can receive direct geniculate afferents (discussed in Stone et al., '79) including many S-type cells with small receptive fields in layer 4. An imbalance in stimulation of the two eyes, such as that produced by MD, influences the outcome of this competition so that afferents from one eye increase the region of cortex which they activate (Hubel et al., '77; Shatz and Stryker, '78; Le Vay et al., '80). As a result of such an imbalance in stimulation, cells in the deprived A laminae are placed at a competitive disadvantage and, because of this, shrink (Guillery and Stelzner, '70). This shrinkage may reflect changes in the axonal arborizations of deprived cells (Guillery and Stelzner, '70; Hubel et al., '77; Shatz and Stryker, '78; Le Vay et al., '80). We suggest that a similar mechanism is unlikely to be operating in area 19 since there is no evidence for any segregation of afferents from the two eyes in normal or deprived cats. Moreover, ocular dominance is unrelated to laminar location and receptive field width in area 19. Finally, within the limits of our technique, the somas of the relay cells in the deprived C laminae projecting to area 19 show no evidence of shrinkage, so there is no evidence at present of any retrograde changes produced by interactions between LGNd afferents from the two eyes onto cells in area 19.

Binocular interactions in area 19 may be generated in a number of ways. First, they may be due to convergence of geniculocortical afferents. For the reasons mentioned above, this seems unlikely. Second, it is possible that binocular interactions in area 19 are generated by the sizable projection this area receives from the pulvinar complex (Berson and Graybiel, '78; Leventhal et al., '80). Unfortunately, little is known about the function of the pulvinocortical projections. Third, binocularity in area 19 may be generated by connections among cortical cells much as they are in the lateral suprasylvian vision area (Spear and Baumann, '79).

Area 19 does not appear to receive inputs from X-type relay cells; it receives much of its W-cell input and all of its Y-cell input indirectly, probably via area 17 (Dreher et al., '80). Thus, the absence of cells responsive to the deprived eye in those layers of area 17 which project to area 19, layers 2 and 3 (Gilbert and Kelly, '75), is consistent with the hypothesis that the effect of MD on cells in area 19 is indirect and reflects changes produced in other cortical visual areas by binocular competition. Most cells in layers 2 and 3 of area 17 exhibit mixed on-off responses so the increase in the proportion of S-type cells we observed in area 19 of MD cats also supports the idea that, in area 19, MD affects corticocortical connections more than geniculocortical ones. A direct test of this idea will require study of binocularity in area 19 of normal and MD cats in which area 17 has been inactivated.

LITERATURE CITED

Berson DM, Graybiel AM (1978). Parallel thalamic zones in the LP-pulvinar complex of the cat identified by their afferent and efferent connections. Brain Res 147:139-148.

Dreher B, Leventhal AG, Hale PT (1980). Geniculate input to cat visual cortex: a comparison of area 19 with areas 17 and 18, J Neurophysiol 44:804-826.

Duysens J, Orban GA, Van Der Glas HW, De Zegher FE (1982). Functional properties of area 19 as compared to area 17 of the cat. Brain Res 231:279-291.

Gilbert CD (1977). Laminar differences in receptive field properties of cells in cat primary visual cortex. J Physiol (Lond) 268:391-421.

Gilbert CD, Kelly JP (1975). The projections of cells in different layers of the cat's visual cortex. J Comp Neurol 183:81-106.

Guillery RW, Stelzner DJ (1970). The differential effects of unilateral lid closure upon the monocular and binocular segments of the dorsal lateral geniculate nucleus. J Comp Neurol 139:413-422.

Hale PT, Dreher B, Leventhal AG (1978). A comparison of the receptive field properties of cells in areas 17, 18, and 19 of the cat. Proc Aust Physiol Pharmacol Soc 9: 193P.

Hirsch HVB, Leventhal AG (1978). Functional modification of the developing visual system. In M. Jacobson (Ed) Handbook of Sensory Physiology, Vol IX, Development of Sensory Systems, Springer, Heidelberg.

Hoffmann KP, Stone J, Sherman SM (1972). Relay of receptive field properties in dorsal lateral geniculate

nucleus of the cat. J Neurophysiol 35:518-531.
Hubel DH, Wiesel TN, LeVay S (1977). Plasticity of ocular dominance columns in monkey striate cortex. Philos Trans R Soc Lond [Biol] 278:377-409.
Hubel DH, Wiesel TN (1962). Receptive fields, binocular interaction, and functional architecture in the cat's visual cortex. J Physiol (Lond) 160:106-154.
Hubel DH, Wiesel TN (1965). Receptive fields and functional architecture in two non-striate visual areas (18 and 19) of the cat. J Neurophysiol 28:229-289.
Kalil R (1980). A quantitative study of the effects of monocular enucleation and deprivation on cell growth in the dorsal lateral geniculate nucleus of the cat. J Comp Neurol 189:438-524.
LeVay S, Wiesel TN, Hubel DH (1980). The development of ocular dominance columns in normal and visually deprived monkeys. J Comp Neurol 191:1-51.
Leventhal AG, Hirsch HVB (1980). Receptive-field properties of different classes of neurons in the visual cortex of normal and dark-reared cats. J Neurophysiol 43:1111-1132.
Leventhal AG (1982). Morphology and distribution of retinal ganglion cells projecting to different layers of the lateral geniculate nucleus in normal and Siamese cats. J Neurosci 2:1024-1042.
Leventhal AG, Keens J, Törk I (1980). The afferent ganglion cells and cortical projections of the retinal recipient zone of the cat's pulvinar complex. J Comp Neurol 194:535-554.
Movshon JA, Van Sluyters RC (1981). Visual neural development. Ann Rev Psychol 32:477-522.
Murakami DM, Wilson PD (1980). Monocular deprivation affects cell morphology in laminae C and C_1 in cat lateral geniculate nucleus. Neuroscience Abstracts 6:789.
Rapaport DH, Dreher B, Rowe MH (1982). Lack of binocularity in cells of area 19 of cat visual cortex following monocular deprivation. Brain Research 246:319-324.
Rowe MH, Stone K (1977). Naming of neurones. Brain Behav Evol 14:185-216.
Shatz CJ, Stryker MP (1978). Ocular dominance in layer IV of the cat's visual cortex and the effects of monocular deprivation. J Physiol (Lond) 281:267-283.
Sherman SM, Hoffmann KP, Stone J (1972). Loss of a specific cell type from the dorsal lateral geniculate nucleus in visually deprived cats. J Physiol (Lond) 35:532-541.

Stone J, Dreher B (1973). Projection of X- and Y-cells of the cat's lateral geniculate nucleus to areas 17 and 18 of the visual cortex. J Neurophysiol 36:551-567.

Wiesel TN, Hubel DH (1963). Single-cell responses in striate cortex of kittens deprived of vision in one eye. J Neurophysiol 26:1003-1017.

Wiesel TN, Hubel DH (1965). Comparison of the effects of unilateral and bilateral eye closure on cortical unit responses in kittens. J Neurophysiol 28: 1029-1040.

Wilson PE, Rowe MH, Stone J (1976). Properties of relay cells in the cat's lateral geniculate nucleus. A comparison of W cells with X and Y cells. J Neurophysiol 38:1193-1209.

Development of Visual Pathways in Mammals, pages 363–380

EXPERIMENTAL AMBLYOPIA

DP Crewther and SG Crewther

National Vision Research Institute, 386 Cardigan St., Carlton, Vic 3053, and
School of Optometry, University of New South Wales, Kensington, NSW 2033.

INTRODUCTION

Amblyopia is defined as optically uncorrectable loss of vision, usually in one eye, without any known pathology (Hess, Campbell, Greenhalgh 1978). The condition affects up to 5% of the human population (Duke-Elder, Wybar 1973). It is usually associated with strabismus (misalignment of the eyes) or with anisometropia (unequal refraction of the eyes), or both, although a small percentage of amblyopes suffer from deprivation amblyopia following congenital cataract. Amblyopia is also associated with other symptoms which suggests that the neural circuitry involved is not confined purely to the primary visual pathway. For example, many amblyopes exhibit exaggerated fixation eye movements and non-central fixation of their amblyopic eyes, indicating an involvement of the oculomotor system in amblyopia, either in a principal or a corollary sense.

Recent psychophysical experiments involving contrast sensitivity analysis have added greatly to the characterization of amblyopia giving a more complete picture of the nature of the visual loss than simple acuity measurements. For example, on the basis of contrast sensitivity functions, the visual dysfunction occurring in strabismic amblyopia has been divided into two classes, one in which the contrast sensitivity is reduced only at high spatial frequencies and the other in which there is lowered sensitivity at all frequencies (Hess, Howell 1977). However, perception of gratings of a single spatial frequency often does not give a true indication of the loss of vision in terms of perception of a complex visual scene. For instance, letter acuity is often considerably worse than would be predicted from the grating acuity deficit. On the other hand, Vernier grating acuities show a strong correlation with Snellen acuities over a wide range of frequencies, and under such a Vernier hyperacuity task, a distinction has been drawn between two classes of amblyopia (Levi, Klein 1982). Anisometropic amblyopes demonstrate hyperacuity even at spatial frequencies approaching their grating resolution, while strabismic amblyopes have reduced hyperacuity at high spatial frequencies, indicating an inability to maintain fine positional information.

Psychophysical evidence has long suggested that amblyopia in humans involves a central neural abnormality. Amblyopic patients show relatively minor disturbances of functions such as increment threshold and dark adaptation, which are classical indicators of retinal function (Wald, Burian 1944; Burian, von Noorden 1974; Duke-Elder, Wybar 1973), while they suffer gross disturbances of form vision and a loss of binocular function. Both of these latter symptoms are typical of high order visual processing such as occurs beyond the lateral geniculate nucleus, i.e. in visual cortex. However there have been recent reports of abnormal pattern electroretinograms (postulated to be indicative of retinal ganglion cell function) in strabismic amblyopes (Arden, Vaegan, Hogg, Powell, Carter 1980).

Neurophysiological understanding of these visual abnormalities has been greatly enhanced by the use of animal models. The discovery that the developing visual system was susceptible to monocular occlusion (Wiesel, Hubel 1963a) has led to many experiments involving post-natal modifications to the visual environment of the animal. Analogously to the clinical observation that amblyopia ia a developmental disorder, experiments in which monocular deprivation was initiated at different ages in kittens indicated that there is a period, early in the animals life, when the brain is very susceptible to external experience. This period is now referred to as the critical or sensitive period (Hubel, Wiesel 1970). Other experimental rearing conditions relevant to clinical visual disorders quickly followed. Congenital strabismus was mimicked by section of the medial rectus muscle, causing an exodeviation of the eye (Hubel, Wiesel 1965). Concomitant strabismus was induced optically by rearing kittens with goggles containing prisms (Bennett,

Smith, Harwerth, Crawford 1980; Van Sluyters, Levitt 1981). Anisometropia was artificially created by the rearing of kittens wearing goggles containing spherical lenses of varying powers (Eggers, Blakemore 1978) and also by the wearing of contact lenses (Crewther, Cleland, Crewther 1982). Chronic paralysis of accommodation in cats by the daily installation of atropine (Ikeda, Tremain 1978) also has the effect of producing blurred images of proximal objects on the retina of the penalized eye, due both to the inability to accommodate and the reduced depth of field caused by the dilated pupil, and also leads to a profound amblyopia (Crewther, Cleland, Crewther 1982) (see Fig 1).

Chronic astigmatism in kittens has been produced by rearing with cylindrical lenses held in goggles (Blasdel, Mitchell, Muir, Pettigrew 1977). This is similar to the experiment of Hirsch and Spinelli (1976) in which the kittens could view only horizontal contours through one eye and vertical contours through the other and also the single orientation drum-rearing experiments of Blakemore, Cooper (1975). Cortical studies showed that there was a reduction in the percentages of cells coding orientations orthogonal to the orientation of exposure - effectively, an orientational or meridional amblyopia.

MECHANISMS OF AMBLYOPIA

In those cases where functional acuity measurements have been made in these model disorders, the symptoms resemble the comparable human disorder. However , in spite of this large literature, our understanding of the fundamental causes or mechanisms of amblyopia is not much further advanced than at the turn of the century. Much of the history of research into amblyopia has been concerned with a fundamental conflict between two theories (for a review, see Haik 1962). According to the first theory advocated by Donders, squint causes an amblyopia which depends on a loss of vision due to either habitual suppression or lack of use - "amblyopia ex anopsia." The second theory considers that amblyopia is a congenital defect which precedes and causes the squint. Chavasse developed Donder's theory to give rise to the idea of learned inhibition which becomes more fixed with age, leading to an obligatory inhibition or amblyopia. This theory also states that the earlier is the age at which inhibition begins to operate, the more profound will be the amblyopia, hence the term "amblyopia of arrest."

It seems that a unitary neural mechanism for the loss of visual acuity that characterizes amblyopia is unlikely to emerge. However, experimental amblyopia can probably be described by two principles, either of which would lead to lowered visual acuity during development. The first relies on direct binocular competition at the level of individual neurons in visual cortex. Active and unequal competition during the period in which the visual system is receptive to modification, causes the loss or suppression of access of one eye's terminals to synaptic sites in visual cortex. In a sense, the driving force in this case is the presynaptic imbalance. The sampling of inputs from the less priviledged eye is then too coarse or the input on single cells is too weak for high spatial frequencies to be coded. The second principle is almost the converse of the first. Rather than amblyopia being caused by active and unequal binocular competition, postsynaptic cortical inactivity or lack of susceptibility during development may well have the same effect - the result being a slowed rate of development and a lack of maturation of receptive field properties. Of necessity, the inactivation of a cortical region or the process by which it remains visually responsive but loses its plasticity, is of extra-visual origin.

LOCUS OF DEPRIVATION AMBLYOPIA

The purest example of the first principle is the neural manifestation of monocular deprivation in visual cortical areas. However, it is convenient to treat this together with the group of experiments involving the degradation of the retinal image to one eye of an animal for a period of time. This group then includes monocular deprivation, induced anisometropia and atropine penalization. Behaviourally, cats appear blind when first exposed to visual stimuli through an eye deprived of form vision by eyelid suture from soon after birth (Wiesel, Hubel 1963a; Ganz, Fitch 1968; Dews, Wiesel 1970; Giffin, Mitchell 1978). This behavioural deficit is accompanied by equally substantial effects in the physiological properties of cells in the visual cortex. With the exception of some cells in layer IV (Shatz, Stryker 1978), virtually all cells in Area

17 of monocularly deprived cats can be excited only by stimuli presented through the experienced eye (Wiesel, Hubel 1963a). Similarly, a profound amblyopia is observed in anisometropic kittens raised wearing a contact lens of high refractive power and also in kittens reared with atropine penalization (Crewther, Cleland, Crewther 1982). Lowered cut-off spatial frequencies have been observed in striate cortical cells of anisometropic cats driven by the eye which had been subject to prolonged image degradation (Eggers, Blakemore 1979). Thus, the locus of these deprivation amblyopias is at least as peripheral as the striate cortex.

The situation in the lateral geniculate nucleus is somewhat less clear. It is certainly true that there is an effect of monocular deprivation on the growth rate of cells in the laminae of the LGN that receive projection from the deprived eye. In fact, the cross-sectional areas of such cells may be smaller by as much as 25% - 40% in comparison to cells in laminae that receive input from the visually experienced eye (Wiesel, Hubel 1963b; Kupfer, Palmer 1964; Guillery, Stelzner 1970; Guillery 1972; Garey, Fisken, Powell 1973; Dursteler, Garey, Movshon 1976). Whether there are any functional changes in the geniculate, particularly with respect to the coding of high spatial frequencies is more controversial. On the one hand, a reduction in the cell acuities or cut-off spatial frequencies of cells receiving input from the deprived eye has been reported (Maffei, Fiorentini 1976; Sireteanu, Hoffmann 1979), while on the other hand, normal resolving power has been reported for deprived geniculate cells of the X-cell pathway (Derrington, Hawken 1981; Friedlander, this vol). In spite of this controversy at the level of the geniculate, there is general agreement that retinal ganglion cells are unaffected, both anatomically and physiologically, by monocular deprivation. This was indicated by the studies of Sherman and Stone (1975) and Kratz, Mangel, Lehmkuhle and Sherman (1979) who recorded from the fibres comprising the optic nerve, and also by direct recording from the region of the area centralis of the monocularly deprived retina of monocularly deprived cats (Cleland, Mitchell, Crewther, Crewther 1980). Cellular cut-off spatial frequencies of up to 10 cycles/deg were found in the area centralis of the monocularly deprived eye of cats deprived from eye opening until at least 12 months of age. Similarly, normal ganglion cell cut-off spatial frequencies have been reported in kittens reared with optical defocus (Crewther, Cleland, Crewther 1982). Thus it would appear that retinal ganglion cells develop normally in spite of a severely degraded retinal image over a long period of time. This would tend to implicate binocular competition as the chief cause of deprivation amblyopia, as the most peripheral organ, the retina, has the capacity to function normally. Also, the behavioural amblyopia resulting from monocular deprivation could then in part be termed a "population" amblyopia arising from the sparseness of sampling of visual information in the cortical sheet.

EFFECTS OF LACK OF VISUAL SYNERGY

There are a number of rearing conditions in which the visual synergy of the retinal inputs has been disrupted. Among these are surgical strabismus, whether by tenotomy or by myectomy of an extraocular muscle, optical strabismus caused by raising animals with goggles containing prisms, surgical or optical cyclotorsion and unilateral eye immobilization. Substantial deficits have been observed in the binocularity of striate cortex of cats and monkeys with strabismus induced by tenotomy of one of the extraocular muscles (Hubel, Wiesel, 1965; Baker, Grigg, von Noorden, 1974; Crawford, von Noorden, 1979), and are probably the cause of the breakdown of stereopsis in human strabismus (Crawford, von Noorden, Meharg, Rhodes, Smith, Miller 1982). In contrast to monocular deprivation, in which the domination by the visually experienced eye seems to be ubiquitous amongst visual cortical areas, the percentage of binocular cells in some cortical areas of strabismic cats is relatively unchanged (von Grunau 1981).

Behavioural evaluation of the acuities of strabismic cats has been carried out (Jacobson, Ikeda 1979; Cleland, Crewther, Crewther, Mitchell, 1982). Jacobson and Ikeda found that their esotropic cats (strabismus by myectomy) were amblyopic and associated the presence of an amblyopia with an inability to fixate with the strabismic eye. Similarly, using the Mitchell jumping stand technique, both tenotomized and myectomized strabismic cats have been shown without exception to possess an amblyopia of approximately two octaves of spatial frequency (Cleland, Crewther, Crewther, Mitchell 1982; Crewther, Cleland, Crewther1982) in both esotropic and exotropic cats. Functional amblyopia in exotropic cats has also been observed using evoked potential techniques (von Grunau, Singer 1980).

The depth of amblyopia induced in artificial strabismus

(by tenotomy or myectomy) is very similar to that induced by anisometropia through optical defocus or by atropine penalization. This is demonstrated in Fig. 1 where the behavioural performance of four classes of amblyopic cats are shown. Monocular acuity through the normal eye is compared with that through the experimental eye.

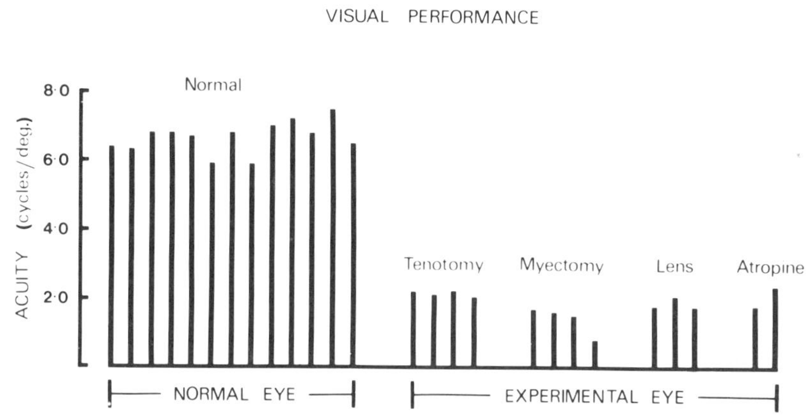

Fig. 1. A profound amblyopia is shown in both strabismic (by tenotomy and myectomy) and anisometropic (lens-reared and atropine-penalized) cats. The monocular acuity for the normal and experimental eyes (in order) of thirteen cats were measured using the jumping stand technique (Mitchell, Giffin , Timney 1978).

Traditionally it has been considered that the loss of binocularly driven cells in the visual cortex of strabismic cats is due to the lack of synergy of visual inputs (because the visual axes are not parallel) leads, over the developmental period, to a situation where the cortical cells cultivate only monocular inputs. Also, neighbouring cortical cells tend to cultivate inputs from the same eye so that the ocular dominance bands in strabismic cats are more prominent than in normals. The results of rearing kittens (Smith, Bennett, Harwerth, Crawford 1979; Van Sluyters, Levitt 1981) or monkeys (Crawford, von Noorden 1980) with an optical strabismus seem to indicate that visual lack of synergy is sufficient to cause the loss of binocularity. On the other hand, lack of visual synergy during development does not necessarily impose a loss of binocularly driven neurons on striate cortex. For example, approximately normal percentages of binocular cortical cells were reported in cats in which both eyes had been surgically rotated soon after birth (Crewther, Crewther, Peck, Pettigrew 1981). These animals possessed a large strabismus - a rotation of 90 deg around the optic nerve for each eye yields a non-alignment of the visual axes of approximately 50 deg.

The symmetrical, predominantly monocular, ocular dominance distributions commonly observed in experimental strabismus could be attributed to the occurrence of alternating fixation, a situation rarely associated with the presence of an amblyopia. Indeed, where strabismic amblyopia has been reliably induced, the reported ocular dominance distributions strongly favour the non-amblyopic eye (Tremain, Ikeda 1982). However, these animals also possess a retinal abnormality (as is discussed in the next section), which should lead to a situation of competitive advantage for the nondeviating eye at the cortical level.

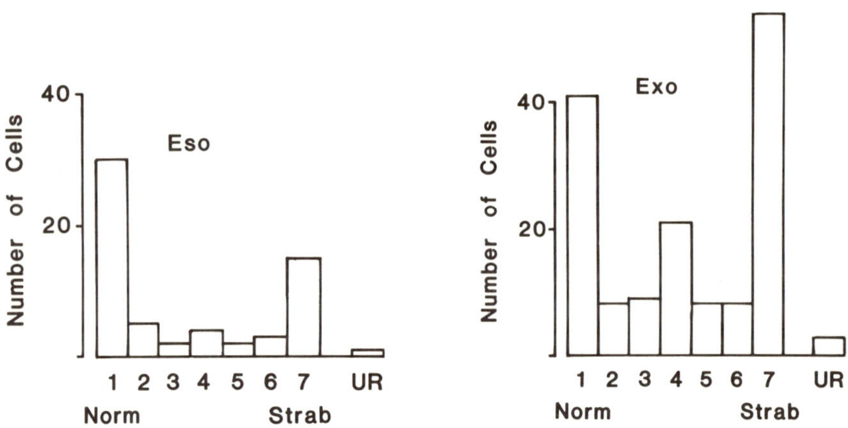

Fig. 2. Ocular dominance distributions from four tenotomized strabismic cats, three exotropic and one esotropic. Approximately equal numbers of cells were recorded from left and right Area 17 and the results displayed in terms of dominance with respect to the normal and amblyopic (strab) eyes.

Where the retinal spatial frequency coding is normal and and amblyopia has been demonstrated (Cleland, Crewther, Crewther, Mitchell 1982), the cortical ocular dominance histograms show approximately equal proportions of cells driven by the amblyopic eye as by the normal eye (see Fig 2). A significant degree of contralateral dominance was also observed in the distributions. This means that the hemisphere contralateral to the deviating eye was dominated by this (amblyopic) eye.

SITE OF STRABISMIC AMBLYOPIA

Ikeda and colleagues have suggested that experimental amblyopia due to bilateral hypermetropia, anisometropia and strabismus could have a common cause: habitual exposure of the fovea or area centralis to defocused images during the sensitive period of early post-natal life. They further proposed that peripheral parts of the visual system are susceptible to degraded images in that normal maturation cannot proceed without sharply focused images. The empirical support for this theory comes from experiments on esotropic kittens in which the strabismus was produced by myectomy of the lateral rectus muscle (and usually the superior oblique, as well). These cats could not fixate with the deviating eye, leading to an inappropriate accommodation of that eye (Ikeda, Wright 1974). Physiological recordings from cells in the LGN showed reduced cut-off spatial frequencies and lowered sensitivities, especially for X-cells (Ikeda, Wright 1976). Furthermore an "arrest of development" was shown; the acuities of cells receiving input from the area centralis being more depressed, the earlier the strabismus operation was carried out. The acuities of cells coding central visual field recorded from these adult strabismic cats were approximately the same as those recorded from normal kittens at an age at which trhe strabismus was performed. A morphological effect in the lateral geniculate nucleus of these strabismic cats was also observed (Ikeda, Plant, Tremain 1976). Cells in the laminae receiving input from the strabismic eye showed a large shrinkage with respect to the laminae receiving input from the unoperated eye. To examine whether the defect could be even more peripheral, Ikeda and Tremain (1979) recorded directly from retinal ganglion cells and showed that the acuities of area centralis retinal ganglion cells of the strabismic eye were depressed by up to two octaves of spatial frequency (results supported by Chino, Shansky, Hamasaki 1980).

In seeming contradiction to this finding, Cleland et al (1982) found that strabismic kittens which had been shown behaviourally to possess a severe amblyopia, had no acuity deficit in the retinal ganglion cells of the amblyopic eye. Also, the cells in the LGN of these animals showed no significant interlaminar differences in cross-sectional areas. However, the strabismus in these kittens was produced by disinsertion of the muscle rather than removal, i.e., by tenotomy rather than by myectomy. Also, in order to promote as normal eye movements as possible following recovery from the strabismus surgery, the muscle was allowed to reinsert by itself no suture being used to hold the strabismus in place. Direct comparison of the retinal physiology and geniculate morphology of cats with strabismus produced by tenotomy and myectomy showed there to be two classes of experimental strabismic amblyopia (Crewther, Cleland, Crewther 1982).

The most peripheral neural locus of the class of strabismic amblyopia in which the retina functions normally, is still unknown. The resolution of individual neurons from striate cortex of tenotomized strabismic cats was found to be indistinguishable for the two eyes, comparable to acuities from cortical cells of normal cats (Blakemore, Eggers 1978). However, the behavioural performance of these cats was not reported, and hence the presence or otherwise of an amblyopia in these animals is unknown.

THE SUPPRESSION OF PLASTICITY

There are several examples in the literature of cases where the normally expected plastic effects of a particular rearing procedure do not occur. A common feature of these examples is that immobilization or paralysis of the whole animal or of the eye is involved. Freeman and Bonds (1979) showed that a marked reduction in the proportion of binocular cells in striate cortex was found if kittens were stimulated monocularly while anaesthetised and held in a stereotaxic apparatus. If, however, the kittens were paralysed as well as anaesthetised, the same changes were not found unless an eye was moved mechanically. It appears that neither eye movement nor visual stimulation alone is sufficient to induce plastic changes. Similarly, in six week old vision deprived (dark-reared) kittens Buisseret, Gary-Bobo and Imbert (1978) showed that six hours of exposure to a normal visual environment was sufficient for the development of refined orientation tuning in cortical cells. However, if kittens were paralysed during this period of visual exposure, then the receptive field properties remained immature, akin to those of visually inexperienced control kittens. Thus paralysis seems effective in reducing the degree of cortical plasticity. In the same fashion, reduced or restricted eye-movements might be expected to play a major role in the abnormal maturation of the visual system. An example arises in the rearing of kittens with ocular rotation of one eye and lid-suture of the other (Singer, Tretter, Yinon 1982). Ocular dominance histograms obtained from the adult animals did not show the expected total dominance of the visually experienced (rotated) eye over the deprived eye, although some shift was found. The degree of mobility of the rotated eye was reported to be far less than normal. The significance of the lack of motility is revealed by the total dominance of striate cortex by the rotated eye of similarly prepared cats (Crewther, Crewther, unpublished results) in which the rotated eye showed unimpeded eye movements.

NON-VISUAL AFFERENTS AND VISUAL DEVELOPMENT

The direct influence of the extra-ocular muscle afferents on the visual system of cats has been substantiated by several experiments, for example, responses in visual cortex to stretch of the extra-ocular muscles (Buisseret, Maffei 1977). Also, interactions between muscle stimulation and visual stimulation of cells in the superior colliculus and LGN have been observed (Donaldson, Long 1980; Donaldson, Dixon 1981). Evidence that this input is indeed afferent input rather than a corollary discharge has been provided by Buisseret and Maffei (1977) who showed that the effects recorded in the visual cortex disappeared when the eye muscles were locally anaesthetised with xylocaine.

The actual function of the eye muscle input is still somewhat of a mystery, the main evidence coming from experiments in which that input is manipulated early in development. Perhaps the most plausible hypothesis for how the extra-ocular muscle input affects cortical plasticity is that it operates a central gating which modulates cortical excitability. Several recent studies indicate that retinal responses to contours are only a necessary, but not a sufficient condition for inducing such modifications (Freeman, Bonds, 1979; Buisseret, Gary-Bobo, Imbert 1978; Buisseret, Gary-Bobo, 1979; Rauschecker, Singer 1979; Singer, Rauschecker 1982). Singer (1982) has shown that unilateral destruction of the nuclear complex of the medial thalamus leads to a sensory neglect in the contralateral visual hemifield and an impairment of experience dependant modifications in the visual cortex on the side of the lesion. Other lesion and stimulation studies suggest that the mesencephalic reticular formation has an ascending projection which contributes to the gating of cortical plastic changes (Singer, Rauschecker 1982).

AMBLYOPIA OF NEGLECT

The second proposed mechanism for the development of an amblyopia, namely, a loss of plasticity in cortex, which prevents the normal maturation of receptive fields, and hence the development of high acuity coding by visual cortex receives support from the experiments, referenced above, involving paralysis or immobilization early in development. But if such a mechanism is to have much credence in explaining how one eye of a kitten made strabismic by surgical means becomes amblyopic, there remain two difficulties. Firstly, although the section of a muscle tendon will cause postoperatively a loss of motility of the eye and an interruption to the afferent information relayed to the visual system, the muscle will reattach itself within a matter of one to two weeks after the surgery, often before the peak of the sensitve period for cortical modification. The fact that lasting visual cortical effects result would suggest that there may be competition between the oculomotor afferents which feed into the visual system, and plasticity in that system may follow a different time course from that existing in visual cortex. Once fixation is controlled by one eye, reversal of any visual cortical effects by environmental modification may be difficult. The second difficulty is that the non-visual afferents from the eye do not have a partial decussation of the nature of the optic nerve fibres. This means that unilateral interruption to the extraocular afferent input will initially produce a hemianopia, rather than effects restricted to one eye or the other. This objection may be obviated if close reciprocal connections exist between afferents from agonist and antagonist muscles of the two eyes, the initial field loss initiating a loss of fixation and the cessation of normal maturation.

Arden GB,Vaegan, Hogg CR, Powell DJ, Carter RM (1980). Pattern ERGs are abnormal in many amblyopes. Trans Ophthal Soc UK 100: 453.

Baker FH, Grigg P, von Noorden GK (1974). Effects of visual deprivation and strabismus on the response of neurons in the visual cortex of the monkey, including studies on the striate and prestriate cortex in the normal animal. Brain Res 66: 185.

Bennett MJ, Smith EL, Harwerth RL, Crawford MLJ (1980). Ocular dominance, eye alignment and visual acuity in kittens reared with an optically induced squint. Brain Res 193: 33.

Blakemore C, Cooper GF (1970). Development of the brain depends on the visual environment. Nature 228: 477.

Blakemore C, Eggers HM (1978). Effects of artificial anisometropia and strabismus on the kitten's visual cortex. Arch Ital Biol 116: 385.

Blasdel GG, Mitchell DE, Muir DW, Pettigrew JD (1977). A physiological and behavioural study in cats of the effect of early visual experience with contours of a single orientation. J Physiol 265: 615.

Buisseret P, Gary-Bobo E (1979). Development of visual cortical orientation specificity after dark-rearing: role of extraocular proprioception. REF

Buisseret P, Gary-Bobo E, Imbert M (1978). Ocular motility and recovery of orientational properties of visual cortical neurones in dark-reared kittens. Nature 272: 816.

Buisseret P, Maffei L (1977). Extraocular proprioceptive projections to the visual cortex. Exp Brain Res 28: 421.

Burian HM , von Noorden GK (1974) Binocular Vision and Ocular Motility. Theory and Management of Strabismus, chapter 14. St. Louis; CV Mosby

Chino YM, Shansky MS, Hamasaki DI (1980). Development of receptive field properties of retinal ganglion cells in kittens raised with a convergent squint. Exp Brain Res 39: 313.

Cleland BG, Crewther DP, Crewther SG, Mitchell DE (1982). Normality of spatial resolution of retinal ganglion cells in cats with strabismic amblyopia. J Physiol 326: 235.

Cleland BG, Harding TH, Tulunay-Keesey U (1979). Visual resolution and receptive field size examination of two kinds of cat retinal ganglion cell. Science 205: 1015.

Cleland BG , Levick WR (1974). Brisk and sluggish

concentrically organized ganglion cells in the cat's retina. J Physiol 240: 421.
Cleland BG, Mitchell DE, Crewther SG, Crewther DP (1980) Visual resolution of retinal ganglion cells in monocularly-deprived cats. Brain Res 192: 261.
Cooper RD and Pettigrew JD (1979). The retinothalamic pathway in the cat with a note on the major meridians of the cat's eye. J Comp Neurol 187: 285.
Crawford MLJ, von Noorden GK (1979). The effects of short-term experimental strabismus on the visual system in Macaca mulatta. Invest Ophthal Vis Sci 18: 496.
Crawford MLJ, von Noorden GK (1980). Optically induced concomitant strabismus in monkeys. Invest Ophthal Vis Sci 19:1105.
Crawford MLJ, von Noorden GK, Meharg LS, Rhodes JW, Harwerth RL, Smith EL, Miller DD (1982). Binocular neurons and binocular function in monkeys and children. Invest Ophthal Vis Sci REF
Crewther DP, Crewther SG, Pettigrew JD (1978) A role for extraocular afferents in post-critical period reversal of monocular deprivation. J Physiol 282:181.
Crewther SG, Cleland BG, Crewther DP (1982). Is amblyopia a retinal ganglion cell phenomenon? Perception 11: 30A
Crewther SG, Crewther DP, Peck CK, Pettigrew JD (1981). Visual cortical effects of rearing kittens with monocular or binocular cyclotorsion. J Neurophysiol 44: 97.
Derrington AM, Hawken MJ (1981) Spatial and temporal properties of cat geniculate neurones after prolonged deprivation. J Physiol 314:107.
Dews PB, Wiesel TN (1970) Consequences of monocular deprivation on visual behaviour in kittens. J Physiol 206:437.
Donaldson IML, Dixon RA (1981). Excitation of units in the lateral geniculate and contiguous nuclei by stretch of the extraocular muscles. Exp Brain Res 38: 245
Donaldson IML, Long AC (1980). Interaction between extraocular proprioceptive and visual signals in the superior colliculus of the cat. J Physiol (Lond) 298, 85.
Duke-Elder S , Wybar K (1973). Ocular motility and strabismus. System of Ophthalmology VI: Ed. Duke-Elder S. London: Henry Kimpton p 301.
Dursteler MR, Garey LJ, Movshon JA (1976) Reversal of the morphological effects of monocular deprivation in the kitten's lateral geniculate nucleus. J Physiol 261:189.
Eggers HM, Blakemore C (1978). Physiological basis of

anisometropic amblyopia. Science 201: 264.
Freeman RD, Bonds, AB (1979). Cortical plasticity in moncularly deprived immotilized kittens depends on eye movement. Science 206: 1093.
Freeman RD, Olson CR (1979) Is there a "consolidation" effect for monocular deprivation. Nature 282:404.
Ganz L, Fitch M (1968) The effect of visual deprivation on perceptual behaviour. Exp Neurol 22:639.
Garey LJ, Dursteler MR (1975) Reversal of deprivation effects in the lateral geniculate nucleus of the cat. Neurosci Lett 1: 19.
Garey LJ, Fisken RA, Powell TPS (1973) Effects of experimental deafferentation on cells in the lateral geniculate nucleus of the cat. Brain Res 52: 363.
Giffin F, Mitchell DE (1978) The rate of recovery of vision after early monocular deprivation in kittens. J Physiol 274: 511.
von Grunau MW (1982). Comparison of the effects of induced strabismus on binocularity in area 17 and LS area in the cat. Brain Res 246: 325.
von Grunau MW , Singer W (1980). Functional amblyopia in kittens with unilateral exotropia. II. Correspondence between behavioural and electrophysiological assessment. Exp Brain Res 40: 305.
Guillery RW (1972) Binocular competition in the control of geniculate cell growth. J Comp Neurol 144: 117:127.
Guillery RW, Stelzner DJ (1970) The differential effects of unilateral lid closure upon the monocular and binocular segments of the dorsal lateral geniculate nucleus in the cat. J Comp Neurol 139:411.
Hirsch HVB, Spinelli DN (1970). Visual experience modifies distribution of horizontally and vertically oriented receptive fields in cats. Science 168: 869.
Hubel DH, Wiesel TN (1965). Binocular interaction in striate cortex of kittens reared with artificial squint. J Neurophysiol 28: 1041.
Hubel DH, Wiesel TN (1970) The period of susceptibility to the physiological effects of unilateral eye closure in kittens. J Physiol 206: 419.
Ikeda H (1980). Visual acuity its development and amblyopia. JR Soc Med 73: 546.
Ikeda H, Plant GT and Tremain KE (1977). Amblyopia occurs in retinal ganglion cells in cats reared with convergent squint without alternating fixation. Exp Brain Res 35: 559.
Ikeda H, Tremain KE (1978). Amblyopia resulting from

penalization: neurophysiological studies of kittens reared with atropinization of one or both eyes. Brit J Ophthal 62: 21.

Ikeda H, Tremain KE (1979). Amblyopia occurs in retinal ganglion cells in cats reared with convergent squint without alternating fixation. Exp Brain Res 35: 559.

Ikeda H, Tremain KE (1982). REF

Ikeda H, Wright MJ (1974). Is amblyopia due to inappropriate stimulation of the 'sustained' pathway during development? Br J Ophthal 58: 165.

Ikeda H, Wright MJ (1976). Properties of LGN cells in kittens reared with convergent squint a neurophysiological demonstration of amblyopia. Exp Brain Res 25: 63.

Jacobson SG, Ikeda H (1979). Behavioural studies of spatial vision in cats reared with convergent squint: is amblyopia due to arrest of development? Exp Brain Res 34: 11.

Kalil RE (1980) A quantitative study of the effects of monocular enucleation and deprivation on cell growth in the dorsal lateral geniculate nucleus of the cat. J Comp Neurol 189: 483.

Kratz KE, Mangel SC, Lehmkuhle S, Sherman SM (1979) An effect of early monocular lid suture upon the development of X-cells in the cat's lateral geniculate nucleus. Brain Res 172: 545.

Kupfer C, Palmer P (1964) Lateral geniculate nucleus: histological and cytochemical changes following afferent denervation and visual deprivation. Exp Neurol 9: 400.

Lehmkuhle S, Kratz KE, Mangel SC, Sherman SM (1978) Effects of early monocular lid suture on spatial and temporal sensitivity of neurones in dorsal lateral geniculate nucleus of the cat. J Neurophysiol 43: 542.

Levick WR (1972) Another tungsten microelectrode. Med Electron Biol 10: 510.

Maffei L, Fiorentini A (1976) Monocular deprivation in kittens impairs the spatial resolution of geniculate neurones. Nature 264: 754.

Mitchell DE, Cynader M, Movshon JA (1977) Recovery from the effects of monocular deprivation in kittens. J Comp Neurol 176: 53.

Mitchell DE, Giffin F, Timney B (1977). A behavioural technique for the rapid assessment of the visual capabilities of kittens. Perception 6: 181.

Mitchell DE, Giffin F, Wilkinson F, Anderson P, Smith ML (1976). Visual resolution in young kittens. Vision Res 16: 363.

Movshon JA (1976a) Reversal of the physiological effects of monocular deprivation in the kitten's visual cortex. J Physiol 261: 125.

Movshon JA (1976b) Reversal of the behavioural effects of monocular deprivation n the kitten. J Physiol 261: 175.

Movshon JA, Blakemore C (1974) Functional reinnervation in kitten visual cortex. Nature 251: 504.

Mower GD, Christen WG (1982) Effects of early monocular deprivation on the acuity of lateral geniculate neurons in the cat. Dev Brain Res 3: 475.

Peck CK, Blakemore C (1975) Modification of single neurons in the kitten's visual cortex after brief period of monocular visual experience. Exp.Brain Res 22: 57 .

Pettigrew JD, Garey LJ (1974) Selective modification of single neuron properties in the visual cortex of kittens. Brain Res 66: 160.

Ramachandran VS, Ary M (1981) Evidence for a "consolidation" effect during changes in the ocular dominance of cortical neurons in kittens. Neuroscience Abst 7: 220.

Rusoff AC, Dubin MW (1988). Development of receptive field properties of retinal ganglion cells in kittens. J Neurophysiol 40: 1188.

Sanderson KJ (1971) The projection of the visual field of the lateral geniculate and medial interlaminar nuclei in the cat. J Comp Neurol 143: 101 .

Shapley RM, So YT (1980) Is there an effect of monocular deprivation on the proportions of X and Y cells in the cat's lateral geniculate nucleus? Exp Brain Res 39: 41.

Shatz CJ, Lindstrom SH, Wiesel TN (1977) The distribution of afferents representing the right and left eyes in the cat's visual cortex. Brain Res 131: 103.

Shatz CJ, Stryker MP (1978) Ocular dominance in layer IV of the cat's visual cortex and the effects of monocular deprivation. J Physiol 282: 267.

Sherman SM, Stone J (1973). Physiological normality of the retina in visually deprived cats. Brain Res 60: 224.

Sireteanu R, Hoffman KP (1979) Relative frequency and visual resolution of X- and Y- cells in the LGN of normal and monocularly deprived cats: interlaminar differences. Exp.Brain Res 34: 591.

Smith EL, Bennett MJ, Harwerth RL, Crawford (1979). Binocularity in kittens reared with an optically induced squint. Science 204: 875.

Stone J (1981). The Wholemount Handbook. Maitland Publications (Sydney).

Stone J, Fukuda Y (1974). Properties of cat retinal

ganglion cells: a comparison of W- cells with X- and Y-cells. J Neurophysiol 37: 722.

Van Sluyters RC, Levitt FB (1980). Experimental strabismus in the kitten. J Neurophysiol 43: 686.

Tong L, Spear PD, Kalil RE, Callahan EC. Loss of retinal X-cells in cats with neonatal or adult visual cortical damage. Science 217: 72.

Wald G, Burian HM (1944). The dissociation of form vision and light perception in strabismic amblyopia. Am J Ophthal 27: 950.

Wiesel TN, Hubel DH (1963a) Single cell responses in striate cortex of kittens deprived of vision in one eye. J Neurophysiol 26: 1003.

Wiesel TN, Hubel DH (1963b) Effects of visual deprivation on morphology and physiology of cells in the cat's lateral geniculate body. J Neurophysiol 26: 978.

Development of Visual Pathways in Mammals, pages 381–392

THE EFFECTIVENESS OF REVERSE OCCLUSION AS A MEANS OF PROMOTING VISUAL RECOVERY IN MONOCULARLY DEPRIVED KITTENS.

Donald E. Mitchell and Kathryn M. Murphy

Psychology Department
Dalhousie University
Halifax, Nova Scotia B3H 4J1, Canada

Of a number of approaches to study of the role of experiential factors in visual development, possibly the most insight has been obtained from investigations of development in animals whose early visual input was biased in certain specific ways. One of the most widely employed strategies of this sort has been that of monocular deprivation where one eye is deprived of normal patterned visual input by eyelid closure (for recent reviews see Movshon, Van Sluyters 1981; Mitchell, Timney 1983). This form of early visual deprivation exerts widespread effects on the development of the central visual pathways of cats and monkeys and can so seriously impair the vision of the deprived eye as to render it effectively blind. The effects of early monocular deprivation on the visual cortex are particularly profound, such that subsequently virtually all cells outside of layer IV can only be excited through the nondeprived eye.

Severe as these initial deficits are, substantial behavioural and physiological recovery can occur if normal visual input is restored to the deprived eye sufficiently early. The extent of recovery is enhanced (particularly in monkeys) if, at the time that visual input is restored to the deprived eye, the nondeprived eye is occluded, a procedure referred to as reverse occlusion (Blakemore, Van Sluyters 1974; Mitchell, Cynader, Movshon 1977; Blakemore, Garey, Vital-Durand 1978). If this procedure is initiated on monocularly deprived kittens at 4 or 5 weeks of age, the ocular dominance distribution of cells in area 17 can be shifted almost completely towards the deprived eye in only

2 to 3 weeks (Movshon 1976). This rapid reversal of the physiological effects of monocular deprivation in the visual cortex is accompanied by equally fast recovery of vision in the deprived eye (Mitchell, Cynader, Movshon 1977; Giffin, Mitchell 1978). Because of its close resemblance to conventional patching therapy for human amblyopia, we have investigated the permanence of the visual recovery promoted by reverse occlusion in monocularly deprived kittens. Surprisingly, this recovery proved to be extremely labile. Upon restoring binocular visual input the vision of the deprived eye dropped rapidly, erasing much of the gains in vision made during the period of reverse occlusion.

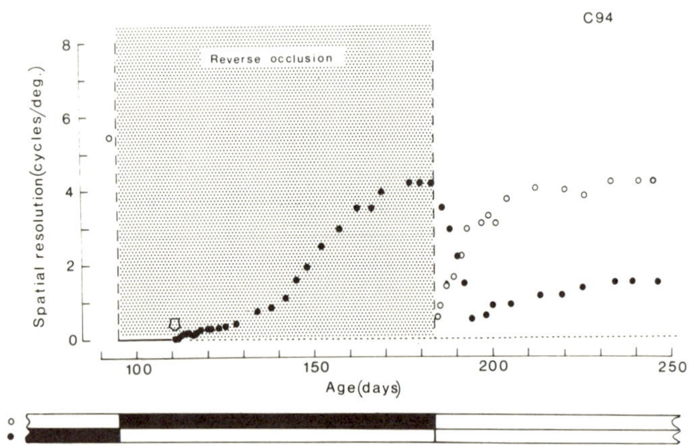

Fig. 1. Changes in the visual acuity of the deprived (filled symbols) and nondeprived eye (open symbols) of a monocularly deprived kitten during and following an 89 day period of reverse occlusion (shaded region) imposed at 94 days of age. Throughout the period of time indicated by the horizontal line the animal appeared blind with its deprived eye. The open arrow indicates the day on which the animal first showed signs of vision with this eye. The visual history of the animal is summarized below in schematic form.

This phenomenon was observed in kittens that were monocularly deprived by eyelid closure from near birth and reverse occluded at 4 to 19 weeks of age for various periods of time. On termination of reverse occlusion the lids of the nondeprived eye were parted to restore normal visual input to both eyes. Regular measurements of the visual acuity of the deprived eye were made on a jumping stand (Mitchell, Giffin, Timney 1977) during the period of reverse occlusion and of both eyes afterwards. During the latter period measurements of the acuities of the deprived and nondeprived eyes were made on alternate days with one eye occluded with an opaque contact lens occluder. Measurements of the contrast sensitivity functions (CSF) of the two eyes were made on some animals on conclusion of the measurements of visual acuity. A modified jumping stand in which the stimuli were presented on the face of a large oscilloscope monitor was employed for this purpose. Details of the testing procedure and the modified jumping stand are provided in other preliminary accounts of this work (Mitchell, Murphy, Kaye 1983; Mitchell, Murphy, Kaye 1984). In the descriptions that follow the two eyes are defined in terms of their status during the initial period of monocular deprivation.

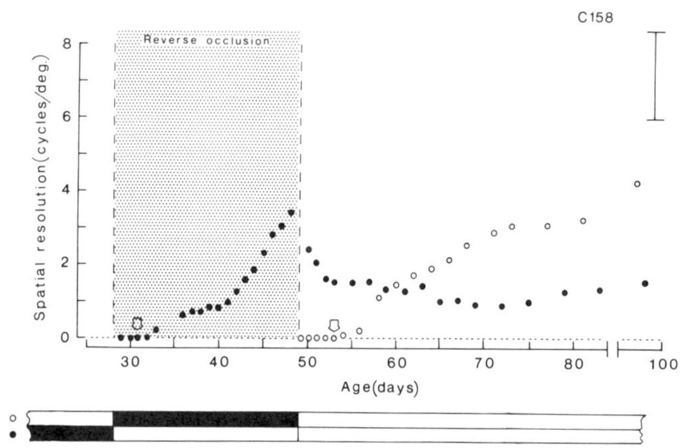

Fig. 2. Changes in the visual acuity of the deprived and nondeprived eye of a monocularly deprived kitten during and following a 21 day period of reverse occlusion imposed at 28 days of age. Symbols as in Fig. 1.

Findings from two representative animals are displayed in Figs. 1 and 2. The monocularly deprived animal of Fig. 1, C94, was reverse occluded at 94 days of age for 89 days. During the period of reverse occlusion the vision of the deprived eye improved gradually from apparent blindness to

4.2 cycles/degree. On termination of reverse occlusion, the vision of this eye dropped rapidly while at the same time the acuity of the nondeprived eye (that had been depressed by the period of reverse occlusion) improved. However, note that the acuity of the nondeprived eye never recovered to the level (5.45 cycles/degree) it had achieved prior to the period of reverse occlusion.

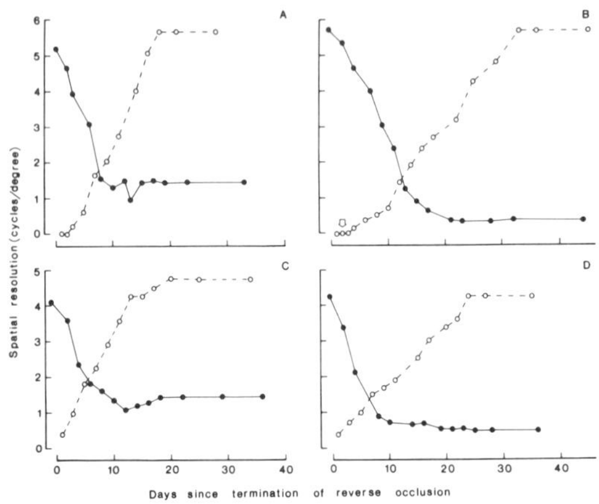

Fig. 3. Changes in the visual acuities of the deprived and nondeprived eyes of four monocularly deprived kittens following termination of a period of reverse occlusion and restoration of visual input to both eyes. The individual deprivation conditions were: A. Reverse occlusion for 38 days imposed at 35 days of age; B. Reverse occlusion for 44 days imposed at 42 days of age; C. Reverse occlusion for 91 days imposed at 103 days of age; D. Reverse occlusion for 102 days imposed at 101 days of age. Symbols as in Fig. 1.

The results shown in Fig. 2 are from an animal, C158, that was monocularly deprived to only 4 weeks of age and then reverse occluded for 3 weeks, a period of time that is sufficiently long for the physiological effects of the initial period of monocular deprivation in the visual cortex to be completely reversed (Movshon 1976). Although the vision of the deprived eye improved considerably to 3.4 cycles/degree during the short period of reverse occlusion, it dropped rapidly to 0.9 cycles/degree in only 3 weeks once both eyes were allowed normal visual input. As with C94, this drop in the visual acuity of the deprived eye was accompanied by a reciprocal improvement in the vision of the nondeprived eye. However, again the vision of this eye never recovered to normal levels (6.4 to 8 cycles/degree for the particular testing conditions employed here).

Similar reciprocal changes in the vision of the deprived and nondeprived eyes following termination of reverse occlusion have been observed under a wide variety of deprivation conditions. Four representative examples are displayed in Fig. 3, which for economy of space show only the results of acuity measurements made after termination of reverse occlusion. As with the animals of Figs. 1 and 2, the acuity of the nondeprived eye of each of these 4 animals did not quite recover to normal levels.

The changes in the vision of the two eyes that occur on termination of reverse occlusion were very surprising, particularly so for those animals that were reverse occluded early at 4 or 5 weeks of age for periods of time that were sufficiently long for the physiological effects of the initial period of monocular deprivation to be completely reversed in layers of the visual cortex other than layer IV (Movshon 1976). In terms of the distribution of ocular dominance of visual cortical cells, the state of affairs at the end of the period of reverse occlusion would be the mirror image of the situation at the beginning. Whereas all cells would be dominated by the nondeprived eye at the outset of the period of reverse occlusion, at the end cells would be dominated by the deprived eye. Consequently, it might be thought that the events that occur on restoring normal visual input to both eyes following reverse occlusion would be very similar to those that occur when binocular visual input is restored after monocular deprivation. However, this was certainly not the case. On restoring binocular visual input to a monocularly deprived kitten, the subsequent improvement in the vision of the deprived eye is not made at the expense of the vision of the nondeprived eye which continues to improve throughout the recovery period (Mitchell, Cynader, Movshon 1977; Giffin, Mitchell 1978). By contrast, following termination of a period of reverse occlusion, the improvement in the vision of the recently occluded nondeprived eye is made at the expense of the vision of the deprived eye as manifested by the reciprocal changes in acuity illustrated in Figs. 1, 2 and 3.

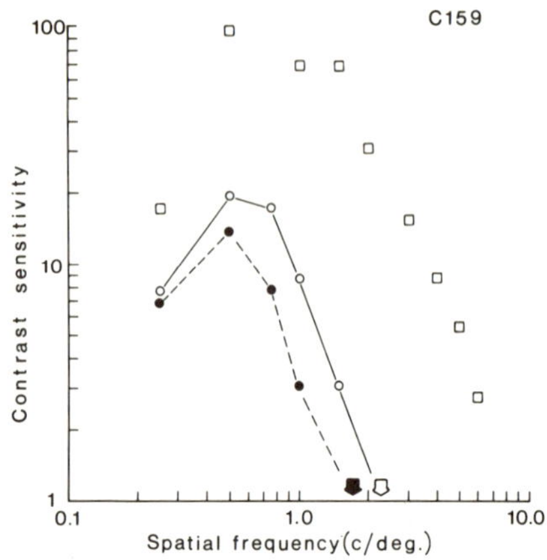

Fig. 4. Contrast sensitivity functions for the deprived (filled circles) and nondeprived (open circles) eye of a monocularly deprived kitten that had been subjected to a 9 day period of reverse occlusion at 5 weeks of age. The measurements were made between 2 and 3 months after termination of the period of reverse occlusion. The open square symbols depict the monocular contrast sensitivity function of a normal cat of comparable age. The open and filled arrows indicate the visual acuities measured earlier on the jumping stand using square-wave gratings of the same luminance.

In an effort to arrive at an understanding of the physiological mechanisms that underlie the behavioural events observed upon termination of reverse occlusion, we have begun (Murphy, Mitchell, in preparation) a systematic study of the visual changes that occur following various periods of reverse occlusion imposed on monocularly deprived kittens at either 4, 5 or 6 weeks of age. Preliminary findings from animals in this series that were reverse occluded for periods of time that were either very brief or very long were particularly surprising. Representative results from an animal of the former category that was reverse occluded for only 9 days at 5 weeks of age, are shown in Fig. 4. This figure shows contrast sensitivity functions of the two eyes measured 3 months after termination of the short period of reverse occlusion. Despite the fact that reverse occlusion was terminated when the animal was only 6.5 weeks of age, the contrast sensitivity functions for the two eyes, while not very different, were both extremely poor in comparison to that of a normal animal of comparable age (open square symbols). At the other extreme, it might be thought that following long periods of reverse occlusion that extend through much of the sensitive period (Mitchell 1981), the deprived eye would retain normal or nearly normal visual acuity. Surprisingly our preliminary findings suggest that such long periods of reverse occlusion do not always afford the degree of protection that might be expected. In Fig. 5 are shown results from a monocularly deprived kitten that was reverse occluded at 5 weeks for 12 weeks. Despite the fact that it was over 4 months old at the end of the period of reverse occlusion, the acuity of the deprived eye began to drop soon after normal visual input was restored to the other eye. Much of the decline in the vision of the deprived eye occurred before the animal exhibited any signs of vision through its other eye! It remains to be seen whether the visual gains made during a long period of reverse occlusion are retained if it is imposed even earlier, at either 3 or 4 weeks of age.

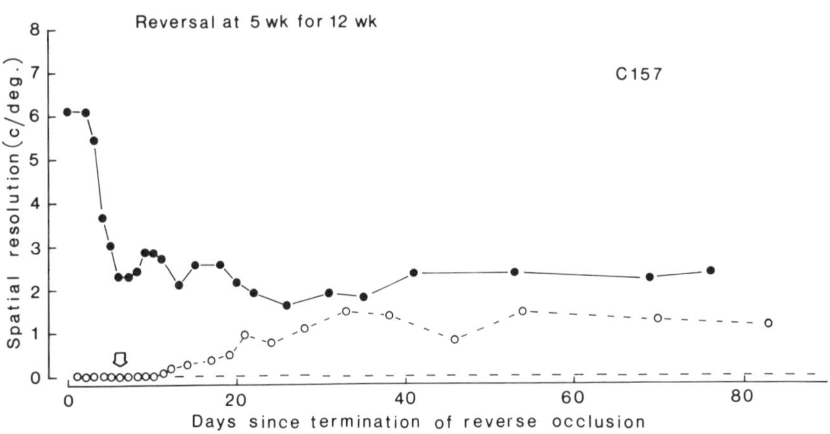

Fig. 5. Changes in the visual acuity of the deprived and nondeprived eye of a kitten following termination of 12 weeks of reverse occlusion imposed at 5 weeks of age. Symbols as in Fig. 1.

Mechanism(s) of the Visual Changes that follow Reverse Occlusion

One possible explanation for the drop in the acuity of the deprived eye that is observed soon after visual input is restored to the other eye is that it represents the onset of strabismic amblyopia in this eye due to a misalignment of the visual axes. While it is highly likely that the visual axes will exhibit a degree of misalignment that may be both large and variable as soon as binocular visual input is restored, the speed of the visual changes taken in conjunction with the age at which they can occur, make this explanation unlikely (Mitchell, Murphy, Kaye 1983). In addition, this argument would be thought to apply equally well when binocular visual input is restored following monocular deprivation since the visual axes would likely be just as misaligned in this situation.

The very different nature of the events that follow termination of reverse occlusion as compared to those subsequent to a period of monocular occlusion, suggest that the mechanism(s) that underlie the shifts of cortical ocular dominance during reverse occlusion may not be the same as those responsible for the shifts that occur during the initial period of monocular deprivation. In any event, the behavioural results imply that the shifts of cortical ocular dominance that are observed during reverse occlusion (Movshon 1976) may not be permanent, particularly so for those cells having high spatial resolution. Preliminary data obtained from the one animal recorded to date, a kitten that was reverse occluded at 6 weeks for about 6 weeks (Fig. 2B), suggest that this is in fact the case (Mitchell, Murphy, Kaye 1983). Only 37 (32.5%) of 114 cells recorded in both hemispheres of this animal were dominated by the deprived eye, a proportion that is much lower than that (at least 65%) reported by Movshon (1976) from two animals from which cortical recordings were made immediately on termination of either 24 or 63 days of reverse occlusion imposed at 6 weeks of age.

The speed of the visual changes that occur on termination of reverse occlusion in animals that are well over 4 months of age at the time (e.g. C94; Fig. 1) suggest that they do not reflect events that occur in layer IV of the visual cortex which appears to be immutable to the effects of monocular occlusion after 8 weeks of age (Shatz,

Stryker 1978). This raises the possibility that some of the changes in cortical ocular allegiance that occur in area 17 during reverse occlusion may be mediated by a process of tonic inhibition (Kratz, Spear, Smith 1976) whereby synaptic

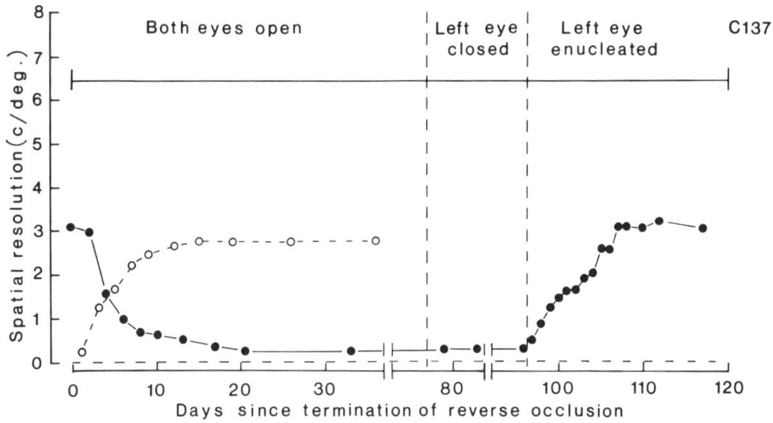

Fig. 6. Changes in the visual acuity of the deprived and nondeprived eye of a cat following termination of 58 days of reverse occlusion imposed at 135 days of age. After 76 days, when the vision of the two eyes had stabilized, the nondeprived eye was occluded once more for 20 days during which time the vision of the deprived eye did not improve. The nondeprived eye was then removed after which the vision of the other eye improved dramatically. The horizontal line indicates that acuity of the nondeprived eye prior to reverse occlusion.

connections from the nondeprived eye are rendered ineffective during reverse occlusion by tonic inhibition arising from activity in the pathway from the other (deprived) eye. Behavioural evidence for such a mechanism can be obtained from examination of the consequences for the vision of one eye of enucleating the other (e.g. Smith 1981; Kratz, Lehmkuhle 1983). Preliminary data from one animal on which this manipulation was performed suggest that mechanisms of this nature are involved during reverse occlusion. This animal (Fig. 6) was reverse occluded for 58 days at 135 days of age during which time the vision of the deprived eye improved to 3.05 cycles/degree. On termination of reverse occlusion there was a rapid and reciprocal change in the acuity of the two eyes such that one month later the acuity of the deprived eye had dropped to 0.25 cycles/degree while the nondeprived eye had improved to 2.75 cycles/degree, a value that fell far short of the level it had attained (6.4 cycles/degree) prior to reverse occlusion. After 76 days the nondeprived eye was occluded once more for 20 days during which time the deprived eye did not improve whatsoever. The left eye was enucleated at this point after which there was a rapid and substantial improvement in the vision of the deprived eye to 3.2 cycles/degree in only 2 weeks. The speed and extent of the changes in the vision of the deprived eye after removing the nondeprived eye invite study of the opposite situation. In particular, would enucleation of the deprived eye several weeks after termination of reverse occlusion result in improvement of the vision of the nondeprived eye to normal levels?

Clinical Implications

The procedure of reverse occlusion that was employed here on kittens is analogous to conventional patching therapy for treatment of amblyopia in humans whereby the good eye is covered with a patch for several hours each day in order to force the child to employ its amblyopic eye. It is therefore of interest that similar reciprocal changes in the visual acuity of the two eyes to those reported here have been reported in young human infants following termination of patching therapy for severe amblyopia (Jacobson, Mohindra, Held 1981; Odom, Hoyt, Marg 1981; Marg 1982). Discovery of a similar phenomenon in kittens allows investigation of the underlying anatomical and physiological mechanisms which may eventually lead to the development of treatment regimens that promote permanent improvement of the vision of the amblyopic eye without compromising the vision of the other eye.

REFERENCES

Blakemore C, Van Sluters RC (1974). Reversal of the physiological effects of monocular deprivation in kittens: further evidence for a sensitive period. J Physiol Lond 237:195.

Blakemore C, Garey LJ, Vital-Durand F (1978). The physiological effects of monocular deprivation and their

reversal in the monkey's visual cortex. J Physiol Lond 283:223.
Giffin F, Mitchell DE (1978). The rate of recovery of vision after early monocular deprivation in kittens. J Physiol Lond 274:511.
Jacobson SG, Mohindra I, Held R (1981). Development of visual acuity in infants with congenital cataracts. Brit J Ophthalmol 65:727.
Kratz KE, Lehmkuhle S (1983). Spatial contrast sensitivity of monocularly deprived cats after removal of the non-deprived eye. Behav Brain Res 7:261.
Kratz KE, Spear PD, Smith DC (1976). Postcritical-period reversal of effects of monocular deprivation on striate cortex cells in the cat. J Neurophysiol 39:501.
Marg E (1982). Is the animal model for stimulus deprivation amblyopia in children valid or useful? Am J Optom Physiol Opt 59:451.
Mitchell DE (1981). Sensitive periods in visual development. In Aslin RN, Alberts J, Petersen M (eds): "The Development of Perception, Vol II," New York: Academic Press, p 3.
Mitchell DE, Cynader M, Movshon JA (1977). Recovery from the effects of monocular deprivation in kittens. J Comp Neurol 176:53.
Mitchell DE, Giffin F, Timney B (1977). A behavioural technique for the rapid assessment of the visual capabilities of kittens. Perception 6:181.
Mitchell DE, Murphy KM, Kaye MG (1983). Labile nature of the visual recovery promoted by reverse occlusion in monocularly deprived kittens. Proc Nat Acad Sci U.S.A. (in press).
Mitchell DE, Murphy KM, Kaye MG (1984). The permanence of the visual recovery that follows reverse occlusion of monocularly deprived kittens. Invest Ophthalmol Vis Sci (submitted).
Mitchell DE, Timney B (1983). Postnatal development of function in the mammalian visual system. In Darian-Smith I, Mountcastle VB (eds): "Handbook of Physiology Section I: The Nervous System, Vol 3. Sensory Processes," Bethesda: American Physiological Society (in press).
Movshon JA (1976). Reversal of the physiological effects of monocular deprivation in the kitten's visual cortex. J Physiol Lond 261:125.
Movshon JA, Van Sluyters RC (1981). Visual neuronal development. Ann Rev Psychol 32:477.
Odom JV, Hoyt CS, Marg E (1981). Effect of natural deprivation and unilateral eye patching on visual acuity

of infants and children. Arch Ophthalmol 99:1412.

Shatz CJ, Stryker MP (1978). Ocular dominance in layer IV of the cat's visual cortex and the effects of monocular deprivation. J Physiol Lond 281:267.

Smith DC (1981). Functional restoration of vision in the cat after long-term monocular deprivation. Science 213:1137.

Development of Visual Pathways in Mammals, pages 393–404

RECOVERY FROM THE CORTICAL EFFECTS OF VISUAL DEPRIVATION IN THE KITTEN

Richard C. Van Sluyters and Rafael Malach

School of Optometry
University of California
Berkeley, CA 94720

The widespread change in functional connectivity of the striate cortex that occurs when normal binocular vision is disrupted in a young kitten is one of the most impressive and well-documented examples of central nervous system plasticity in higher mammals. This ability to modify binocular interactions in primary visual cortex is also found in other predatory animals with forward-facing eyes, and may stem from their need to compensate for the subtle optical and positional changes that take place during normal development of the eyes in order to sustain highly precise cortical processing of binocular visual signals.

Soon after the initial demonstration of a massive shift in cortical ocular dominance was reported (Hubel, Wiesel 1963), interest turned to the question of the permanence of these changes in cortical function. If cortical plasticity is a positive adaptive feature of the developing visual system, then it would seem likely that the mechanism underlying this process should be flexible enough to be capable of reversing the changes in cortical function that occur as a result of a temporary alteration of visual input. Wiesel and Hubel (1965; Hubel, Wiesel 1970) were the first to address this issue when they subjected 10-day-old kittens to periods of monocular deprivation (MD) ranging from 4 wks to 4 mos, and then opened the deprived eye and allowed an extensive period of binocular vision. While the deprived eye did regain some functional cortical connections in these kittens, there was almost no improvement in the proportion of striate cortex cells with functional binocular input. Since this initial work, there have been several studies in which variations of this rearing strategy have been employed to investigate recovery from the cortical effects of MD (e.g., Mitchell, Cynader, Movshon 1977; Blasdel, Pettigrew 1978; Olson, Freeman 1978; Freeman, Olson 1982), but none has been successful in demonstrating complete reversal of the loss of binocularity caused by MD. The results of these studies led to the speculation that once cortical binocularity was lost, it never could be fully recovered.

However, if cortical plasticity really is designed to cope with the subtle environmental demands brought about by growth of the head and eyes, then the dramatic shifts in cortical ocular dominance that researchers produce in the laboratory by disrupting binocular vision completely for prolonged periods of time, actually must represent the outcome of pushing this process to its limit. Such massive disruptions, and the gross changes in cortical function they lead to, would seem to occur rarely, if ever, in the wild. In this sense, it might be unreasonable to expect the visual system to be able to recover fully from the effects of a prolonged period of abnormal visual experience. Indeed, evidence that the potential for recovery of binocularity is greater following shorter periods of deprivation was obtained in some of the earlier studies mentioned above.

In this chapter, we discuss the outcome of two sets of experiments in which we assessed the potential for recovery of cortical binocularity in kittens exposed to strabismus or monocular deprivation. In both these studies we employed a very brief period of deprivation -- in the range of those that could possibly be encountered in the wild -- to disrupt binocularity, and found that the cortex can recover fully from the effects of such a limited period of abnormal visual experience.

RECOVERY FROM STRABISMUS

Our first study of the potential for recovery of binocularity in striate cortex was carefully designed to maximize the chances for full recovery (Levitt, Van Sluyters 1982b). Thus, we chose to use strabismus, rather than MD, to break down binocularity, since the loss of binocular cells in strabismic kittens is not accompanied by the marked asymmetry in ocular dominance that is found in MD kittens. It seemed to us that the potential for binocularity to reappear would be higher in animals whose cortex contained two populations of monocular neurons than in animals in which one eye was functionally disconnected from the cortex. Since the traditional method for inducing ocular misalignment, extra-ocular muscle tenotomy, is not a reversible procedure, we used Fresnel prisms mounted in lightweight goggles to induce a condition very much like concomitant strabismus in man (Van Sluyters 1977). In this way we could study recovery from strabismus simply by removing the prism goggles.

We also attempted to expose our kittens to the briefest episode of strabismic vision that would cause a marked loss of cortical binocularity. Based on our earlier study of the sensitive period for strabismus (Levitt, Van Sluyters 1982a), we knew that exposing normally reared, 4-wk-old kittens to 2 days of ocular misalignment was sufficient to produce a loss of binocularity as severe as that found in kittens exposed to prolonged periods of strabismus. In order to minimize sampling problems due to between-animal variability, we decided to record from each kitten twice -- once immediately after exposure to strabismus, and a second time following a period of recovery.

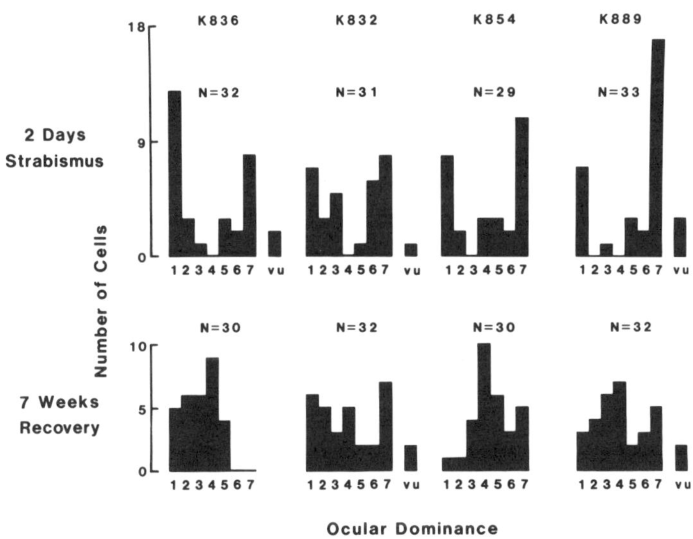

Fig. 1. Recovery of binocularity in kittens exposed to 2 days of strabismus followed by 7 wks of binocular vision. Top row of ocular dominance histograms obtained prior to

recovery, and bottom row after. Histograms directly above and below one another are from successive recording sessions in the same kitten (redrawn from Levitt, Van Sluyters 1982b).

The top row of ocular dominance histograms in Fig. 1 shows the loss of binocular cells caused by exposing four normally reared, 4-wk-old kittens to a total of 12 hrs of strabismic vision, given in 2-hr sessions spread evenly over a period of 2 days. Between exposure sessions, these kittens were kept in total darkness with their mothers and on the morning of the third day they were prepared for single-unit recording using techniques and procedures described in detail elsewhere (Levitt, Van Sluyters 1982b). The proportion of binocular cells in these kittens was severely reduced (only about 30%) as a result of this exposure regimen. However, as indicated by the bottom row of histograms in Fig. 1, when these kittens were allowed to recover from the initial recording session, exposed to a period of normal binocular vision lasting 7 wks and recorded from once again, cortical binocularity recovered to an average level (about 80%) that was no different from that found in a similar sample of cells from six normally reared kittens (81% binocular).

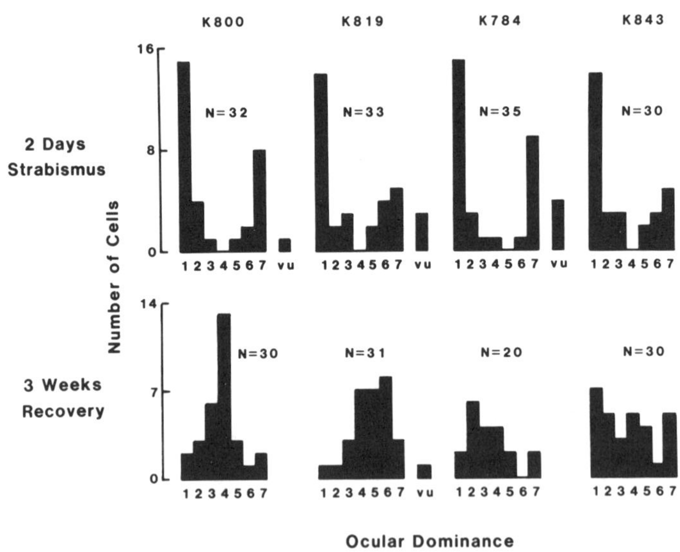

Fig. 2. Recovery of binocularity in kittens exposed to 2 days of strabismus followed by 3 wks of normal binocular vision

(redrawn from Levitt, Van Sluyters 1982b) . Same conventions as Fig. 1.

Fig. 2 presents data obtained from four additional kittens that were exposed to a similar 2-day period of optical strabismus and then allowed to recover for a period lasting only 3 wks. Once again, there was no evidence of the marked loss of binocularity caused by the initial period of stabismus (top row of histograms) when these kittens were studied a second time only 3 wks later (bottom row of histograms).

This first study demonstrated that cortical binocularity can be recovered fully in kittens in which it has been reduced significantly by exposure to a brief episode of strabismus at the peak of the sensitive period. Our next study was designed to determine whether this finding in strabismic kittens could be extended to include kittens exposed to a brief period of MD.

RECOVERY FROM MONOCULAR DEPRIVATION

As pointed out earlier, the results of previous studies of recovery of binocularity in MD kittens suggest that full recovery is not possible. If that were the case, it would indicate that the cortical effects of strabismus and MD differ in a very fundamental way. On the other hand, none of these previous studies used a period of deprivation as brief as 2 days and, as mentioned earlier, there was a tendency for recovery to be greater in those studies in which the period of MD was briefest. Thus, we began our second experiment (Malach, Ebert, Van Sluyters 1983) by examining recovery of binocularity in MD kittens exposed to periods of deprivation and binocular vision identical to those used in our strabismus experiment.

The top row of ocular dominance histograms in Fig. 3 shows the loss of binocular cells and the shift in ocular dominance caused by exposing six normally reared, 4-wk-old kittens to a total of 12 hrs of MD, given in 2-hr sessions spread evenly over a period of 2 days. As indicated by these receptive field data obtained from the hemisphere ipsilateral to the deprived eye, the proportion of binocular striate cortex cells in these kittens was severely reduced (to about

23%) as a result of this exposure regimen. These kittens were allowed to recover from the initial recording session, and were exposed to a period of normal binocular vision lasting 3 wks. When they were recorded from a second time, the level of cortical binocularity had recovered considerably (67%), but was still somewhat lower than that in normally reared kittens.

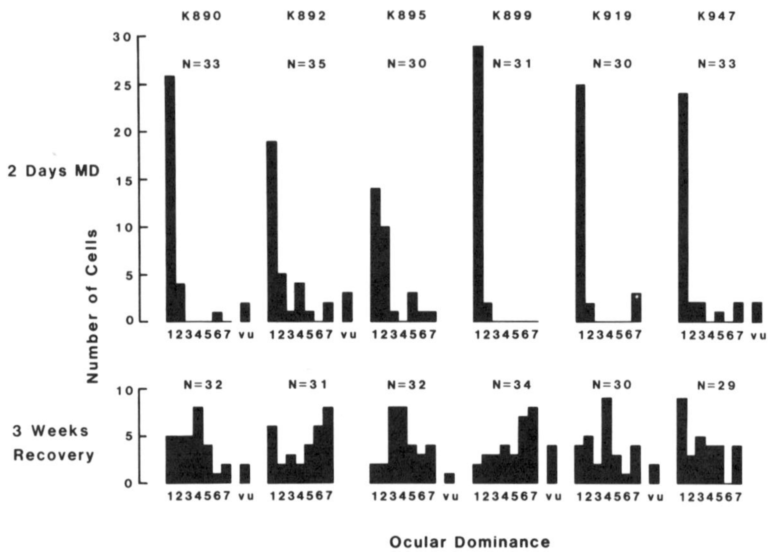

Fig. 3. Recovery of cortical binocularity in kittens exposed to 2 days of MD followed by 3 wks of normal binocular vision (redrawn from Malach, Ebert, Van Sluyters 1983). Same conventions as Fig. 1.

Four of these kittens, and a fifth kitten that was not recorded from immediately after the initial 2-day period of MD (K898), were recorded from once again after experiencing a further 4-wk period of normal binocular vision. Fig. 4 presents data obtained from these five kittens at all three stages of this experiment. Inspection of the individual histograms in this figure suggests the level of binocularity improved slightly during the second recovery period, and averaging the data indicated that it had increased from 67% to 71%, a level not significantly different from that found in normally reared kittens.

In the final part of this study we examined the speed with which binocularity recovered in kittens exposed to 2 days of MD. Would 2 days of binocular vision substantially reverse the effects of 2 days of MD? This question was answered by exposing MD kittens to a 2-day recovery period designed to match the initial deprivation regimen exactly (three 2-hr exposure sessions each day). These kittens were recorded from only one time, immediately after the 2-day recovery period, since we felt it was unwise to subject them to the rigors of two recording sessions separated by only 2 days.

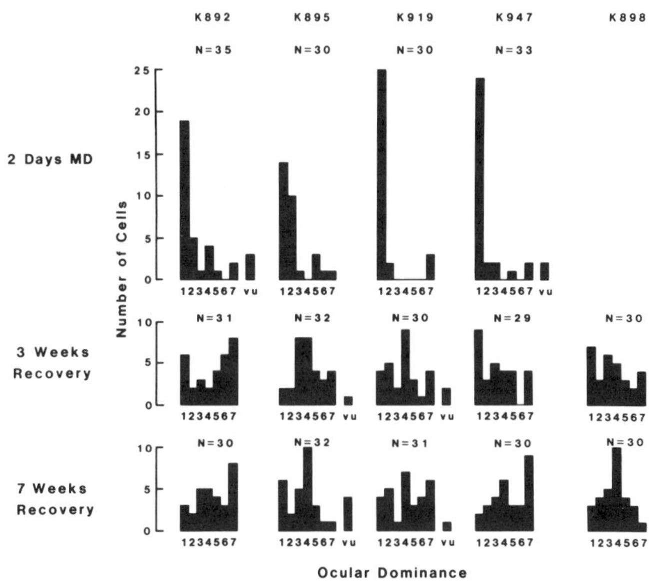

Fig. 4. Recovery of cortical binocularity in kittens exposed to 2 days of MD followed by periods of normal binocular vision lasting a total of 3 wks (middle row of histograms) or 7 wks (bottom row; redrawn from Malach, Ebert, Van Sluyters 1983). Same conventions as Fig. 1.

The top row of ocular dominance histograms in Fig. 5 reveals that 2 days of binocular vision were sufficient to promote a substantial recovery of functional input from the deprived eye. A total of 55% of cortical cells are now binocular, compared to only 23% in kittens recorded immediately after the 2-day period of MD (see Fig. 3). The bottom row of histograms in this figure gives the data from four additional kittens that were allowed only 1-day of binocular vision (three 2-hr sessions) following the 2-day period of MD. These data indicate that the expansion of cortical input from the deprived eye during the earliest stage of recovery occurs exclusively as a result of an increase in the level of cortical binocularity (from 23% to 43%).

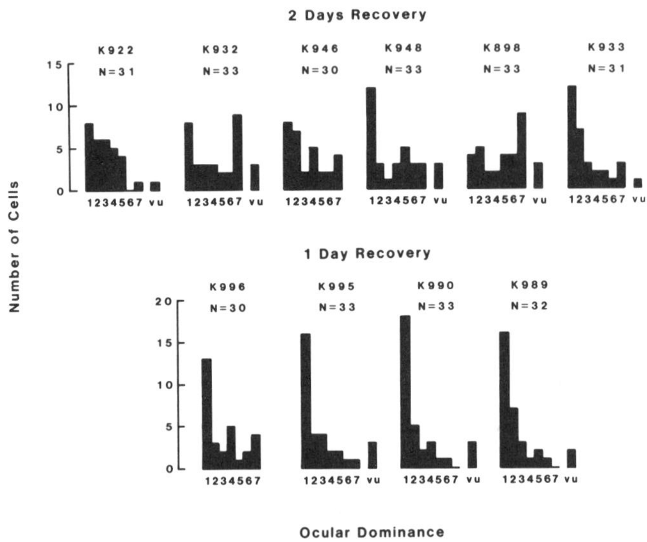

Fig. 5. Recovery of cortical binocularity in kittens exposed to 2 days of MD followed by 2 days (top row of histograms) or 1 day (bottom row) of binocular vision (redrawn from Malach, Ebert, Van Sluyters, 1983).

AN OVERVIEW OF RECOVERY FROM STRABISMUS AND MONOCULAR DEPRIVATION.

Fig. 6 summarizes the time course of recovery from MD and strabismus, and thus allows direct comparison of these two phenomena. Despite slight differences, the similarities that can be observed in both the initial breakdown of binocularity and the time course of its recovery are quite remarkable. These similarities provide further support for the notion that the cortical effects of MD and strabismus are manifestations of the same developmental process.

When our data are considered in the context of previous experiments, one clear theme emerges: the change in cortical binocular processing caused by exposure to a brief period of deprivation differs significantly from that caused by prolonged deprivation. The effects of brief deprivation can be recovered from completely, while prolonging an episode of deprivation leads to additional changes that can never be recovered from fully.

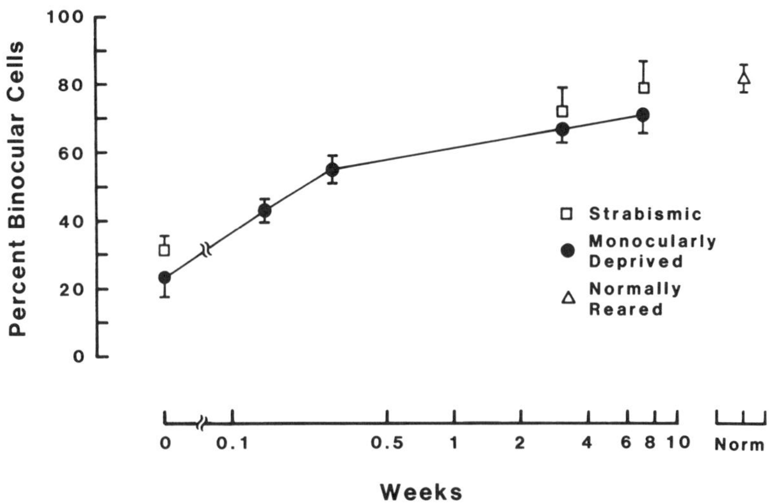

Fig. 6. Change in cortical binocularity as a function of the duration of the recovery period in MD and strabismic kittens (redrawn from Levitt, Van Sluyters 1982b; Malach, Ebert, Van Sluyters 1983).

Recent studies of cortical connections in MD animals suggest a plausible morphological basis for the differences between the effects of brief versus prolonged deprivation. It now appears that subliminal cortical input from the deprived eye is abundant following short-term but not long-term MD (e.g., Kratz, Spear, Smith 1976; Van Sluyters 1978; Blakemore, Hawken, Mark 1982). In addition, studies of thalamo-cortical projections in deprived animals suggest that the period of deprivation required to produce gross morphological changes in axonal arborizations is much longer than that required to modify physiological responses (Shatz, Lindstrom, Wiesel 1977; Shatz, Stryker 1978). On the basis of these findings, it is tempting to speculate that the initial stage of deprivation involves a change only in synaptic efficacy, leaving the overall distribution of cortical connections undisturbed. This rapid change in synaptic gain can be adjusted in either direction depending on the kitten's subsequent visual experience. Prolonging the period of deprivation triggers a slower second stage of the process that involves actual dislocation of synapses and the retraction of axonal arbors. This latter stage is somewhat reversible, in the sense that opening the deprived eye can still lead to functional reinnervation, but now binocularity is permanently disrupted since the expansion of cortical input from the deprived eye is always accompanied by the retraction of that from the other eye (Hubel, Wiesel 1970; Blakemore, Van Sluyters 1974). With regard to the initial change in synaptic efficacy, it is interesting to note recent proposals for a specific mechanism, involving rapid modulation of the electrical properties of dendritic spines, that could perform just such a task (Crick 1982; Koch, Poggio 1983). In light of this hypothesis, it would be extremely valuable to compare the morphology of dendritic spines receiving synaptic input from a briefly deprived eye with that of spines connected to the non-deprived eye.

Although the actual neural process underlying recovery from the changes in cortical function caused by brief deprivation remains to be described, our results and those of others reveal several of its characteristics. The rapid re-establishment of binocularity that occurs immediately following a brief period of deprivation indicates there is little or no competitive interaction between inputs from the two eyes even during the earliest stage of recovery. In addition, functional connections from the deprived eye reappear and spread quite rapidly, despite the fact that it is at an overwhelming disadvantage when it comes to activating cortical cells. These findings rule out the class of "Hebbian" models of cortical plasticity that regards the ability to drive cortical cells as a prerequisite for inducing synaptic changes. Instead, it appears that, as long as the two eyes receive equal visual stimulation, the degree to which functional connections from the deprived eye reappear depends to a large extent on the overall level of activity in these connections.

In conclusion, our results suggest that the changes in cortical function caused by visual deprivation stem from a process that passes through at least two distinct stages, and that as long as synaptic connections remain within the first stage of this process, complete recovery from the effects of deprivation is possible. Whether a corresponding two-stage process is present in the human visual system, and how it is reflected in the potential for the neonate to recover fully from the effects of abnormal visual experience, remains to be seen.

ACKNOWLEDGEMENTS

FB Levitt deserves credit for much of the data from strabismic kittens, and R Ebert collaborated on the MD experiments. We are indebted to J Fiorillo, P Lee and C Morgan for their excellent technical assistance. This research was supported by grants from the Alfred P. Sloan Foundation, the National Eye Institute (EY02193) and the National Science Foundation (BNS8200083) to RCVS, and a Core Facilities Grant from the National Eye Institute (EY03176).

Present address of R Malach: Dept of Psychology, Bldg E25-618, MIT, Cambridge, MA 02139

REFERENCES

Blakemore C, Hawken MJ, Mark RF (1982). Brief monocular deprivation leaves subthreshold synaptic input on neurones of the cat's striate cortex. J Physiol 327:489.

Blasdel GG, Pettigrew JD (1978). Effect of prior visual experience on cortical recovery from the effects of unilateral eyelid suture in kitten. J Physiol 272:601.

Blakemore C, Van Sluyters RC (1974). Reversal of the physiological effects of monocular deprivation in kittens: further evidence for a sensitive period. J Physiol 237:195.

Crick F (1982). Do dendritic spines twitch? Trends NeuroSci 5:44.

Freeman RD, Olson CJ (1982). Brief periods of monocular deprivation in kittens: effects of delay prior to physiological study. J Neurophysiol 47:139.

Hubel DH, Wiesel TN (1970). The period of susceptibility to the physiological effects of unilateral eye closure in kittens. J Physiol 206:419.

Koch C, Poggio T (1983). Electrical properties of dendritic spines. Trends NeuroSci 6:80.

Kratz KE, Spear PD, Smith DC (1976). Postcritical-period reversal of effects of monocular deprivation on striate cortex cells in the cat. J Neurophysiol 39:501.

Levitt FB, Van Sluyters RC (1982a). The sensitive period for strabismus in the kitten. Devel Brain Res 3:323.

Levitt FB, Van Sluyters RC (1982b). Recovery of binocular function in kitten visual cortex. J Neurophysiol 48:6

Malach R, Ebert R, Van Sluyters RC (1983). Recovery from the effects of brief monocular deprivation in the kitten. J Neurophysiol (submitted for publication).

Mitchell DE, Cynader M, Movshon JA (1977). Recovery from the effects of monocular deprivation in kittens. J Comp Neurol 176:53.

Olson CR, Freeman RD (1978). Monocular deprivation and recovery during sensitive period in kittens. J Neurophysiol 41:65.

Shatz CJ, Lindstrom S, Wiesel TN (1977). The distribution of afferents representing the right and left eyes in the cat's visual cortex. Brain Res 131:103

Shatz CJ, Stryker MP (1978). Ocular dominance in Layer IV of the cat's visual cortex and the effects of monocular deprivation. J Physiol 281:267.

Van Sluyters RC (1977). Artificial strabismus in the kitten. Invest Ophthalmol Visual Sci, ARVO Abst Suppl 16:40.

Van Sluyters, RC (1978). Reversal of the physiological effects of brief periods of monocular deprivation in the kitten. J Physiol 284:1.

Wiesel TN, Hubel DH (1963). Single-cell responses in striate cortex of kittens deprived of vision in one eye. J Neurophysiol 26:1003.

Wiesel TN, Hubel DH (1965). Extent of recovery from the effects of visual deprivation in kittens. J Neurophysiol 28:1060.

Development of Visual Pathways in Mammals, pages 405–423

MONOCULAR DEPRIVATION AND BINOCULAR DEPTH PERCEPTION IN KITTENS

Brian Timney

Department of Psychology
University of Western Ontario
London, Ontario, Canada N6A 5C2

It has been known at least since the work of Wheatstone (1838) that the disparate images falling on the two retinae were sufficient to provide an observer with a vivid sensation of depth. However, it was not until 1967 that a possible physiological mechanism for stereopsis was discovered in the cat (Barlow, Blakemore, Pettigrew, 1967). Barlow et al. reported that the optimum stimulus for the majority of cells which they studied was one which fell on slightly disparate positions on the two retinae. This would correspond to a target lying in front of or behind the fixation plane. Since the initial report other investigators have confirmed and extended Barlow et al's. original findings (e.g. Nelson, Kato, Bishop, 1977; Nikara, Bishop, Pettigrew, 1968; Poggio, Fischer, 1977).

Although there is now quite a lot of information concerning disparity-tuned neurones in the cat, there is a paucity of data on their origin. This is somewhat surprising when one considers the vast amount of effort which has gone into the study of other aspects of visual development. Several authors (e.g. Blakemore, 1979; Pettigrew, this symposium) have pointed out that the progressive changes in size and separation of the eyes which occur as a result of an animal's growth, would necessitate a constant recalibration of any mechanism designed to obtain accurate distance information on the basis of retinal disparity. It would seem likely therefore, that the disparity-sensitive system must rely heavily upon normal visual experience in order to achieve its optimal level of adult functioning.

There is some physiological evidence to suggest that this is the case. Pettigrew (1974) has described the development of disparity-sensitive neurones in kittens raised with normal visual experience or ones which had been binocularly deprived by lid suture. Prior to the fourth week of life there were few differences between the normal and deprived kitten cortex. In both, neurones were broadly tuned for disparity, showing binocular facilitation over a wide range. During and after the fourth week, however, there was a progressive sharpening of the disparity tuning curves for the normal kittens and a comparable decline in responsiveness for the deprived animals. Figure 1 shows the binocular response, plotted as a function of disparity, for the most selective neurones encountered in normal kittens of different ages. It may be seen that the most

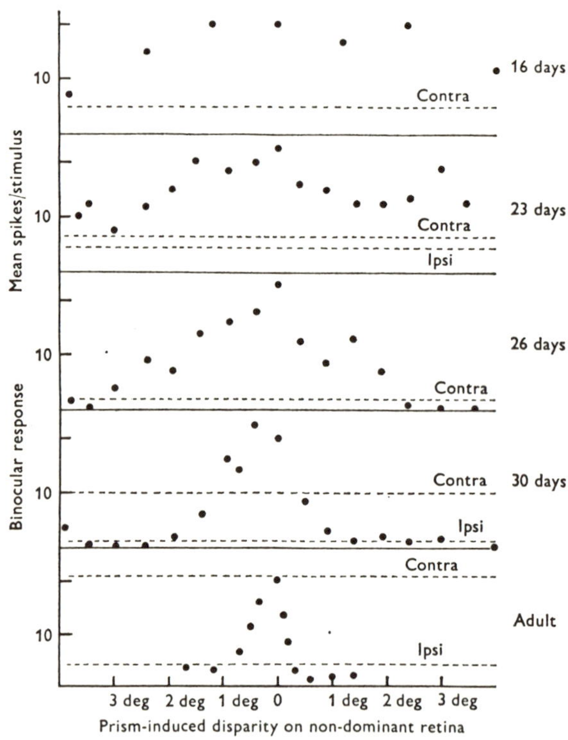

Fig. 1. Development of disparity-selectivity in the most selective cortical neurones of normally-reared kittens. Single unit responses are plotted as a function of the disparity of the optimum stimulus for a given neurone (from Pettigrew, 1974).

selective neurones are acquiring adult-like characteristics during the fifth week. At the same age, binocularly deprived kittens possessed no disparity-specific neurones.

Pettigrew's (1974) data suggest that normal visual experience in the first few weeks is necessary for the development of disparity-selectivity. However, the precise role of visual experience and its age-dependence remain unexplored. There are, of course, many studies which have examined the breakdown of binocularity as a consequence of deprivation (see Mitchell, Timney, in press; Movshon, Van Sluyters, 1981 for reviews). But these do not address the present issue directly, since it is quite possible for cells to retain their binocularity while losing their disparity-selectivity. On the other hand, a breakdown of binocularity necessarily implies a loss of disparity-sensitivity.

An example of the lack of correspondence between cortical binocularity and the potential use of retinal disparity depth cues may be found in the work of Mitchell and his colleagues. Kaye, Mitchell and Cynader (1982) reported that dark-reared kittens, which had a relatively high proportion of binocular neurones, nevertheless fared poorly in a binocular depth discrimination task. These results suggest that the kittens were unable to use retinal disparity cues, presumably because the binocular neurones were not disparity selective. The data of Pettigrew (1974) are certainly consistent with such an interpretation.

In the present chapter I shall describe a series of behavioural studies designed to examine the role of anomalous visual experience on the development of binocular depth perception in kittens. In these studies, kittens were monocularly deprived for varying periods, beginning at different ages and then later, some time after their eyelids were parted, the kittens' ability to make binocular and monocular depth discriminations was measured. Before presenting the results of these experiments I shall describe the procedure employed to obtain the data and the rationale underlying its use.

MEASUREMENT OF DEPTH THRESHOLDS

All of the data were gathered using a much modified

version of the jumping stand technique first described by Lashley (1930). The apparatus is shown in Figure 2 and has been described in detail elsewhere (Mitchell, Kaye, Timney, 1979). Here, I shall give just a brief overview. The main component of the apparatus is a box containing a large parallel balance. The 'pans' of the balance are large square pieces of Plexiglass on which are placed translucent acetate sheets covered with a random array of dots of varying sizes. By moving a handle, either the left or right 'pan' may be brought up flush against a sheet of transparent Plexiglass which covers the whole top surface of the box. The vertical separation between the two dotted surfaces may be changed by moving the whole balance unit up or down. To minimize the number of monocular depth cues which are available when the box is viewed from above, an opaque mask may be placed on the transparent Plexiglass

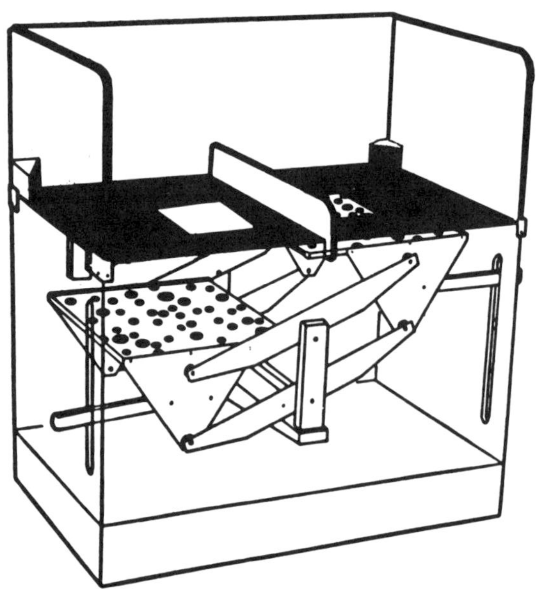

Fig. 2. Jumping stand used to measure monocular and binocular depth thresholds. The kittens are trained to jump from a raised platform towards the side of the box where the dotted target surface is closer. See text for details. (adapted from Timney, 1981)

such that only a small central region of the dots is visible. The kitten views these targets from a platform, whose height may be varied from almost flush with the landing surface up to about 80 or 90 cm.

Depth discrimination is measured by having the kittens jump from the platform toward the side of the landing box with the closer dotted surface. Initially, this is done with the opaque mask removed in order to provide the maximum number of depth cues. Most kittens will perform this task with little or no training; that is, even from the first trial they will tend to jump towards the closer surface. The kittens are rewarded with a small amount of commercial baby food for making a correct decision. Errors are signalled by a loud, high-frequency tone.

Once the discrimination is well-established, the opaque mask is replaced to minimize the number of monocular cues and threshold determinations are made. This is done by progressively reducing the separation between the dotted target surfaces until the kitten can no longer make the discrimination. Typically, an estimate of threshold is made first using a psychophysical staircase procedure, then a more accurate estimate is made using the method of constant stimuli. For the constant stimuli procedure five target separations are chosen to bracket the initial threshold estimate. The kitten is then run in blocks of five trials at each of these separations, which are presented in a random order. Within a session, a total of 15 or 20 trials are generally given at each separation and several sessions are run in order to obtain 50 or more trials per datum point. From the percentage of correct jumps a psychometric function may be derived and threshold is interpolated at the 70% correct level.

All animals are run under two viewing conditions: binocular, and monocular, when an opaque occluder is placed over one eye. Normally, the deprived eye is occluded. This is particularly important for cats who may have reduced acuity in the deprived eye as a consequence of a prolonged deprivation period. The principle underlying the procedure is the same as that for the Howard-Dolman apparatus (Howard, 1919). It is assumed that under both viewing conditions there are monocular depth cues present. However, when both eyes are open, additional, uniquely binocular cues become available. Of these, the most important is retinal disparity. The use of disparity information permits very fine depth judgements to be made, in contrast to most of the monocular cues which allow only quite coarse discriminations. Thus, if the kitten does much better when tested with both eyes open, the presence of stereopsis may be inferred.

It is possible to argue that superior binocular performance may result from some form of binocular summation, or the fact that the cat may be distracted by the presence of the occluder in the monocular condition. However, there are several lines of evidence to argue against this. The strongest counterargument against a simple binocular summation explanation is the finding that binocular depth performance may decline to the level of monocular, even though there is no reason to expect a visual deficit in either eye (e.g. Blake, Hirsch, 1975; Kaye et al., 1982). The failure of an animal to show a strong binocular superiority may be taken as evidence for a breakdown of stereopsis.

DEPTH THRESHOLDS IN NORMAL CATS

When normally-reared cats are tested on the jumping stand, typically they do very well binocularly, but performance declines radically when they are tested with one eye covered. In fact many cats, after being trained with both eyes open, act as if they no longer understand what is required of them when they are first tested monocularly. Their initial few jumps are made hesitantly and they are often clumsy. Often, they are unable to discriminate even the largest target separation of 20 cm. It should be noted though, that some cats do very well monocularly, although monocular thresholds are never as good as binocular.

Figure 3 shows a summary of thresholds based upon data from 10 normal cats. Threshold is expressed in angular units of retinal disparity, which is appropriate and convenient for binocular viewing, since it permits comparisons between cats with different interocular separations jumping from different heights. Monocular thresholds have also been expressed in units of retinal disparity, which obviously can have no functional significance, but the calculations were made in terms of the disparity that would have been present had both eyes been open. This provides a convenient comparison unit for the binocular results.

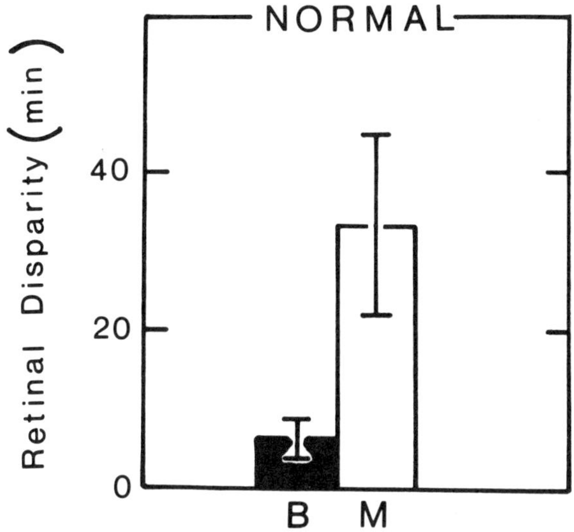

Fig. 3. Mean binocular and monocular depth thresholds of 10 normally-reared cats. Vertical bars indicate ±1 standard error of the mean.

The average binocular threshold is 6.8 min/arc disparity, although the best cats are able to discriminate separations of less than 4 min. This is equivalent to a separation of about 2 or 3 cm from a viewing distance of 70 cm. The monocular thresholds are much poorer than binocular, an average of 32.2 min/arc., but the reader should note the large between-animal variability. Some cats did very well monocularly, with thresholds only two or three times worse than those obtained binocularly. Others were scarcely able to do the task at all with one eye covered. An example of the differences in monocular thresholds between normal cats is given in Figure 4.

EARLY MONOCULAR DEPRIVATION

There is now an abundance of data on the effects of monocular deprivation upon cortical ocular dominance (see Movshon, Van Sluyters, 1981). Simply put, periods of monocular deprivation imposed at the time of natural eye opening lead to a breakdown in the functional connections between the cortex and the deprived eye. Some limited

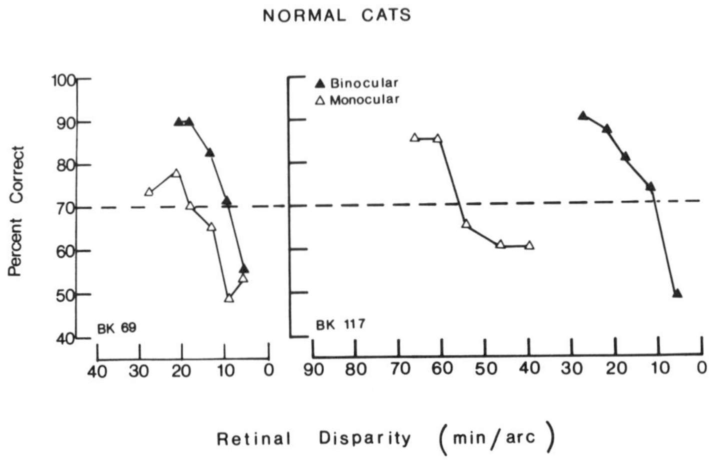

Fig. 4. Individual psychometric functions for two cats which were tested binocularly and monocularly. The data were collected over several sessions and the percentage of correct jumps to the closer surface is plotted as a function of the retinal disparity present under binocular viewing conditions. Threshold is defined as the disparity value where the function crosses the 70% correct level (from Timney, 1983).

recovery of binocularity is possible if the eye is opened prior to the end of the sensitive period (Cynader, Timney, Mitchell, 1981; Hubel, Wiesel, 1970; Mitchell, Cynader, Movshon, 1977) and extensive behavioural recovery of visual acuity is possible, especially if the deprivation period if short (Giffin, Mitchell, 1978).

If it is the case that stereopsis is subserved by disparity-selective binocular neurones, then conditions which lead to a breakdown in binocularity should also abolish the ability to use retinal disparity as a depth cue. There is limited evidence to support this argument. Blake and Hirsch (1975) raised kittens under a regime of alternating monocular occlusion then tested the kittens' monocular and binocular depth thresholds. They found that the two thresholds for each animal did not differ, in contrast to normal control kittens whose binocular thresholds were much better. Physiological recordings from the experimental kittens showed a loss of binocular neurones. A similar behavioural result was obtained by

Packwood and Gordon (1975) who also tested for stereopsis using random element stereograms. In addition, Packwood and Gordon tested Siamese cats, who are known to have a reduced complement of binocular neurones, and failed to find evidence for stereopsis.

These findings are supplemented by those shown in Figure 5, which presents data from a cat which was deprived from the time of natural eye opening until 90 days of age. The main part of the figure shows individual session thresholds obtained shortly after the deprived eye was opened. The inset contains the final psychometric functions obtained after time had been allowed for recovery

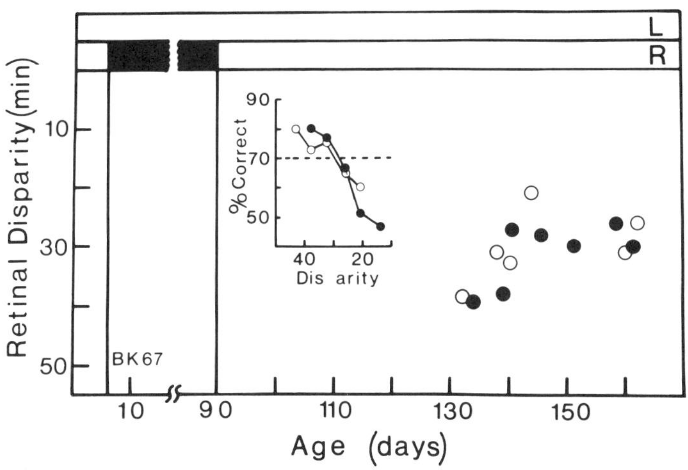

Fig. 5. Monocular (open symbols) and binocular (filled symbols) thresholds data from a cat which was monocularly deprived from the time of natural eye opening until day 90, as shown above the figure. The main part of the figure shows threshold estimates obtained in individual sessions after the deprived eye was opened. The inset shows final threshold data obtained using the method of constant stimuli.

of vision in the deprived eye (cf. Giffin, Mitchell, 1978). It may be seen that binocular and monocular thresholds are virtually identical, indicating that this cat has lost the ability to use binocular depth cues. Such a result is not unexpected, given the paucity of binocularly excitable neurones found following three months of monocular deprivation (e.g. Wiesel, Hubel, 1963).

Of more interest are the consequences of shorter deprivation periods in which the behavioural recovery of visual acuity is extensive (Giffin, Mitchell, 1978) and also a reasonable number of binocular neurones remain (Mitchell et al, 1977). It should be stressed again at this point that although disparity-selective neurones are by their nature binocular, the presence of binocular neurones does not necessarily imply disparity selectivity.

Figure 6 shows the results of monocular deprivation lasting until 30, 35, 40 or 50 days. At the end of each deprivation period the kitten's sutured eye was opened and it was allowed binocular vision until its depth vision was tested several months later. Although no physiological recordings were made of these kittens, the available published data suggest that there should be at least a partial recovery of binocularity (Hubel, Wiesel, 1970; Mitchell et al., 1977). It is evident from Figure 6 that only the shortest period of deprivation, 30 days, permits stereoacuity to reach more or less normal levels. Even 35 days of deprivation is sufficient to disrupt binocular depth thresholds permanently. Data from other animals have confirmed this finding (Timney, 1983).

The reader should note that the age at which the kittens becomes susceptible to the effects of monocular deprivation coincides with the time at which disparity selective neurones are just achieving their adult-like characteristics (Pettigrew, 1974). It is also the age at which the most rapid development of stereoacuity is occurring (Timney, 1981). To some extent these results are similar to those found for the effects of monocular deprivation on visual acuity; that is, susceptibility to deprivatation runs parallel to the acuity development function (Giffin, Mitchell, 1978; Mitchell, Wilkinson, Anderson, Smith, 1976). However, the results differ in that there appears to be no recovery of binocular depth vision, in contrast to the extensive recovery observed for acuity. It appears that, once lost, the ability to use binocular depth cues does not return.

It is possible that the lack of recovery is not due to an irreversible breakdown in disparity-selective mechanisms, but is instead a secondary consequence of the

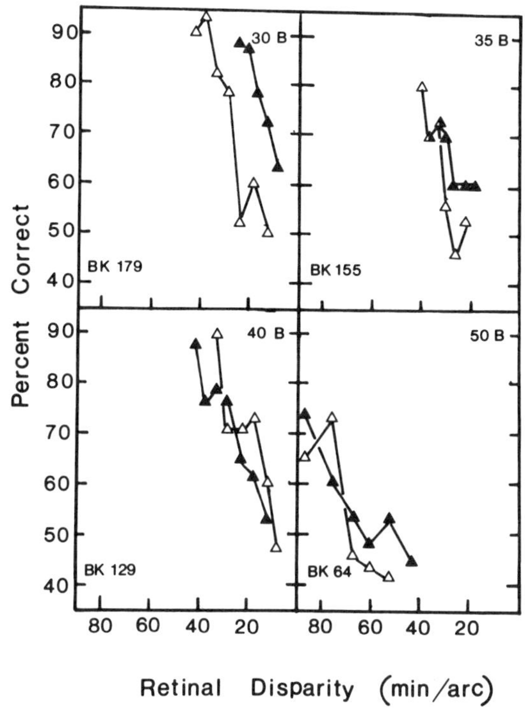

Fig. 6. Monocular (open symbols) and binocular (filled symbols) data from four kittens which were deprived from the time of natural eye opening until days 30, 35, 40 and 50 respectively.

strabismus that sometimes follows a period of monocular deprivation (Sherman, 1972). While this explanation cannot be ruled out, observation of the cats used in this study showed that there were only a few instances where a pronounced strabisumus was evident. Among cats which were raised under identical conditions for periods longer than 35 days those which apppeared orthophoric did not do any better than those which had a manifest deviation. A similar finding has been reported by Kaye et al. (1982) for dark-reared cats and by Crawford, Von Noorden, Meharg, Rhodes, Harwerth, Smith and Miller, (1983) for monkeys reared with optically induced strabismus, suggesting that post-deprivation strabismus may not be the primary factor.

LATE MONOCULAR DEPRIVATION

The results of the early deprivation study suggested that there is a sensitive period of susceptibility which begins during the fifth week of life. Unfortunately, this study was unable to provide any information about the end of the sensitive period, since all periods beyond 30 days had equally severe effects with no recovery. In order to obtain information about the end of the sensitive period, a second approach was employed. Kittens were allowed normal experience until the age of four months, during which time depth thresholds were obtained. Then they were deprived for one, two or three months while monocular testing was continued. Finally, the deprived eye was opened and depth testing was begun almost immediately.

The results are shown in Figure 7, which plots data from individual testing sessions. It is quite obvious that even three months of monocular deprivation begun at four months of age has no effect on binocular depth discrimination. Even a few hours after the deprived eye has been opened the cats did as well as they did before the eye was closed. Thus it appears that the sensitive period is over completely by the end of the fourth month.

PROFILE OF THE SENSITIVE PERIOD - BRIEF MONOCULAR DEPRIVATION

The final study to be described here represents a first step towards the construction of a profile of the sensitive period. The strategy employed was the same as that used by Olson and Freeman (1980), who imposed 10 day deprivation periods at different ages then examined cortical ocular dominance. In the present study, kittens were allowed normal visual experience except for a 10 day period when the eyelids of one eye were sutured closed, then at a later age depth thresholds were obtained. So far, we have data on several groups of kittens deprived at 10 day intervals from day 30 until day 70. The typical pattern of results obtained is illustrated in Figure 8. This shows data from four kittens deprived at 30, 40, 50 and 60 days. Deprivation between days 30 and 40 results in binocular thresholds which are poorer than those found for most normal cats, but nevertheless are slightly better than their own monocular thresholds. Any binocular superiority

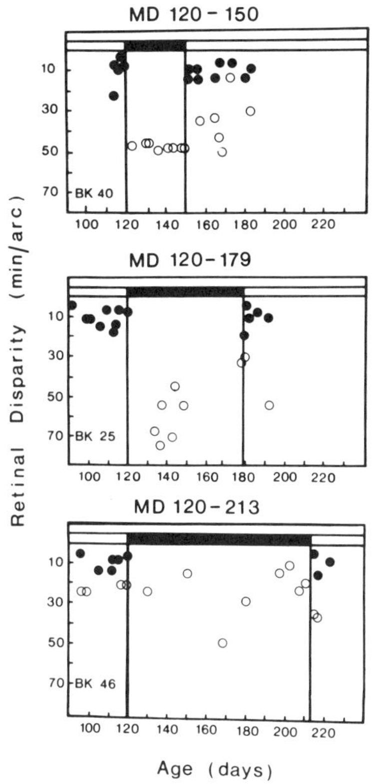

Fig. 7. Monocular (open symbols) and binocular (filled symbols) thresholds obtained in individual sessions from three cats which were reared normally until the age of four months then were monocularly deprived for one, two or three months, as shown above the figure. Post-deprivation thresholds were obtained on the same day or the day following eye-opening (from Timney, 1983).

appears to be abolished completely by a 40 to 50 day deprivation period, but thereafter the deprivation effect becomes progressively smaller.

In order to obtain a profile of the period of susceptibility to brief deprivation, an index was constructed, consisting of the ratio of monocular to binocular thresholds (cf. Henson, Williams, 1980). In

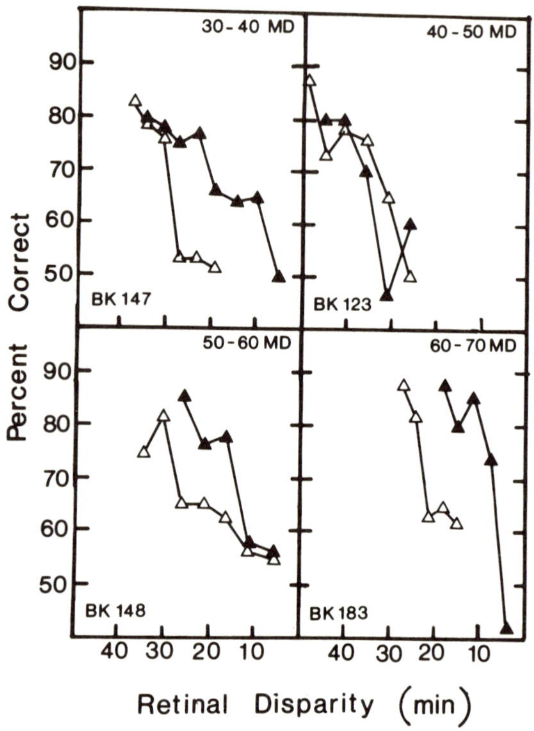

Fig. 8. Monocular (open symbols) and binocular (filled symbols) data from four kittens which were monocularly deprived for 10-day periods as indicated at the top right of each panel.

normal cats this ratio is approximately six, indicating that binocular thresholds are better than monocular by that factor. A ratio of one indicates no difference between the thresholds. Figure 9 summarizes these ratios for all kittens tested so far. It can be seen that deprivation beginning on either day 30 or day 40 has the most severe effects. Deprivation applied at later ages is less disruptive, so that the average monocular/binocular ratio after a 70-80 day deprivation is close to normal levels. It is worth pointing out here that the large variability associated with this latter condition is due primarily to variations in monocular thresholds and does not result from large differences in binocular performance. From these data one may conclude that the sensitive period begins

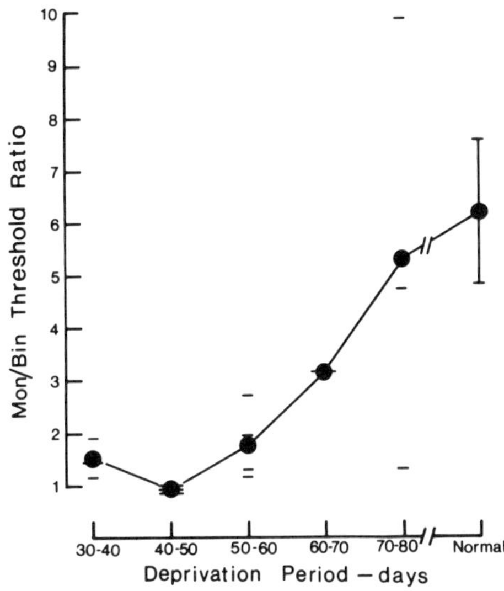

Fig. 9. Monocular/binocular threshold ratios for kittens whichwere monocularly deprived for 10-day periods as shown on the abscissa. Filled symbols represent the average ratio for all kittens with the same deprivation period and the horizontal bars represent the ratio for individual kittens. The vertical bar around the point for normal cats indicates ±1 standard error of the mean, based upon the 10 animals shown in Figure 3.

around the end of the first month and seems to be essentially over by the age of 2.5 months.

CONCLUSIONS

In the preceding pages I have attempted to bring together a number of studies being conducted in my lab which we directed towards the question of the susceptibility of binocular depth perception to abnormal early visual experience. The results obtained so far paint a picture of a system which is highly susceptible to disruption over a relatively short period early in life. For the kitten, the peak of susceptibility to monocular deprivation seems to occur sometime between days 30 and 50, with a fairly rapid decline after that. Our late deprivation studies show that by 90 or 120 days binocular depth perception is resistant to a month or more of monocular deprivation, and these results, taken in conjunction with those from the short-term deprivation studies suggest that the sensitive period is over completely no later than the beginning of the fourth month.

It is instructive to compare the results of the present studies with those which have been concerned with the development of visual acuity in kittens and the effects of monocular deprivation. There are a number of differences. For example, compared with the very rapid development of binocular depth vision in kittens, which improves markedly between days 35 and 45 and has reached adult-like levels no later than the end of the third month (Timney, 1981), the development of visual acuity is relatively slow (Giffin, Mitchell, 1978; Mitchell et al. 1976). Acuity improves gradually over the first few months of life, to reach full adult levels only after the end of the fourth month. Another difference between resolution acuity and stereoacuity is in the extent of recovery found after periods of monocular deprivation. Following monocular deprivation imposed at the time of eye-opening there is initially a period of blindness in the deprived eye, which varies with the duration of deprivation. Afterwards, there is a gradual recovery of resolution acuity, whose extent again depends upon the deprivation period. Deprivation imposed after a kitten has been allowed some normal visual experience typically has less severe effects, with an initial reduction in acuity followed by extensive recovery (Giffin, Mitchell, 1978). In contrast, it appears that once the ability to use binocular depth cues has been lost, then the probability of recovery is much less than for visual acuity. Periods of deprivation which result in only minor acuity losses can produce a severe and permanent impairment of binocular depth discrimination. This result in cats has been confirmed in the monkey by Crawford et al. (1983), who reported that one month of optically-induced strabismus, imposed during the second month of life, was sufficient to abolish stereopsis permanently.

These differences between visual acuity and binocular depth vision measured behaviourally suggest that the physiological mechanisms underlying them may also be different. It seems likely that the more stringent requirements for binocular processing make this system a much more fragile one during the course of development, one which cannot withstand even minor disruptions. We are now conducting studies to determine the minimum amount of binocular disruption which will lead to a deficit in binocular depth vision.

ACKNOWLEDGEMENTS. I wish to thank Cathy Bryans who assisted in all technical aspects of the experiment and preparation of the chapter. I also thank Linda Toenders for able secretarial help. Supported by grants from the Medical Research Council of Canada (MA7123) and the Natural Sciences and Engineering Research Council of Canada (A7062).

References

Barlow HB, Blakemore C, Pettigrew JD (1967). The neural mechanisms of binocular depth discrimination. J Physiol 193:327.

Blake R, Hirsch HVB (1975). Deficits in binocular depth perception after alternating monocular deprivation. Science 190:1114.

Blakemore C (1979). The development of stereoscopic mechansims in the visual cortex of the cat. Proc R Soc Lond B 204:477.

Crawford MLJ, von Noorden GK, Meharg LS, Rhodes JW, Harwerth RS, Smith EL III, Miller DD (1983). Binocular neurons and binocular function in monkeys and children. Invest Ophthalmol Vis Sci 24:491.

Cynader M, Timney B, Mitchell DE (1980). Period of susceptibility of kitten visual cortex to the effects of monocular deprivation extends beyond six months of age. Brain Res 191:545.

Giffin F, Mitchell DE (1978). The rate of recovery of vision after early monocular deprivation in kittens. J Physiol 274:511.

Henson DB, Williams DE (1980). Depth perception in strabismus. Br J Ophthalmol 64:349.

Howard HJ (1919). A test for the judgement of distance. Amer J Ophthalmol 2:656.

Hubel DH, Wiesel TN (1970). The period of susceptibility to the physiological effects of unilateral eye closure in kittens. J Physiol 206:419.

Kaye M, Mitchell DE, Cynader M (1982). Depth perception, eye dominance and cortical binocularity of dark-reared cats. Develop Brain Res 2:37.

Lashley KS (1930). The mechanism of vision: I. A method for the rapid analysis of pattern vision in the rat. J Genetic Psychol 37:453.

Mitchell DE, Timney, B (in press). Postnatal development of function in the mammalian visual system. In Darian-Smith I ((ed): "Sensory Processes," Baltimore, Maryland: Williams & Wilkins.

Mitchell DE, Cynader M, Movshon JA (1977). Recovery from the effects of monocular deprivation in kittens. J Comp Neurol 176:53.

Mitchell DE, Kaye M, Timney B (1979). Assessment of depth perception in cats. Perception 8:389.

Mitchell DE, Giffin F, Wilkinson FE, Anderson P, Smith ML (1976). Visual resolution in young kittens. Vis Res 16:363.

Movshon, JA, Van Sluyters, RC (1981). Visual neural development. Ann Rev Psychol 32:477.

Nelson JI, Kato H, Bishop PO (1977). Discrimination of orientation and position disparities by binocularly activated neurons in cat striate cortex. J Neurophysiol 40:260.

Nikara T, Bishop PO, Pettigrew JD (1968). Analysis of retinal correspondence by studying receptive fields of binocular single units in cat striate cortex. Exp Brain Res 6:353.

Olson CR, Freeman RD (1980). Profile of the sensitive period for monocular deprivation in kittens. Exp Brain Res 39:17.

Packwood J, Gordon B (1975). Stereopsis in normal domestic cat, Siamese cat and cat raised with alternating monocular occlusion. J Neurophysiol 38:1485.

Pettigrew JD (1974). The effect of visual experience on the development of stimulus specificity by kitten cortical neurones. J Physiol 237:49.

Poggio GF, Fischer B (1977). Binocular interaction and depth sensitivity in striate and prestriate cortex of behaving rhesus monkey. J Neurophysiol 40:1392.

Sherman SM (1972). Development of interocular alignment in cats. Brain Res 37:187.

Timney B (1981). Development of binocular depth perception in kittens. Invest Ophthalmol Vis Sci 21:493.

Timney B (1983). The effects of early and late monocular deprivation on binocular depth perception in cats. Develop Brain Res 7:235.

Wheatstone C (1938). Contributions to the physiology of vision. I. On some remarkable and hitherto unobserved phenomena of binocular vision. Phil Trans Roy Soc 2:371.

Wiesel TN, Hubel DH (1963). Single unit response in striate cortex of kittens deprived of vision in one eye. J Neurophysiol 26:1003.

Development of Visual Pathways in Mammals, pages 425–438

MECHANISMS FOR GENERATING ORIENTATION SELECTIVITY OF DIFFERENT CELL TYPES IN PRIMARY VISUAL CORTEX OF CAT

Helmut V.B. Hirsch

State University of New York at Albany

Albany, New York 12222

DEVELOPMENT OF ORIENTATION SENSITIVE CELLS

Nerve cells in the primary visual cortex, area 17, of an adult cat respond preferentially to lines, bars or edges of a particular orientation (Hubel and Wiesel, 1959, 1962). While some cells in area 17 display this orientation selectivity in young, inexperienced animals, others develop it only if patterned visual stimulation is provided during early postnatal life (Blakemore and Van Sluyters, 1975; Fregnac and Imbert, 1978; Bonds, 1979; Fregnac et al., 1981; McCall, 1983; but see also Hubel and Wiesel, 1963; Pettigrew, 1974; Sherk and Stryker, 1976). Neuronal activity thus can affect postnatal development of at least some cells in the visual system. In this paper I will compare and contrast cells innately selective for orientation and cells that develop experience-induced orientation selectivity in area 17. I will also outline ways in which the orientation selectivity of cells in these two groups might be generated.

Most experimenters studying area 17 in young kittens have described considerable variation in the extent to which different cells respond selectively to stimulus orientation. First, there are cells that are visually unresponsive; they typically have low spontaneous firing rates and are easy to miss as the electrode is advanced through the cortex (Blakemore and Van Sluyters, 1975; Fregnac and Imbert, 1978; Bonds, 1979; McCall, 1983). Second, there are cells that are responsive, but not selective; these cells, which have been called <u>non-oriented</u> or <u>non-specific</u> cells, typically have circular receptive fields and can be activated by lines or spots moving in any direction (Pettigrew 1974; Blakemore and Van Sluyters, 1975; Fregnac and Imbert, 1978; Bonds, 1979; McCall, 1983). Third, there are cells that are responsive and highly selective; these cells, which have been called orientation selective or specific cells, have a clear preferred orientation (PO) and no reliable response to a stimulus orthogonal to this optimal orientation (Blakemore and Van Sluyters, 1975; Fregnac and Imbert, 1978; Bonds, 1979; McCall, 1983). There are also cells intermediate in selectivity between the latter two groups; these cells, which have been called orientation biased or immature cells, have a preference for a particular stimulus orientation, but the range of orientations over which the cells respond, and the variability of their response is larger than in the adult cat (Pettigrew 1974; Blakemore and Van Sluyters, 1975; Fregnac and Imbert, 1978; Bonds, 1979; McCall, 1983). Some workers have also identified pure direction selective cells (Pettigrew 1974; Blakemore and Van Sluyters, 1975) which respond to moving spots and have both a preferred and null direction (Barlow et al., 1964); these cells are as selective to spots as they are to lines and bars.

The orientation selective cells in very young kittens have a number of distinctive features, in addition to their selectivity for stimulus orientation. They are mostly simple cells (Blakemore and Van Sluyters, 1975; Beckman and Albus, 1982), probably monosynaptically activated by thalamic afferents (Beckman and Albus, 1982) and are concentrated in the deeper layers of area 17, especially throughout layer IV (Blakemore and Van Sluyters, 1975). Furthermore, the distribution of the POs of the orientation selective cells is not uniform, but is biased towards 0° and 90°(Fregnac and Imbert, 1978; Fregnac et al., 1981). In addition, the orientation selective cells (and possibly also the orientation biased cells) are monocular, while the non-specific cells are binocular (Blakemore and Van Sluyters, 1975; Fregnac and Imbert, 1978; McCall, 1983); most are dominated by the contralateral eye (Fregnac and Imbert, 1978; but see Blakemore and Van Sluyters, 1975 and McCall, 1983, for failures to find contralateral eye dominance among specific cells). Finally, the specific cells have the smallest receptive fields (Fregnac and Imbert, 1978; McCall, 1983).

The properties of orientation selective cells in very

young kittens are remarkably similar to those of orientation selective cells found in cats reared from birth to 7-10 months of age in total darkness (Leventhal and Hirsch, 1977, 1980). These latter cells also mostly prefer orientations near 0° or 90°, are monocular and dominated by the contralateral eye, have small receptive fields and are concentrated in and around layer IV. There is, however, one important difference between the orientation selective cells in the two types of cats: they differ in the velocities of movement to which they will respond. In the dark-reared adult cats the orientation selective cells respond only to stimuli moving slowly (low cutoff velocity), whereas the non-selective cells respond readily to rapidly moving stimuli (high cutoff velocity) (Leventhal and Hirsch, 1977, 1980). In fact, the orientation sensitive cells in dark-reared cats have response properties indicative of a direct excitatory input, probably from one particular cell type in the dorsal lateral geniculate nucleus (lgn), the so called X-cells, while the non-selective cells have response properties indicative of an excitatory input from a second cell type in the lgn, the so called Y-cells, and/or from other cortical cells in area 17 (Leventhal and Hirsch, 1977, 1980; Dreher et al., 1980; Mustari et al., 1982). In kittens there is no evidence for a difference in cutoff velocity between cells which are selective for orientation and those which are not (McCall, 1983), however, the significance of such negative findings in young kittens is difficult to evaluate since many of the differences that exist between X- and Y-cells in adult cats, cannot be demonstrated in the lgn and retina of young kittens (Daniels et al., 1978; Hamasaki and Sutija, 1979; Friedlander, 1982). It may be very difficult to distinguish types of excitatory afferents at this stage of development.

Because of the similarities between the first cells that are orientation selective in very young kittens and the cells that remain orientation selective despite prolonged dark-rearing, and because of the evidence that dark-rearing may "freeze" development of orientation selectivity in area 17 (Blakemore and Van Sluyters, 1975; Bonds, 1979), Leventhal and Hirsch (1980) suggested that the orientation selective cells in young, inexperienced kittens are likely to be activated directly by excitatory afferents from the thalamus, and, like the majority of cells in area 17, are likely to receive their afferent input from X-cells. The non-oriented (and possibly the orientation biased) cells, on the other hand, are likely to be neurons that receive much of their excitatory input from Y-cells and/or from other cortical cells in area 17. The distribution of afferents from a third group of cells in the lgn, the so called W-cells is poorly understood. Given the relative maturity of W-cells in the lgn of very young kittens W-cells may have a very important role in the postnatal development of area 17 (Daniels et al.,1978; Friedlander, 1982).

The characteristics of cells in area 17 of a normally reared cat change dramatically after about the third week of life (Pettigrew 1974; Blakemore and Van Sluyters, 1975; Fregnac and Imbert, 1978; Bonds, 1979; McCall, 1983; but see Hubel and Wiesel, 1963 and Sherk and Stryker, 1976). The majority of these changes fail to take place if the animals are deprived of normal patterned visual stimulation either by being kept in darkness (Blakemore and Van Sluyters, 1975; Fregnac and Imbert, 1978; Bonds, 1979) or by having the lids of both eyes sutured together (Pettigrew, 1974; Blakemore and Van Sluyters,1975). Neither normal orientation selectivity, nor the neuronal mechanisms thought to be involved in stereopsis, develop in visually deprived animals.

CORTICAL EFFECTS OF EARLY SELECTIVE VISUAL EXPOSURE

The neuronal activity produced by visual exposure affects development of orientation selectivity of at least some cells in area 17. We can now ask whether in the very young kitten this activity affects only the development of orientation preferences of those cells that lack adult levels of orientation selectivity, i.e. the non-oriented and orientation bias cells, or whether it also affects the orientation preferences of those cells that are already orientation selective (Hirsch and Spinelli, 1971; Rauschecker and Singer, 1981). We have attempted to ask two questions: first, are the cells responsive to horizontal and vertical lines as affected by stimulation during development as the cells responsive to diagonal lines, and, second, are there demonstrable differences in the extent to which cell types are affectd by postnatal stimulation? Differences between the cells with POs near horizontal and vertical, and the cells with POs near the diagonals have been demonstrated previously (Leventhal and Hirsch, 1975), and we will focus here on variations in experience-sensitivity between cell types. As a first step, we have raised kittens from birth in a controlled visual environment (Hirsch and Spinelli, 1970, 1971), looked for effects that are specific to the stimulation provided, and then determined if the effects could be seen in those cells that were likely to have been orientation selective at the beginning of the rearing (Hirsch et al., 1983).

Cats were raised wearing goggles within which each eye could view lines of only one orientation, stripe-rearing. This rearing introduces a significant overall bias in the distribution of the POs of the cells recorded from area 17: POs were normalized to the orientation presented during rearing, a mean preferred orientation was computed for each cat, and circular statistics (Zar, 1974) were used to show that the distribution of these means differed significantly from a uniform distribution ($r=0.725$; V test; $p < .001$; for details see Hirsch et al., 1983).

Although stripe-rearing produces a significant overall effect on the distribution of POs of cells in area 17, it does not, as Figure 1 shows, significantly affect the dis-

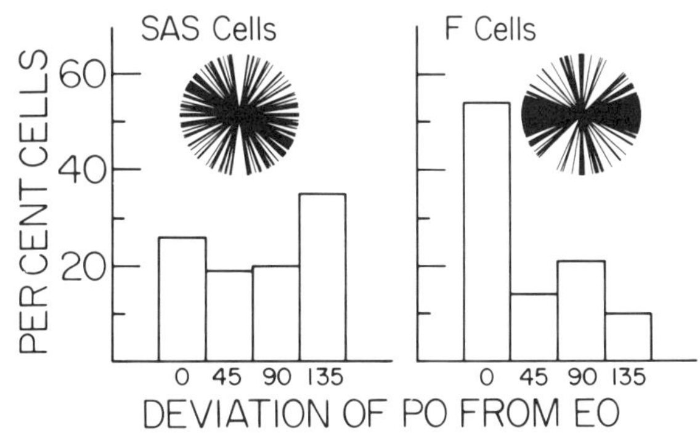

Figure 1. Histogram showing the distribution of POs of SAS and F cells. POs computed as deviations from the exposed orientation (EO) which has been normalized to 0°. SAS cells and F cells have each been divided into four groups: cells in bins labelled 0 have POs within 22.5° of EO; cells in bins labelled 45 have POs within 22.5° of EO+45°, cells in bins labelled 135 have POs within 22.5° of EO+135°. The polar plots above each histogram indicate the normalized POs for each cell. Reproduced with permission from Hirsch et al., J. Physiol. (Lond.) 337, 241-255, 1983, Figure 4.

tribution of the POs of the so-called Small and Slow or SAS cells, cells with the small receptive fields and low cutoff velocities characteristic of neurons receiving excitatory input from lgn X-cells (Hirsch et al., 1983). As Figure 1 shows, stripe-rearing does affect the distribution of the POs for cells with high cutoff velocity, the so called Fast or F cells, which are likely to receive their excitatory input from lgn Y-cells and/or from other cortical cells (Leventhal and Hirsch, 1980; Dreher et al., 1980; Mustari et al., 1982).

To quantify the difference between SAS and F cells we determined for each cat the mean orientation for SAS cells separately from the mean orientation for F cells. We then computed two over-all mean orientations, one for the SAS cells and a second for F cells. For the former group, there was no consistent bias across individual stripe-reared cats (r=0.169; V test; not significant). In contrast, for F cells there was a statistically significant bias across individual stripe-reared cats, and this bias was centered about the exposed orientation (r=0.364; V test, $P < 0.0005$)(see Hirsch et al., 1983 for details).

To summarize, the data suggest that there is a group of cells in area 17 whose members develop and maintain their POs in the absence of visual stimulation (Leventhal and Hirsch, 1977, 1980), and that the distribution of the POs of these cells is not modified by stripe-rearing (Hirsch et al., 1983). The characteristics of the cells in this group suggest that they are likely to receive their excitatory input directly from cells in the lgn (Leventhal and Hirsch, 1977, 1980; Dreher et al., 1980; Bullier and Henry, 1979; Mustari et al., 1982), and that like most cells in area 17, they receive their afferents from lgn X-cells (Leventhal and Hirsch, 1980; Dreher et al., 1980; Mustari et al., 1982). Other groups of cortical cells, including those with high cutoff velocities, which are likely to be higher order cells and/or receive their excitatory input from lgn Y-cells (Leventhal and Hirsch, 1980; Dreher et al., 1980; Bullier and Henry, 1979; Mustari et al., 1982), are not selective for orientation in dark reared cats (Leventhal and Hirsch, 1977, 1980) and the distribution of their POs peaks at the orientation presented during the rearing (Hirsch et al., 1983).

Having described the differences in experience-sensitivity among physiologically identified cell groups in area 17 (Hirsch et al., 1983), we will next consider the evidence for differences in experience-sensitivity among morphologically identified cell groups. Tieman and Hirsch (1982) examined the dendritic morphology of layer IV stellate cells, and of the basal dendrites of layer III pyramidal cells in normal cats and in cats exposed to lines of one orientation (either 0°or 90°). Animals were exposed using the same goggles to control visual exposure as were used by Hirsch et al. (1983). The effects of the selective exposure were confirmed physiologically, and blocks of visual cortex were impregnated by the Golgi-Kopsch procedure and cut tangential to the surface of the brain. Complete neurons from layers III and IV were drawn using a camera lucida, and the orientation of the dendritic fields analyzed quantitatively. In normally reared cats dendritic fields of both layer IV stellate cells and layer III pyramidal cells were elongated and the distributions of the orientations of the dendritic fields were uniform (Figure 2; left column). In cats exposed to lines of only one orientation, however, the distributions of the orientations of layer III pyramidal cells, but not of layer IV stellate cells, were shifted (Figure 2; middle and right column). The direction of the shift for the layer III pyramidal cells varied with the stimulus orientation to which the cats had been exposed during rearing. In cats reared viewing vertical lines, the dendritic fields of the layer III pyramidal cells were oriented perpendicular to the representation of the vertical meridian, and in cats reared viewing horizontal lines the dendritic fields were oriented parallel to the representation of the vertical meridian.

Tieman and Hirsch (1982) thus found that stripe-rearing alters the distribution of dendritic field orientations for layer III pyramidal cells, but not for layer IV stellate cells. These results, when viewed together with those of Coleman et al. (1981), who report little effect of stripe-rearing on stellate cells, suggest that some cortical cells, namely the layer IV stellate cells, are less sensitive to the effects of selective exposure than are other cortical cells, such as the layer III pyramidal cells.

In order to permit integration of the results of physiological and morphological studies of the effects of stripe-rearing, Tieman and Hirsch (1982) suggested a relationship between cell classes identified on physiological and morphological grounds. They suggested that the layer IV stellate cells might contribute heavily to the population of SAS cells and that the layer III pyramidal cells might contribute heavily to the population of F cells.

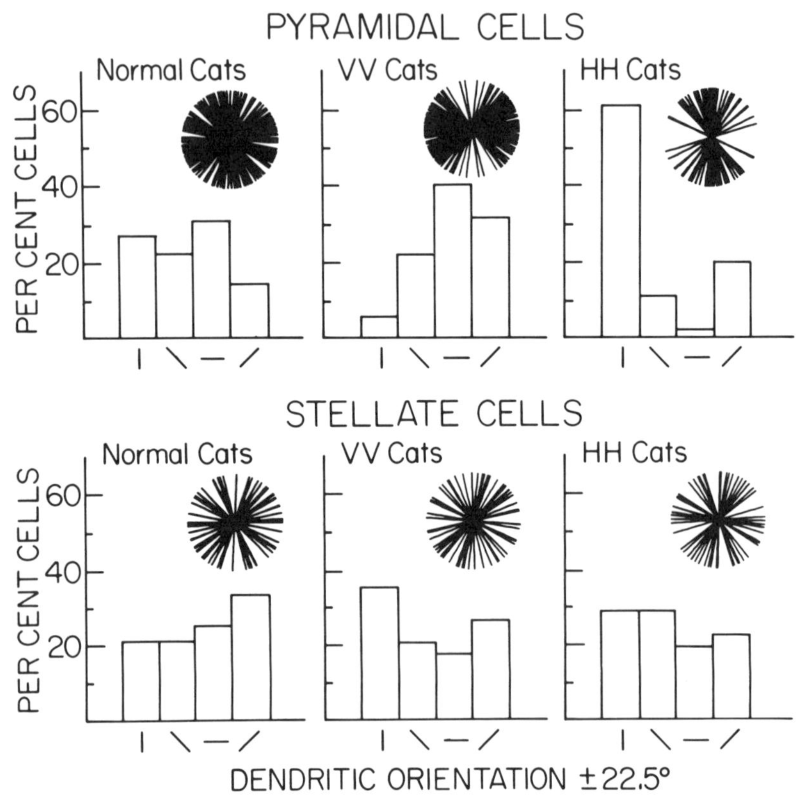

Figure 2. Dendritic field orientations of layer III pyramidal cells and layer IV stellate cells from area 17 of normal cats, stripe-reared cats exposed to 90° (VV Cats) or to 0° (HH Cats). The data from all cats in each condition have been grouped together. The histograms plot the percentage of cells having dendritic field orientations within 22.5° of 0°,90°, 45° and 135°. The inserts are the polar plots of these distributions. Reproduced with permission from Tieman and Hirsch, J. comp. Neurol., 211, 353-362, 1982, Figure 7.

They based this correspondence on the following points: (1) F cells are more readily modified by stripe-rearing than are SAS cells (Hirsch et al., 1983). (2) SAS cells predominate in layer IV (Leventhal and Hirsch, 1980). (3) Medium diameter axons (presumably axons of X cells) are distributed to most of layer IV (Bullier and Henry, 1979) and SAS cells are thought to receive X cell input (Dreher et al., 1980; Leventhal and Hirsch, 1980; Mustari et al., 1982), whereas the distribution of large diameter axons (presumably axons of Y cells) matches the area in which the basal dendrites of layer III pyramidal cells arborize (Bullier and Henry, 1979), and F cells are thought to receive Y cell input (Dreher et al., 1980; Leventhal and Hirsch, 1980; Mustari et al., 1982). (4) SAS cells tend to have smaller receptive fields than F cells (Dreher et al., 1980; Leventhal and Hirsch, 1980; Mustari et al., 1982), and the dendritic fields of layer IV stellate cells are smaller than those of layer III pyramidal cells.

To summarize, we have demonstrated the following points. (1) The POs of innately orientation selective cells are very robust, being affected neither by prolonged periods of dark-rearing (Leventhal and Hirsch, 1977, 1980) nor by prolonged exposure to lines of only one orientation (Hirsch et al., 1983). These cells are likely to receive much of their excitatory input directly from the lgn; like most cells in area 17 this input is probably from lgn X-cells, and this physiologically characterized cell group probably corresponds to layer IV stellate cells. (2) Cells which do not display adult levels of selectivity in young kittens do not develop POs if the young kitten is deprived of patterned visual stimulation (Pettigrew 1974; Blakemore and Van Sluyters, 1975; Leventhal and Hirsch, 1977, 1980; Fregnac and Imbert, 1978; Bonds, 1979), but if the animal does receive visual stimulation they will develop POs and the distribution of these POs is affected in a consistent, systematic fashion by the orientation of the stimuli to which the animals are exposed (Hirsch et al., 1983). These cells are likely to be second order cortical neurons and/or receive their excitatory afferent input from either Y- or W-cells in the lgn; this physiologically characterized cell group (or at least a subset of it) probably corresponds to the layer III pyramidal cells.

A MODEL FOR GENERATION OF ORIENTATION SELECTIVITY

I will conclude by outlining briefly how orientation sensitive cells in the retina and the lgn may be involved in the genesis in area 17 of the kitten of both the innate selectivity of some cells and the capability of other cells to develop a PO as a consequence of visual exposure: the orientation sensitive cells in the retina and lgn might provide the framework needed in order for a "self-organizing" system (e.g. Malsburg, 1973; Malsburg and Cowan, 1982) to produce a population of cells which are orientation selective, and whose POs are distributed in the orderly fashion characteristic of area 17 of normal cats.

Most cells in the retina (Levick and Thibos, 1982) and lgn (Daniels et al., 1977; Vidyagassar and Urbas, 1982) display a PO. The distribution of these POs is not uniform in either the retina or the lgn: cells located near one another tend to have similar POs and except for cells with receptive fields close to the area centralis, PO is correlated with the orientation relative to the horizontal meridian of a line connecting the center of the receptive field with the center of the area centralis, the polar angle of the receptive field (Levick and Thibos, 1982). A possible morphological substrate for the orientation bias of retinal ganglion cells has been demonstrated by Leventhal and Schall (1983): the dendritic fields of most retinal ganglion cells are elongated, and the distribution of the major axis of these elongated fields is related in the same fashion to polar angle as is the physiologically measured orientation bias of the retinal ganglion cells. Finally, there is rather compelling evidence that elongation of the dendritic fields of retinal ganglion cells, and the POs presumed to result from this, may provide the foundation relative to which the distribution of the POs of cells in area 17 is determined: the distribution of POs of most cells in area 17, but especially of simple cells, is correlated in the same fashion with polar angle as the major axis of the elongated dendritic fields of the retinal ganglion cells (Leventhal, 1983).

Thus, the orientation bias of retinal ganglion cells could be relayed to cells in the lgn, and then to cells in area 17. Cells in the lgn display an orientation bias in very young kittens (Albus et al., 1983), so that they could provide orientation biased input to cells in area 17, at a very early stage of development. By controlling the number of orientation biased lgn afferents converging onto an individual cell in area 17 it is possible to create either a cell that has a PO or a cell that is not selective for orientation, but that has the capacity to develop experience-induced selectivity for orientation (Leventhal, 1983). Cells that receive very restricted input, e.g. the small stellate cells in layer IV, would have an orientation selectivity comparable to that of the lgn afferents. Cells that receive a very extensive input from a large area on the retina, e.g. the basal dendrites of layer III pyramidal cells, would not be orientation selective because they are activated by such a heterogeneous population of afferents that they respond well to all orientations. However, by permitting stimulation-induced neuronal activity to modify the ability of individual afferents to activate their target cells, visual exposure can lead to the development of POs of cells like the layer III pyramidal cells. This might be achieved in the following fashion.

Suppose that geniculo-cortical synapses onto dendrites such as those of the layer III pyramidal cells are modified according to the principles suggested by Hebb (1949), i.e. that the effectiveness of a given presynaptic input increases relative to all other inputs, whenever its activity is coincident in time with firing of the postsynaptic cell (for details see Malsburg, 1973; Rauschecker and Singer, 1981; Malsburg and Cowan, 1982). When lines of a given orientation move through the receptive field of a cell that receives an extensive afferent input, those lgn afferents that have a preference for the orientation being presented will be activated more effectively than will afferents with preferences for other orientations. As a result, there will be selective strengthening of synapses from afferents that have POs matching the orientation of lines presented during early postnatal life. This ultimately establishes a PO for the postsynaptic cell. In this way we could account for the aquisition of POs by non-selective cells, e.g. the layer III pyramidal cells, in stripe-reared cats.

In summary, I have suggested that layer IV stellate cells have innately determined POs because they receive a very restricted input from a set of orientation biased lgn cells that all have similar POs. On the other hand, in order for the layer III pyramidal cells to display a PO it is necessary that experience modify synaptic input to these cells. Following these synaptic rearrangements, the POs of the pyramidal cells also reflect directly the orientation bias of the cells comprising their afferent input. Since the orientation bias of cells in area 17 of adult cats is considerably sharper than the bias of retinal ganglion cells (Levick and Thibos, 1982), it is likely that there must also be sharpening of orientation tuning within the cortex. This sharpening might involve intracortical inhibitory connectivity (e.g. Sillito, 1975): it has been suggested that development of intracortical inhibitory connectivity is affected by the neuronal activity produced by early visual experience (Leventhal and Hirsch, 1980; Leventhal, 1983).

Acknowledgements:
I thank A.G. Leventhal, for having shared his unpublished data freely with me; H. Ghiradealla, A.G. Leventhal and D.G. Tieman for critical discussions and for trying to help me to express myself clearly. P. Caruccio helped with the preparation of the manuscript and figures. Support provided by NSF Grant BNS-8217540.

References:

Albus K, Wolf W, Beckman R (1983). Orientation bias in the response of kitten LGNd neurons to moving light bars. Develop Brain Res 6:308.

Barlow HB, Hill RM, Levick WR (1964). Retinal ganglion cells responding selectively to direction and speed of image motion in the rabbit. J Physiol (Lond) 173:377.

Beckman R, Albus K (1982). The geniculocortical system in the early postnatal kitten: an electrophysiological investigation. Exp Brain Res 47:49.

Blakemore C, Van Sluyters RC (1975). Innate and environmental factors in the development of the kitten's visual cortex. J Physiol (Lond) 248:663.

Bonds AB (1979). Development of orientation tuning in the visual cortex of kittens. In Freeman RD (ed): "Developmental Neurobiology of Vision," New York: Plenum Press, p31.

Bullier J, Henry GH (1979). Laminar distribution of first-order neurons and afferent terminals in cat striate cortex. J Neurophysiol 42:1271.

Coleman PD, Flood DG, Whitehead MC, Emerson RC (1981). Spatial sampling by dendritic trees in visual cortex. Brain Res 214:1.

Daniels JD, Norman JL, Pettigrew JD (1977). Biases for oriented moving bars in lateral geniculate nuecleus neurones of normal and stripe-reared cats. Exp Brain Res 29:155.

Daniels JD, Pettigrew JD, Norman JL (1978). Development of single-neuron responses in kitten's lateral geniculate nucleus. J Neurophysiol 41:1373.

Dreher B, Leventhal AG, Hale PT (1980). Geniculate input to cat visual cortex: a comparison of area 19 with areas 17 and 18. J Neurophysiol 44:804.

Fregnac Y, Imbert M (1978). Early development of visual cortical cells in normal and dark-reared kittens: relationship between orientation selectivity and ocular dominance. J Physiol (Lond) 278:27.

Fregnac Y, Trotter Y, Bienenstock E, Buisseret P, Gary-Bobo E, Imbert M (1981). Effects of neonatal unilateral enucleation on the development of orientation selectivity in the primary visual cortex of normally and dark-reared kittens. Exp Brain Res 42:453.

Friedlander, MJ (1982) Structure of physiologically classified neurons in the kitten dorsal lateral geniculate nucleus. Nature 300:180.

Hamasaki DI, Sutija VJ (1979). Development of X- and Y-cells in kittens. Exp Brain Res 35:9.

Hebb DO (1949). "Organization of Behavior," New York: John Wiley.

Hirsch HVB, Spinelli DN (1970). Visual experience modifies distribution of horizontally and vertically oriented receptive fields in cats. Science 168:869.

Hirsch HVB, Spinelli DN (1971). Modification of the distribution of receptive field orientation in cats by selective visual exposure during development. Exp Brain Res 12:509.

Hirsch HVB, Leventhal AG, McCall MA, Tieman DG (1983). Effects of exposure to lines on different cell types in striate cortex of cat. J Physiol (Lond) 337:241.

Hubel DH, Wiesel TN (1959). Receptive fields of single neurons in the cat's striate cortex. J Physiol (Lond) 148:574.

Hubel DH, Wiesel TN (1962). Receptive fields, binocular interaction, and functional architecture in the cat's visual cortex. J Physiol (Lond) 160:106.

Hubel DH, Wiesel, TN (1963). Receptive fields of cells in striate cortex of very young, visually inexperienced kittens. J Neurophysiol 26:994.

Leventhal AG, Hirsch HVB (1975). Cortical effect of early selective exposure to diagonal lines. Science 190: 902.

Leventhal AG, Hirsch HVB (1977). Effects of early experience upon orientation sensitivity and binocularity of neurons in

visual cortex of cats. Proc Natl Acad Sci USA 74:1272.
Leventhal AG, Hirsch HVB (1980). Receptive-field properties of different classes of neurons in visual cortex of normal and dark-reared cats. J Neurophysiol 43:1111.
Leventhal AG (1983). Systematic relationship between preferred orientation and receptive field position of neurons in cat striate cortex. J comp Neurol (in press).
Leventhal AG, Schall JD (1983). Structural basis of orientation sensitivity of cat retinal ganglion cells. J comp Neurol (in press).
Levick WR, Thibos LN (1982). Analysis of orientation bias in cat retina. J Physiol (Lond) 329:243.
Malsburg C von der (1973). Self-organization of orientation sensitive cells in the striate cortex. Kybernetic 14:85.
Malsburg C von der, Cowan JD (1982). Outline of a theory for the ontogenesis of iso-orientation domains in visual cortex. Biol Cybernetics 45:49.
McCall MA (1983). The relationship between ocular dominance and other response properties of cortical cells in normal kittens and in monocularly deprived kittens. Unpublished PhD Thesis, State University of New York at Albany, Albany New York.
Mustari MJ, Bullier J, Henry GH (1982). Comparison of response properties of three types of monosynaptic S-cells in cat striate cortex. J Neuophysiol 47:439.
Pettigrew JD (1974). The effect of visual experience on the development of stimulus specificity by kitten cortical neurones. J Physiol (Lond) 237:49.
Rauschecker JP, Singer W (1981). The effects of early visual experience on the cat's visual cortex and their possible explanation by Hebb synapses. J Physiol (Lond) 310:215.
Sherk H, Stryker MP (1976). Quantitative study of cortical selectivity in visually inexperienced kitten. J Neurophysiol 39:63.
Sillito AM (1975). The contribution of inhibitory mechanisms to the receptive field properties of neurones in the striate cortex of the cat. J Physiol (Lond) 250:305.
Tieman SB, Hirsch HVB (1982). Exposure to lines of only one orientation modifies dendritic morphology of cells in the visual cortex of the cat. J comp Neurol 211:353.
Vidyasagar TR, Urbas JV (1982). Orientation sensitivity of cat LGN neurones with and without inputs from visual cortical areas 17 and 18. Exp Brain Res 46:137.
Zar JH (1974). "Biostatistical Analysis," Englewood Cliffs: Prentice-Hall.

Development of Visual Pathways in Mammals, pages 439-462

CRITICAL PERIODS AND THE DEVELOPMENT OF THE RABBIT VISUAL CORTEX

E. Hazel Murphy

The Medical College of Pennsylvania
Department of Anatomy
3200 Henry Avenue, Philadelphia, PA 19129

CRITICAL PERIODS IN THE DEVELOPMENT OF THE MAMMALIAN VISUAL SYSTEM

The development of an organism involves a complex interaction between genetically determined and experiential factors. The concept of a sensitive period, during which experiential influences have a critical effect on subsequent development, has been applied to studies of imprinting for many years (Bateson 1978, Bischof 1983, Hess 1973). The application of this concept to the role of afferent input and sensory experience in the development of the visual system has been more recent (for reviews, see Aslin 1981, Blakemore 1978, Movshon & Van Sluyters 1981, Schneider 1977, Sherman & Spear 1982, Spear 1979).

In this chapter we will consider three areas of research on visual system development in which the concept of the critical period is often invoked: 1) manipulation of the internal environment by early lesions; 2) manipulation of the external environment by early deprivation; 3) morphological indices of development and maturation which normally occur during the time period in which these manipulations have maximum effect.

The first of these, manipulation of the internal environment by early lesions, has provided a substantial body of evidence that, during an early critical period, lesions of the central visual system have behavioral, physiological and anatomical consequences which differ significantly from the consequences of such lesions made after the critical period. The second area of research, manipulation of the external environment, has produced a wealth of evidence that early deprivation results in severe and often permanent disruption of the physiological organization of the primary visual cortex. Finally, studies of ontogenetic development of synapse and spine densities in visual cortex have produced evidence of a developmental trend which is observed repeatedly: this sequence, which we call the peak/decline trend, involves a postnatal increase to densities significantly higher than those observed in the adult, followed by a decline to adult levels. We suggest that the peak period, which occurs during the critical period for sensory experience, represents a period of selection and stabilization of contacts after which experimental manipulations of the internal or external environment are relatively ineffective.

These observations have been made in a variety of mammals. The effects of early deprivation have been most thoroughly documented in the cat. The effects of early lesions, and the ontogenetic sequences of spine and synapse density have been documented in many mammals, including the monkey, cat, rat, hamster and opossum. However, until recently, the rabbit appeared an isolated and anomalous case. Earlier literature indicated that the rabbit showed no behavioral sparing of visual discrimination following early visual cortical lesions (Murphy & Stewart, 1974); that early lesions of the rabbit visual system resulted in only minor alterations in visual system projections (Chow et al 1973, 1981; Stewart & Rieson 1972); and that those reorganized projections which were observed in the rabbit were not functional (Chow et al 1973, 1981). In addition, it was reported that dark rearing had no effects on visual discrimination (VanHof & Kobayashi 1972) and that pattern deprivation and selective orientation deprivation had no effects on visual cortical organization other than some delay in development of some receptive field characteristics in pattern deprived animals (Chow & Spear 1974, Crabtree et al 1981, Grobstein et al 1975, Mize & Murphy 1973).

We were initially interested in using the rabbit to elucidate those characteristics which determined its apparent lack of plasticity, i.e., its invulnerability to deprivation, and its lack of plasticity following neonatal lesions. More recently, however, it has become apparent that (i) the rabbit viusal cortex does show physiological plasticity in response to deprivation; (ii) there is a demonstrable critical period for this plasticity; (iii) there are postnatal changes in spine and synapse density which overlap the critical period and (iv) early lesions of the visual system result in altered projections similar in many respects to those observed in other mammals. The remainder of this chapter will be devoted to a review of these recent findings in the rabbit. For comparison, reference will be made to findings in other mammals. Details on studies involving animals other than rabbits, can be found in several excellent recent reviews (Movshon & Van Sluyters 1981; Schneider 1977; Sherman & Spear 1982; Spear 1979).

EARLY LESIONS OF THE VISUAL SYSTEM

Neonatal Ablation of Visual Cortex. Alteration of projections and of patterns of terminations following CNS lesions during an early critical period has been widely demonstrated in the mammalian visual system. The anatomical consequences of such neonatal lesions differ from those following the same lesion in the adult. Sprouting into adjacent vacated terminal space, failure to retract early exuberant projections, and the formation of anomalous projections have all been observed following early lesions (for review, see Schneider 1977). The behavioral deficits following lesions made in newborn animals are often less severe than those observed in animals lesioned as adults. For example, the ablation of visual cortex in adult cats and monkeys results in severe deficits in visual pattern discrimination, but relatively good discrimination occurs in animals operated as neonates (Murphy et al 1975, Weiskrantz 1963, Spear 1979). However, in rabbits the effects of removal of visual cortex in neonates or adults do not differ. Rabbits, like rats (Tees 1975), show severe deficits in visual discrimination regardless of their age at the time of visual cortex ablation (Murphy & Chow 1974, Murphy & Stewart 1974, Moore & Murphy 1976).

In rats, although behavioral sparing is not observed, there is significant anatomical re-organization following early visual cortical lesions. The lesion causes severe shrinkage and retrograde degeneration of the dorsal lateral geniculate nucleus (LGd) and results in increased projections from the retina to the lateral posterior nucleus and to the nucleus of the optic tract (Cunningham 1972, Cunningham et al 1979, Perry & Cowey 1979, 1982). In cats, increased projections from the retina to the pulvinar nucleus is observed and there is also an anomalous survival of large cells in the LGd of infant operated cats (Labar et al 1981; Murphy et al 1975). In rabbits, early studies using Nauta methods failed to demonstrate such changes in retinofugal projections in infant operated rabbits (Stewart & Riesen 1972). However, we have recently re-examined this issue using autoradiographic methods and have found clear evidence of changes similar to those observed in the rat, i.e., there is an increased projection to the lateral posterior nucleus and to the nucleus of the optic tract in infant operated animals but not in adult operated animals (Wilkes & Murphy, unpublished observations).

Retrograde degeneration sometimes occurs transneuronally in response to CNS lesions, especially when the lesions are made in very young animals. Early visual cortical ablations have been reported to result in transneuronal retrograde degeneration of medium size retinal ganglion cells in cats and monkeys (Pearson et al 1981, Weller et al 1979). These medium size cells are presumed to be X cells which project only to LGd and not to the superior colliculus (SC) (Dreher et al 1976, Stone et al 1979); the surviving cells are presumed to be W and Y cells which project to both LGd and to SC. Thus the intact collateral branch is assumed to provide protection from transneuronal retrograde changes. In the rat and the rabbit (Vaney et al 1981, Linden et al 1983), there is evidence that virtually all ganglion cells project to both LGd and SC, so one can hypothesize that retrograde neuronal degeneration in retinal ganglion cells might not occur in rats and rabbits. This result has been reported for rats (Perry & Cowey 1979b), and we now have evidence that the same is true for rabbits: there is no change in the cell size distribution of retinal ganglion cells of rabbits following early visual cortical ablation and long term survival (Wilkes & Murphy, unpublished observations). In rabbits, the critical period for these effects has not yet been established, but in rats it has been reported that the sensitive period is over by the 5th postnatal day (Perry & Cowey, 1982). Thus, the effects of neonatal ablation of visual cortex appear to be quite similar in rat and rabbit, with respect to the absence of transneuronal retrograde degeneration in the retina, the sites of termination of anomalous projections from the retina and the absence of behavioral sparing of visual discrimination.

Neonatal Enucleation. Following early monocular enucleation there is also extensive re-organization of visual system projections. The ipsilateral projection from the remaining eye to SC, which normally terminates in a very restricted area, often covers much of the vacated space normally innervated by the contralateral eye. This expanded termination in the SC is functionally active in rats, hamsters and opossums (Cunningham & Speas 1975; Finlay et al 1979, Rhoades 1980, Takeuchi et al 1982). However, in rabbits, although some increase in the size of the ipsilateral projection to the superior colliculus is observed in neonatal enucleates, it is smaller in extent than that seen in rats, hamsters and opossums, and it is not functionally active (Chow et al 1973). That this difference between rabbits and the other mammals cited above is not simply a function of the degree of immaturity at birth, is shown by the finding that even in rabbits enucleated prenatally the projection is still restricted in size compared with that observed in the rat, and is still not functional (Chow et al 1981).

Early enucleation has also been reported to result in changes not only in retinotectal projections but also in projections from the remaining eye to the thalamus (Guillery 1972, Kalil 1973, So et al 1982). We have shown recently that retinothalamic projections in neonatally enucleated rabbits are also abnormal (Grigonis et al 1983), and these abnormalities are quite similar to those observed in neonatally enucleated hamsters (So et al 1982): the contralateral projection from the remaining eye fills the entire LGd, including the medial band which is normally innervated by the ipsilateral eye. This projection pattern is one that is observed in the normal newborn hamster (Frost et al 1979, So et al 1978), suggesting that early enucleation prevents the normal developmental transition from the neonatal to the adult projection pattern. It is now apparent that the interpretation of many of the changes in projections which are observed after early lesions depends on a knowledge of the state of projections at the time the lesion is made. There is evidence in many mammals, and in many parts of the visual system, that projections at the time of birth are often more 'exuberant' or widespread than in the mature animal, and that interference with normal development, either by deprivation or by early lesions, may result in a failure of the normal retraction of these exuberant projections. Some of the retinofugal projections in normal newborn rabbits have been described (Ostrach et al 1981), and the existence of some exuberant projections in the newborn rabbit has already been established (Chow et al 1981; Hollander et al 1979). However, further interpretation of the differences between rabbits and other mammals in the response of the visual system to early lesions must await a more detailed knowledge of neonatal retinofugal projections.

Nevertheless, the data summarized above clearly indicate that the rabbit is more similar to the rodent than was previously recognized, in both the behavioral and the anatomical consequences of early ablation of visual cortex, and in the anatomical consequences of early enucleation. A major and still unexplained difference between the rabbit and other mammals studied, is that the expanded ipsilateral retinotectal projection in the enucleated rabbit is not functional. However, an investigation of the functional integrity of the other anomalous retinofugal pathways resulting from early lesions in the rabbit, described above, must be made before any general conclusions can be drawn.

MORPHOLOGICAL INDICES OF MATURATION IN THE VISUAL CORTEX

In species other than the rabbit, two aspects of the development of visual cortical neurons have been shown to be temporally related to the critical period for visual experience: spine density, and synapse density. In both measures, a characteristic postnatal developmental trend has been frequently observed: an increase in density to levels _higher_ than that observed in adults, followed by a _decline_ to adult levels. We have called this pattern the peak/decline trend (Murphy & Magness 1983). The development of spines, which has been shown to be a site of synaptic contact (Gray et al 1959, LeVay 1973) is a major feature of postnatal maturation in cortical neurons (Boothe et al 1979, Juraska & Fifkova 1979, Ruiz-Marcos & Valverde 1968). In monkeys, spine density increases rapidly during the first 8 postnatal weeks. Some but not all, neurons show a decline in spine density to adult levels. A similar peak, followed by a decline in spine density of visual cortical neurons has been observed in rodents (Juraska & Fifkova 1979, Ruiz-Marcos & Valverde 1969). In rabbits, physiological evidence had suggested that visual cortical development was complete by 25 days postnatal (Mathers et al 1974). Early studies of ontogenetic changes in spine density did not describe a peak/ decline trend (Mathers, 1979). However this study did not investigate ages beyond 25 days, at which spine densities have reached adult levels. We have since re-examined this issue and have continued the period of observation into the second postnatal month (Murphy & Magness 1983). In this study, we show that in both stellate cells of lamina IV and in pyramidal cells of lamina V, a peak/decline trend is observed, similar to that observed in primates. Spine density increases above adult levels at the end of the first postnatal month and then decrease to adult levels during the second postnatal month. These data are shown in Fig 1.

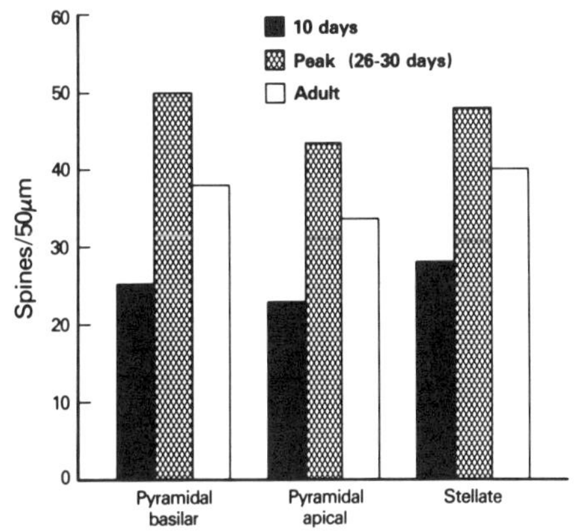

FIGURE 1 Mean spine density of dendrites of Golgi impregnated lamina V pyramidal cells and lamina IV stellate cells in visual cortex of immature and adult rabbits. Basilar pyramidal and stellate density counts are taken from secondary and tertiary dendritic branches. Apical pyramidal density counts are taken from an oblique branch and from the apical dendrite as it traverses lamina IV.

The other index of cortical development which overlaps the visual deprivation critical period in cats is synapse development. In the kitten visual system, in which the critical period extends from the fourth to the twelth week, a peak to supra-adult levels of synapse density has been observed early in the critical period, followed by a decline to adult levels by the end of the critical period (Cragg 1975, Winfield 1983). In rabbits, although the effect is somewhat less striking, a peak/decline trend in synapse density has been reported (Degroot & Vrenson 1978, Muller et al 1981) with the peak occurring towards the end of the first postnatal month. Thus, spine densities and synapse densities of visual cortical neurons in immature rabbits undergo ontogenetic changes in the form of peak/decline trends similar to those observed in cat, monkeys and rodents. If, as we have hypothesized (Murphy & Magness 1983), these peaks do indicate a peak in vulnerability to deprivation, then it should be possible to demonstrate effects of deprivation in rabbit visual cortex during this period.

FUNCTIONAL CONSEQUENCES OF DEPRIVATION

It is now well established that, in the cat, deprivation of visual pattern experience by monocular lid suture during a critical period, which occurs between the fourth and twelth postnatal week, results in a severe and usually permanent disruption of the functional organization of the primary visual cortex (for review, see Sherman & Spear 1982). Other types of visual deprivation are also effective. For example, there is a critical period for artificially induced strabismus, deprivation of movement, and selective deprivation of specific orientations or directions all of which result in changes in visual cortical organization in cats (Berman & Daw 1977, Berman & Murphy 1981, Daw et al 1978, Leventhal & Hirsch 1975).

The critical periods for different deprivations are not identical, and their onset and duration may be experimentally manipulated (Cynader & Mitchell 1980, Mower et al 1983). However, despite these variables in the precise temporal parameters of the critical period, the basic phenomenon of functional alteration following visual deprivation during an early critical period is now well established. The mechanisms underlying the phenomenon are complex, but involve both changes in connectivity (Levay et al 1980, Shatz & Stryker 1978, Winfield 1983), and modifications in the efficacy of synapses (Birchfield & Duffy 1981, Blakemore & Hawken 1982, Blakemore et al 1982, Kratz et al 1976).

Early studies suggested that the functional development of the rabbit visual system was remarkably independent of experience. Both behavioral studies of visual discrimination in rabbits reared in total darkness for the first 7 postnatal months and studies of visual discrimination in lid sutured rabbits indicated no discrimination effects (VanHof & Kobayashi 1972, Chow & Spear 1974). Single cell recordings from primary visual cortex following selective orientation deprivation and following pattern deprivation by lid suture also failed to isolate significant differences between normal and deprived cortex (Chow & Spear 1974, Mize & Murphy 1973). Even in the small binocular segment of primary visual cortex, the disruption of binocularity following monocular lid suture was much less than that observed in cats and monkeys (Van Sluyters & Stewart 1974). Abnormalities are observed in receptive field characteristics in the monocular cortical segment of monocularly deprived cats; abnormalities are observed in binocular cortex of binocularly deprived cats (Buisseret & Imbert 1976, Fregnac et al 1981, Leventhal & Hirsch 1980, Mower et al 1981, Wilson & Sherman 1977). The relative ineffectiveness of monocular suture in rabbits can not be attributed solely to the fact that the rabbit has laterally placed eyes and a predominantly monocular visual system (Hughes 1971, Hughes & Vaney 1982). Subsequent studies in rabbits indicated that lid suture does in fact cause some delay in the development of receptive fields, and that, if the deprivation is extended over a very long time period, some abnormalities of receptive field organization occur (Crabtree et al 1981, Grobstein et al 1975). However, this appears to be a different phenomenon from that observed in the cat, since binocularity in the cat can be disrupted by as little as a few hours of monocular exposure during the critical period (Schechter & Murphy 1976, Olson & Freeman 1975).

A possible explanation for the relative ineffectiveness of lid suture in modifying visual cortical organization in the rabbit might be sought in a consideration of differences in the basic organization of receptive field properties of the cat and rabbit. The physiological organization of the visual cortex of the mature rabbit has been described in detail (Chow et al 1971, Bousfield 1977, Glanzman 1983, Hughes & Vaney 1982; Murphy & Berman 1979, Reuter & Chow 1982). These studies differ on some points, but there is good agreement that, although a significant percentage of rabbit cortical cells have orientation selective cells which can be classified as simple or complex (60-70% in Murphy & Berman 1979), there is also a significant percentage of rabbit visual cortical cells which are not orientation selective but are responsive to moving or transient, non-oriented stimuli. (For a more detailed discussion of differences between cat and rabbit visual cortical receptive fields, see Murphy & Berman, 1979). Examples of the responses of an orientation selective cell and a non-orientation selective cell are shown in Figure 2. This lower percentage of cortical cells with orientation selective receptive fields in rabbits, suggests that the rabbit visual cortex is less specialised for the discrimination of orientation selectivity than cat and monkey visual cortex, in which the vast majority of cells have orientation selective receptive fields and in which orientation selectivity is particularly sensitive to pattern deprivation. We therefore decided to investigate plasticity in rabbit visual cortex by using a selective form of visual deprivation more appropriate to the organization of the rabbit visual cortex. The rabbit visual system is highly tuned for detection of direction of movement at all levels of the system from retinal ganglion cells through the tectum, pretectum, accessory optic nuclei, thalamus and cortex (Ariel & Daw 1982, Barlow & Levick

1965, Collewijn 1975, Levick et al 1969, Reuter & Chow 1982, Semm 1978, Simpson et al 1979, Winterson & Collewijn 1981). Although rabbits have fewer orientation selective visual cortical cells than cats, they are similar to cats in that almost all orientation selective cells also have a preferred direction of movement of the oriented stimulus. An example of such a cell is shown in Fig. 2. We therefore sought a deprivation to which direction selectivity might be vulnerable. It had already been reported that direction selectivity, but not orientation selectivity, is almost totally abolished in cat visual cortical cells following rearing in an environment illuminated only by a stroboscopic light. This environment permits experience of patterned oriented stimuli, but no experience of movement or direction (Cynader & Chernenko 1976, Cynader et al 1973, Kennedy & Orban 1983). We therefore raised rabbits in a stroboscopically illuminated environment.

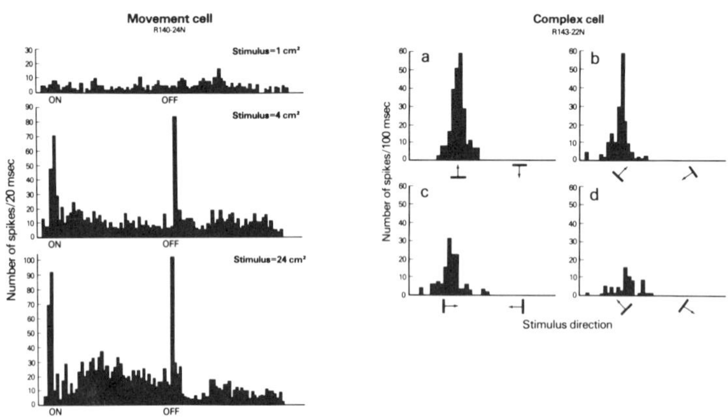

FIGURE 2 On the left, poststimulus time histograms of the response of a movement sensitive cell are displayed. The cell responds strongly to the onset and offset of light, shows internal summation but no surround inhibition. On the right are poststimulus time histograms of the response of a complex cell. The cell has no spontaneous activity, strong direction selectivity and responds most vigorously to a horizontally oriented stimulus.

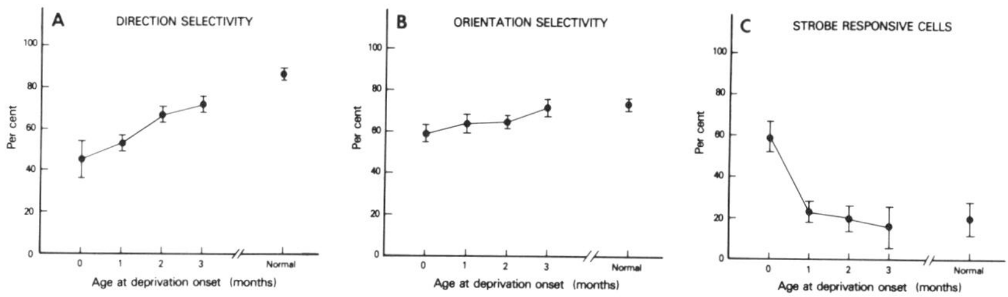

FIGURE 3 The percentage of cells in rabbit visual cortex which are: A) Direction selective; B) Orientation selective, C) Responsive to a stroboscopic stimulus, is shown for normal rabbits and for rabbits reared in a stroboscopically illuminated environment. Rabbits were placed in the strobe environment at birth, or following 1, 2 or 3 months of normal visual experience. Reproduced, with permission, from Pearson HE, Berman N & Murphy EH. Exp Brain Res (1983) 50:367.

The results of this study were clear (Pearson et al 1981). Rabbit visual cortex is sensitive to early environmental experience. The percentage of orientation selective cells was not greatly changed in strobe reared rabbits, but the percentage of those cells which were also direction selective was dramatically reduced from about 90% to 40%. Another major change in cortical responses was that there was a significant increase in the percentage of cells responsive to a stroboscopic stimulus in strobe reared animals.

Our next goal was to establish whether rabbits also have a critical period, and, if so, whether it overlaps the period of supra-adult peaks in spine density and synapse density. Rabbits were placed in the stroboscopically illuminated room at birth or at the ages of 1, 2 or 3 months postnatally. Results are shown in Figure 3. The greatest effect of strobe rearing is seen in animals deprived from birth. However, for direction selectivity, the effects of deprivation are still significant for rabbits deprived after 3 months of normal experience. For orientation selectivity, the effects are smaller for all conditions, and there is no significant difference in orientation selectivity for animals deprived later than 2 months. These results indicate that the critical period for orientation selectivity ends earlier than the critical period for direction selectivity. The critical period for the change in probability of a cell responding to a stroboscopic stimulus ends even earlier, by the end of the first month. These results show not only that there is a critical period for experience of movement in rabbit visual cortex, but also that this critical period differs for different response characteristics. It is also noteworthy that the period of maximum sensitivity, although not precisely defined by these experiments, appears to occur during the first month and end during the second month, i.e., it overlaps the time period during which peak spine and synapse densities are observed in the rabbit visual cortex.

Our results, indicating that direction selectivity is severely reduced by early strobe rearing, suggest that this deprivation is a particularly powerful one for the rabbit visual system in which the detection of direction of movement is so predominant. We suggested earlier that pattern deprivation by lid suture might not be an optimum deprivation for the rabbit visual system since rabbit visual cortex is less specialised for orientation discrimination. However, those earlier studies, which reported no significant differences between normal and lid-sutured rabbits in the organization of cortical receptive field properties, did not report on the percentage of orientation selective cells which were also direction selective. It is possible, therefore, that our results indicate not that strobe rearing is a particularly effective deprivation for the rabbit, but that direction selectivity in the rabbit is particularly susceptible to disruption by deprivation of any sort. We therefore decided to repeat the studies of pattern deprivation by rearing rabbits with monocular lid suture (Murphy 1982).

The results are shown in Figure 4 and reveal a significant loss of direction selectivity in monocularly deprived rabbits compared with normals but no significant change in orientation selectivity. Thus the results indicate that direction selectivity _is_ a characteristic of rabbit visual cortex which is particularly vulnerable to deprivation. However, the results also indicate that strobe rearing is a more powerful deprivation than lid suture, since the loss of direction selectivity in lid sutured rabbits is significantly less than the loss observed in strobe reared rabbits (Fig. 4). Finally, it must be noted that even in strobe reared rabbits, 40% of cells still maintain direction selectivity, whereas in strobe reared cats, the loss of cortical direction selectivity is almost complete. This suggests that there is a population of cells in the rabbit visual cortex which develops or maintains direction selectivity despite this powerful deprivation; such a population of cells is small or absent in the cat.

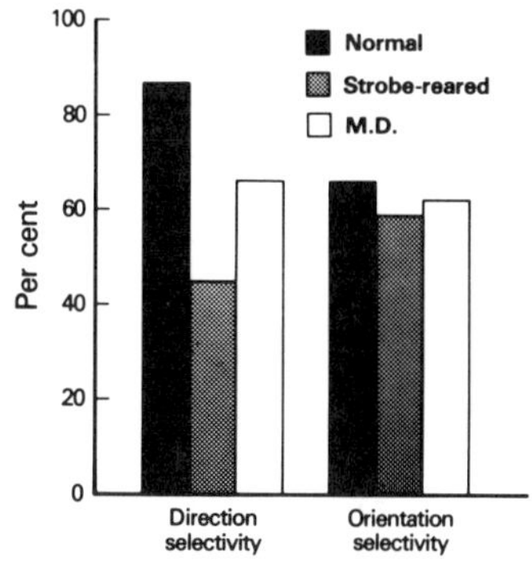

FIGURE 4 The percentage of cells which are direction selective or orientation selective recorded from visual cortex of normal, strobe-reared or monocularly deprived (MD) rabbits.

DEVELOPMENT OF ORIENTATION SELECTIVITY AND DIRECTION SELECTIVITY IN THE NORMAL RABBIT

The question of whether deprivation causes a breakdown of existing organization or a failure to develop is one of continuing interest in the search for the mechanisms involved in normal and abnormal development. In order to answer this question, we have recorded from immature, normally reared rabbits, and have studied the time course of development of orientation selectivity and direction selectivity. Our hypothesis was that, since direction selectivity is more sensitive to early deprivation than orientation selectivity and since the critical period for direction selectivity extends later in ontogeny than the critical period for orientation selectivity, we would expect direction selectivity to develop later than orientation selectivity.

The results of this study are shown in Figure 5, and fit well with our predictions. In rabbits aged 30 days or less, the percentages of both orientation selectivity and direction selectivity are significantly lower than in adults. By 30-50 days, orientation selectivity has reached adult levels, but direction selectivity attains adult levels much more slowly.

The results suggest that, in the rabbit visual cortex, different subpopulations of cells exist, which differ in their rate of development and in their response to deprivation. Some cells maintain direction selectivity despite deprivation; some cells fail to develop direction selectivity in the absence of normal experience, and some cells must lose direction selectivity as a result of early strobe rearing.

It is well established that direction selectivity involves inhibitory mechanisms (Sillito 1977, Sillito et al 1980), which may be among the latest to develop in the mammalian cortex (Komatsu 1983, Winfield 1983). Isolation and identification of those parameters which determine the developmental program of subpopulations of cells and which determine their response to experimental manipulation should prove fruitful in contributing to our understanding of these mechanisms.

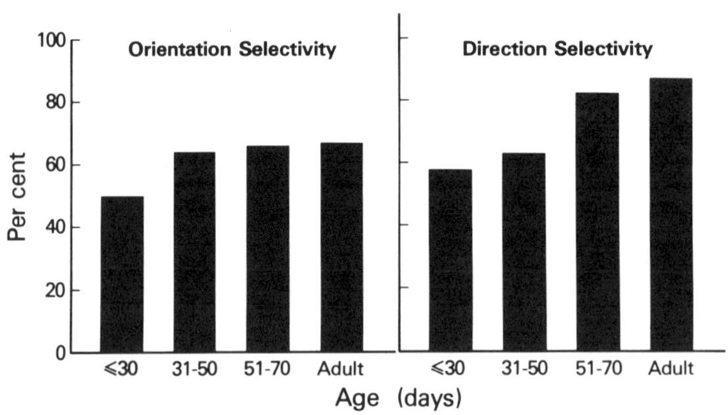

FIGURE 5 The percentage of cells which are orientation selective or direction selective recorded from visual cortex of immature and adult rabbits.

CONCLUSIONS

In summarizing the results of these recent studies of the development of the rabbit visual system, we must note both differences and similarities between the rabbit and other commonly studied mammals. Similarities include: a) rabbits are similar to rodents in the anatomical re-organization which follows early lesions of the visual system, and in the relative absence of behavioral sparing following such lesions; b) the rabbit is similar to all other mammals studied in showing a peak/decline trend postnatally in measures of spine and synapse density, c) the rabbit is similar to the cat in showing some loss of direction selectivity in visual cortical neurons following early rearing in a stroboscopically illuminated environment.

Differences include: a) direction selectivity, but not orientation selectivity, is decreased by early deprivation in the rabbit: in the cat both direction selectivity and orientation selectivity can be disrupted by early deprivation; b) in the rabbit, direction selectivity develops later in ontogeny than orientation selectivity; in the cat, the available evidence indicates that orientation selectivity develops later than directional selectivity (Pettigrew 1974). c) the loss of direction selectivity is much less complete in strobe-reared rabbits than in cats. This suggests that, in rabbits, but not in cats, there is a population of cortical cells whose development and maintenance of direction selectivity is not dependent on experience; d) in rabbits there is evidence that following early lesions, the resultant increased retino-tectal projections are not functional. This has been observed in no other mammal, and deserves further investigation.

Comparative studies of similarities and differences between different mammals, in their normal development and in their response to experimental manipulations of the internal and external environment, should provide insight into some of the rules governing development. Recent evidence, summarised in this review, indicates that the rabbit is similar to other mammals in that its visual system is modified by experimental manipulations. However, there are significant differences in the nature and extent of their modifications. These differences provide the opportunity to address such issues as the identification of characteristic features of deprivation-sensitive and deprivation-insensitive neurons, the relationship between ontogenetic development and susceptibility to deprivation, and the factors which determine the functional integrity of anomalous projections. Thus, the visual system of the rabbit provides a valuable model system for the exploration of some of the current issues in developmental neurobiology.

Ariel M, Daw NW (1982). Pharmacological analysis of directionally sensitive rabbit retinal ganglion cells. J Physiol 324:161.
Aslin RN (1981). Experiential influences and sensitive periods in perceptual development: a unified model. In Aslin et al. (ed): "Development of Perception", New York: Academic Press, p 45.
Barlow HB, Levick WR (1965). The mechanism of directionally sensitive units in rabbit retina. J Physiol 178:477.
Bateson PPG (1981). The control of sensitivity to the environment during development. In Immelman K, Barlow GW, Petrinovich L, Main M (eds): "Behavioral development: The Bielefeld Interdisciplinary Project", New York: Cambridge University Press.
Berman N, Daw NW (1977). Comparison of the critical periods for monocular and directional deprivations in kittens. J Physiol 265:249.
Berman N, Murphy EH (1982). The critical period for alteration in cortical binocularity resulting from divergent and convergent strabismus. Dev Brain Res 2:181.
Birchfield JL, Duffy FH (1981). Role of intracortical inhibition in deprivation amblyopia: reversal by micro-iontophoretic bicuculline. Brain Res 206:479.
Bischof HJ (1983). Imprinting and cortical plasticity: A comparative review. Neurosci & Behav Rev 7:213.
Blakemore C (1978). Maturation and modification in the developing visual system. In Held R, Leibowitz H, Teuber HL (eds): "Handbook of Sensory Physiology, Perception", Berlin, Heidelberg, New York: Springer Verlag, p. 377-436.
Blakemore C, Hawken MJ (1982). Rapid restoration of functional input to the visual cortex of the cat after brief monocular deprivation. J Physiol 327:463.
Blakemore C, Hawken MJ (1982). Brief monocular deprivation leaves subthreshold synaptic input on neurones of the cat's visual cortex. J Physiol 327:489.
Boothe RG, Greenough WT, Lund JS, Wrege K (1979). A quantitative investigation of spine and dendrite development of neurons in visual cortex (area 17) of Macaca nemestrina monkeys. J Comp Neurol 180:277.
Bousfield JD, (1977). Columnar organization and the visual cortex of the rabbit. Brain Res 136:154.
Buisseret P, Imbert M (1976). Visual cortical cells: their developmental properties in normal and dark-reared kittens. J Physiol 225:511.
Chow KL, Baumbach HD, Lawson R. (1981). Callosal projections of the striate cortex in the neonatal rabbit. Exp Brain Res 42:122.

Chow KL, Masland RH, Stewart DL (1971). The receptive field characteristics of striate cortical neurons in the rabbit. Brain Res 33:337.

Chow KL, Mathers LH, Spear PD (1973). Spreading of uncrossed retinal projections in superior colliculus of neonatally enucleated rabbits. J Comp Neurol 151:307.

Chow KL, Ostrach LH, Crabtree JW, Bernegger O, Baumbach HD, Lawson R (1981). Anomalous uncrossed retinal projections fail to activate superior colliculus neurons in rabbits unilaterally enucleated by fetal surgery. J Comp Neurol 196:189.

Chow KL, Spear PD (1974). Morphological and functional effects of visual deprivation on the rabbit visual system. Exp Neurol 42:429.

Collewijn H (1975). Direction-selective units in the rabbit's nucleus of the optic tract. Brain Res 100:489.

Leventhal AG, Hirsch HUB (1975). Effects of early selective exposure to diagonal lines. Science 190:902.

Crabtree JW, Chow KL, Conlee J, Ostrach, LH, Grobstein P (1981). Effects of progressively longer durations of monocular deprivaton on development of visuocortical receptive fields in the rabbit. Neurosci Lett 26:61.

Cragg, BG (1975). The development of synapses in the visual system of the cat. J Comp Neurol 160:279.

Cunningham, TJ (1972). Sprouting of optic projections after cortical lesions. Anat Rec 172:298.

Cunningham, TJ, Speas G (1975). Inversion of anomalous uncrossed projections along the mediolateral axis of the superior colliculus: implications for retino-collicular specificity. Brain Res 88:73.

Cunningham TJ, Huddleston C, Murray M (1979). Modification of neuron number in the visual system of the rat. J Comp Neurol 184:423.

Cynader M, Berman N, Hein A (1973). Cats reared in stroboscopic illumination: effects on receptive fields in visual cortex. Proc Nat Acad Sci, USA. 70:1353.

Cynader M, Chernenko G (1976). Abolition of direction selectivity in the visual cortex of the cat. Science 193:504.

Cynader M, Mitchell, DE (1980). Prolonged sensitivty to monocular deprivation in dark reared cats. J Neurophysiol 43:1026.

Daw NW, Berman N, Ariel M (1978). Interaction of critical periods in the kitten visual cortex. Science 197:565.

De Groot D, Vrenson G. (1978). Postnatal development of synaptic contact zones in the visual cortex of rabbits. Brain Res 147:362.

Dreher B, Fukada Y, Rodieck R W (1976). Identification, classification and anatomical segregation of cells with X-like and Y-like properties in the lateral geniculate nucleus of old-world primates. J Physiol 258:433.

Finlay BL, Wilson KG, Schneider GE (1979). Anomalous ipsilateral retinotectal projections in Syrian hamsters with early lesions: topography and functional capacity. J Comp Neurol 183:721.

Freeman RD, Mallach R, Hartley S (1981). Responsivity of normal kitten striate cortex deteriorates after brief binocular deprivation. J Neurophysiol 45:1074.

Fregnac Y, Trotter Y, Bienenstock E, Buisseret P, Gary-Bobo E, Imbert M (1981). Effect of neonatal unilateral enucleation on the development of orientation selectivity in the primary visual cortex of normally and dark-reared kittens. Exp Brain Res 42:453.

Frost DO, So KF, Schneider GE, Sanderson TO (1979). Postnatal development of retinal projections in Syrian hamsters: a study using autoradiographic and anterograde degeneration techniques. Neurosci 4:1649.

Glanzman, DL (1983). Spatial properties of cells in the rabbit's striate cortex. J Physiol 340:535.

Graham J, Pearson HE, Berman N, Murphy EH (1981). Laminar organization of superior colliculus in the rabbit: a study of receptive field properties of single units. J Neurophysiol 45:915.

Gray EG (1959). Axo-somatic and axo-dendritic synapses of the cerebral cortex: an electron microscope study. J Anat 93:420.

Grigonis AM, Murphy EH, Pearson HE (1983). Retinothalamic projections in the rabbit following monocular enucleation. Neurosci Abstr 9:701.

Grobstein P, Chow KL, Fox PC (1975). Development of receptive fields in rabbit visual cortex: changes in time course due to delayed eye-opening. Proc Nat Acad Sci 72:1543.

Guillery, RW (1972). Experiments to determine whether retinogeniculate axons can form translaminar collateral sprouts in the dorsal lateral geniculate nucleus of the cat. J Comp Neurol 146:407.

Hendrickson A, Dineen JT (1982). Hypertrophy of neurons in dorsal lateral geniculate nucleus following striate cortex lesions in infant monkeys. Neurosci Lett 30:217.

Hess E (1973). Imprinting. New York: van Nostrand Reinhold.
Hughes A (1971). Topographic relationships between the anatomy and physiology of the rabbit visual system. Documenta Opthal 30:33.
Hughes A, Vaney DI (1982). The organization of binocular cortex in the primary visual area of the rabbit. J Comp Neurol 204:151.
Johnston MV, Coyle JT (1981). Development of central neurotransmitter systems. In Ciba Foundation (ed): "The Fetus and Independent Life", London, Pitman: CIBA Foundation Symposium #86.
Komatsu Y (1983). Development of cortical inhibition in kitten striate cortex investigated by a slice preparation. Dev Brain Res 8:136.
Juraska JM, Fifkova E (1979). A Golgi study of the early postnatal development of the visual cortex of the hooded rat. J Comp Neurol 183:247.
Kalil RE (1973). Formation of new retinogeniculate connections in kittens: effects of age and visual experience. Anat Rec 175:353.
Kennedy H, Orban GA (1983). Response properties of visual cortical neurons in cats reared in stroboscopic illumination. J Neurophysiol 49:686.
Kratz, KE, Spear PD, Smith DC (1976). Post-critical period reversal of effects of monocular deprivation on the striate cortex in cats. J Neurophysiol 39:501.
Labar DR, Berman NE, Murphy EH (1981). Short and long-term effects of neonatal and adult visual cortex lesion on the retinal projection to the pulvinar in cats. J Comp Neurol 197:639.
LeVay, S. (1973). Synaptic patterns in the visual cortex of the cat and monkey. Electron microscopy of Golgi preparations. J Comp Neurol 150:53.
LeVay S, Wiesel TN, Hubel DH (1980). The development of ocular dominance columns in normal and visually deprived monkeys. J Comp Neurol 191:1.
Leventhal AG, Hirsch HVB (1980). Receptive field properties of different classes of neurons in visual cortex of normal and dark-reared cats. J Neurophysiol 43:1111.
Levick WR, Oyster CW, Takahashi E (1969). Rabbit lateral geniculate nucleus: sharpener of directional information. Science 165:172.
Linden R, Cowey A, Perry VH (1983). Tectal ablation at different ages in developing rats has different effects on retinal ganglion cell density but not on visual acuity. Exp Brain Res 51:368.

Lund RD, Cunningham TJ, Lund JS (1973). Modified optic projections after unilateral eye removal in young rats. Brain Behav Evol 8:51.

Mathers LH (1979). Postnatal dendritic development in the rabbit visual cortex. Brain Res 168:21.

Mathers LH, Chow KL, Spear PD, Grobstein P (1974). Ontogenesis of receptive fields in the rabbit striate cortex. Exp Brain Res 19:20.

Mize RR, Murphy EH (1973). Selective visual experience fails to modifiy receptive field properties of rabbit striate cortex neurons. Science 180:320.

Moore DT, Murphy EH (1976). Differential effects of two visual cortical lesions in the rabbit. Exp Neurol 53:21.

Movshon JA, Van Sluyters RC (1981). Visual neural development. Ann Rev Psychol 32:447.

Mower GD, Berry D, Burchfield JL, Duffy FH (1981). Comparison of the effects of dark rearing and binocular suture on development and plasticity of cat visual cortex. Brain Res 220:255.

Mower GD, Christen WG, Caplan CJ (1983). Very brief visual experience eliminates plasticity in the cat visual cortex. Science 221:178.

Muller L, Pattiselanno A, Vrenson G (1981). The postnatal development of the presynaptic grid in the visual cortex of rabbits and the effect of dark rearing. Brain Res 205:39.

Murphy EH (1982). Monocular suture disrupts the functional organization of rabbit striate cortex. Neurosci Abstr 8:297.

Murphy EH, Berman N (1979). The rabbit and the cat: a comparison of some features of single cells in the primary visual cortex. J Comp Neurol 188:401.

Murphy EH, Kalil RE (1979). Functional re-organization of cells in the lateral geniculate nucleus following removal of visual cortex in the newborn kitten. Science 206:713.

Murphy EH, Magness R (1983). Development of the rabbit visual cortex: a quantitative Golgi analysis. Exp Brain Res (in press).

Murphy EH, Mize RR, Schechter PB (1975). Visual discrimination following infant and adult ablation of areas 17, 18 and 19 in the cat. Exp Neurol 49:386.

Murphy EH, Stewart DL (1974). Effects of neonatal and adult striate lesions on visual discrimination in the rabbit. Exp Neurol 42:89.

Olson CR, Freeman RD (1975). Progressive changes in kitten striate cortex during monocular vision. J Neurophysiol 38:26.

Ostrach LH, Chow KL, Crabtree JW (1981). Postnatal development of ipsilateral retinotectal and retinogeniculate projections in the rabbit. Invest Opthal Vis Sci Suppl 20:73,

Pearson HE, Berman N and Murphy EH (1981). Stroboscopic rearing reduces direction selectivity in rabbit visual cortex. Dev Brain Res 1:127.

Pearson HE, Berman N, Murphy EH (1983). Critical periods in development for susceptibility to the effects of stroboscopic rearing in the rabbit visual cortex. Exp Brain Res 50:367.

Pearson HE, Labar DR, Payne BR, Cornwell P, Aggarwal N (1981). Transneuronal retrograde degeneration in the cat retina following neonatal ablation of visual cortex. Brain Res 212:470-475.

Pettigrew JD (1974). The effect of visual experience on the development of stimulus specificity by kitten cortical neurones. J Physiol 237:49.

Perry VH, Cowey A (1979). Changes in the retinofugal pathways following cortical and tectal lesions in the neonatal and adult rat. Exp Brain Res 35:97.

Perry VH, Cowey A (1979b). The effects of unilateral cortical and tectal lesions on retinal ganglion cells in rats. Exp Brain Res 35:85.

Perry VH, Cowey A (1982). A sensitive period for ganglion cell degeneration and the formation of aberrant connections following tectal lesions in rats. Neurosci 7:583.

Rakic P (1977). Prenatal development of the visual system in rhesus monkey. Phil Trans Roy Soc Lon B 278:245.

Reuter JH, Chow KL (1982). Velocity tuning of motion sensitive and direction selective cells in the rabbit striate cortex. Behav Brain Res 6:237.

Rhoades RW (1980). Effects of neonatal enucleation on the functional re-organization of the superior colliculus in the golden hamster. J Physiol 301:383.

Ruiz-Marcos A, Valverde F (1969). The temporal evolution of the distribution of dendritic spines in the visual cortex of normal and dark raised mice. Exp Brain Res 8:281.

Schechter PB, Murphy EH (1976). Brief monocular visual experience and kitten cortical binocularity. Brain Res 109:165.

Schneider GE (1973). Early lesions of the superior colliculus: factors affecting the formation of abnormal retinal projections. Brain Behav Evol 8:73.

Schneider GE (1977). Growth of abnormal neural connections following focal brain lesions; constraining factors and functional effects. In Sweet WH, Obrador S, Matrin-Rodriquez JG (eds): "Neurosurgical Treatments in Psychiatry, Pain and Epilepsy." Baltimore: University Park Press , p 5.

Semm P (1978). Antidromically activated direction selective ganglion cells of the rabbit. Neurosci Lett 9:207.

Shatz CJ, Stryker MP (1978). Ocular dominance in layer IV of the cat's visual cortex and the effects of monocular deprivation. J Physiol 282:267.

Sherman SM, Guillery RW, Kaas JH, Sanderson KJ (1974). Behavioral, electrophysiological and morphological studies of binocular competition in the development of the geniculo-cortical pathways of cats. J Comp Neurol 158:1.

Sherman SM, Spear PD (1982). Organization of visual pathways in normal and visually deprived cats. Physiol Rev 62:738-855.

Simpson JI, Soodak RE, Hess R (1979). The accessory optic system and its relation to the vestibulocerebellum. In Granit R, Pompeiano O (eds): "Reflex Control of Posture and Movement", New York: Elsevier/North Holland Press.

So KF, Schneider GE, Frost DO (1978). Postnatal development of retinal projections to the lateral geniculate body in Syrian hamsters. Brain Res 142:343.

So KF, Woo HH, Jen LS (1982). Effect on segregation of retinogeniculate projections in early unilateral enucleated or dark-reared hamsters. Neurosci. Abstr. 8:816.

So KF, Schneider GE, Ayres S (1981). Lesions of the brachium of the superior colliculus in neonate hamsters: correlation of anatomy with behavior. Exp Neurol 72:379.

Spear PD (1979). Behavioral and neurophysiological consequences of visual cortex damage: mechanisms of recovery. In Sprague JM, Epstein AN (eds). "Progress in Psychobiology and Physiological Psychology", New York: Academic Press. 8:45.

Spear PD, Kalil RE, Tong L (1980). Functional compensation in lateral suprasylvian visual area following neonatal visual cortex removal in cats. J. Neurophysiol 43:851.

Stewart D, Riesen A (1972). Adult versus infant brain damage: Behavioral and electrophysiological effects of striatectomy in adult and neonatal rabbits. Adv Psychobiol 1:171.

Stone J, Dreher B, Leventhal AC (1979). Hierarchical and parallel mechanisms in the organization of visual cortex. Brain Res Rev 1:345.

Takeuchi H, Fukuda Y, Hara Y, Hsiao C (1982). Physiological properties of expanded ipsilateral retinocollicular projections in neonatally enucleated albino rats. Brain Res 231:191.

Tees RC (1975). The effects of neonatal striate lesions and visual experience on form discrimination in the rat. Can J Psychol 29:66.

Thompson ID (1979). Changes in the uncrossed retinotectal projection after removal of the other eye at birth. Nature 279:63.

Thompson JM, Woolsey CN, Talbot SA (1950). Visual areas I and II of cerebral cortex of rabbit. J Neurophysiol 13:277.

Tong L, Spear PD, Kalil RE (1981). The critical period for functional compensation in lateral suprasylvian visual area following neonatal visual cortex removal in cats. Neurosc Abstr 7:675.

Vaney DL, Peichl L, Wassle H, Illing RB (1981). Almost all ganglion cells in the rabbit retina project to the superior colliculus. Brain Res 212: 447.

Van Hof MW, Kobayashi K (1972). Pattern discrimination in rabbits deprived of light for seven months after birth. Exp Neurol 35:551.

Van Sluyters R, Stewart DL (1974). Binocular neurons of the rabbit's visual cortex: effects of monocular sensory deprivation. Exp Brain Res 19:196.

Watkins DW, Wilson JR, Sherman SM (1977). Receptive field properties of neurons in binocular and monocular segments of striate cortex in cats raised with binocular suture. J Neurophysiol 41:332.

Weiskrantz L (1963). Contour discrimination in a young monkey with striate cortex ablation. Neurophysiol. 1:145.

Weller RE, Kaas JH (1980). Loss of retinal ganglion cells and altered retinogeniculate projections in monkeys with striate cortex lesions. Invest Opthalmol 19:2.

Weller RE, Kaas JH, Wetzel AB (1979). Evidence for the loss of X-cells of the retina after long term ablation of visual cortex in monkeys. Brain Res. 160:134.

Wiesel TN, Hubel DH (1965). Comparison of the effects of unilateral and bilateral eye closure on cortical unit responses in kittens. J Neurophysiol 28:1029.

Wilson, JR, Sherman SM (1977). Differential effects of early monocular deprivation on binocular and monocular segments of cat striate cortex. J Neurophysiol 40:892. 114.

Winfield DA (1983). The postnatal development of synapses in the different laminae of visual cortex in the normal kitten and in kittens with eyelid suture. Dev Brain Res 9:155.

Winterson BJ, Collewijn H (1981). Inversion of direction-selectivity to anterior fields in neurons of nucleus of the optic tract in rabbits with ocular albinism. Brain Res 220:31.

The assistance of Dr. Antony Grigonis, Ms. Mary Wilkes and Ms. Gloria Zingaro in many of the experiments reviewed above is gratefully acknowledged.

Development of Visual Pathways in Mammals, pages 463-467

DEVELOPMENT OF THE MONKEY'S GENICULO-CORTICAL SYSTEM

Colin Blakemore and François Vital-Durand

University Laboratory of Physiology
Parks Road, Oxford OX1 3PT, U.K. and
Laboratoire de Neuropsychologie Expérimentale
INSERM Unité 94, 16 Avenue du Doyen Lepine,
69500 Bron, France

Monkeys, close as they are to man in genetic make-up and in the general organization of the visual system, would seem to be ideal experimental animals for the study of visual development in relation to human developmental disorders of vision. There have indeed been a variety of reports of the effects of monocular deprivation and artificial strabismus on the properties of neurons in the primate striate cortex (e.g. Baker, Grigg, von Noorden 1974; von Noorden, Crawford 1978; Blakemore, Garey, Vital-Durand 1978; LeVay, Hubel, Wiesel 1980; Blakemore, Vital-Durand, Garey 1981; Swindale, Vital-Durand, Blakemore 1981). These studies were concerned with the ocular dominance of cortical cells and the distribution, in layer 4 of the cortex, of the terminals of afferent axons from the lateral geniculate nucleus (LGN). In these respects the striate cortex of primates seems generally similar in the extent and nature of its plasticity to that of the cat, which has been much more extensively studied (see Movshon, Van Sluyters 1981).

Amblyopia, a very common human clinical disorder associated with deprivation of form vision, defocus of the retinal image or strabismus (squint) early in life, is characterized by a reduction in visual acuity and usually also in contrast sensitivity, even with optimal refractive correction, in the absence of any obvious ocular or retinal pathology. Infant monkeys can certainly become amblyopic, defined by reduced visual acuity, after similar forms of abnormal visual experience: it is therefore of interest to ask how the neural mechanism for the analysis of contrast and fine detail in the retinal image matures in infant monkeys and how the spatial properties of neurons in the visual pathway are affected by early deprivation of normal patterned vision, which rapidly and potently precipitates amblyopia.

Over the past few years, then, we have been studying the receptive fields, and especially their spatial properties, for cells in the LGN and striate cortex of Old world primates (mainly *Erythrocebus patas*, but some vervet, cynomolgus and rhesus monkeys) at various ages, from the day of birth onwards.

LATERAL GENICULATE NUCLEUS

The ability of neurons in the foveal representation at the posterior pole of the LGN to resolve high-contrast drifting gratings increases gradually over the first year or so of life (Blakemore, Vital-Durand 1979, 1981). On the day of birth those cells with the best spatial performance are unable to resolve high-contrast gratings above about 5 c/deg (cycles per degree of visual angle), whereas in the adult, the best X (linear-summating) cells in both parvocellular and magnocellular layers are able to resolve 35 c/deg or more. The time-course of this change is similar to that of the behaviourally-determined postnatal improvement in visual acuity (Teller, Regal, Videen, Pulos 1978; Teller 1981) suggesting that the maturation of some element in the peripheral part of the visual pathway normally limits behavioural acuity during early development. Candidates for this limiting peripheral process or processes include growth of the eyeball (which could cause effective magnification of the image at the photoreceptor level), improvement in the optical quality of the eye and increased packing of receptors in the central fovea. All of these processes do occur and undoubtedly they contribute to the improvement in neural performance, but quantitative analysis suggests that collectively they could account for only a fraction of the change in properties of LGN cells; therefore some kind of refinement of synaptic connexions between optic tract axons and geniculate cells or, more likely, in the retina itself, is likely to be occurring and to be making a substantial contribution to the maturational improvement.

One might imagine, then, that conditions leading to a failure of normal visual acuity would somehow interfere with this peripheral maturation. This seems not to be the case, at least for deprivation of form vision. We examined LGN cells in the foveal representation of monkeys that had been deprived of patterned vision by closure of the lids of one eye (ipsilateral to the LGN studied) from within a few hours after birth until as late as seven months of age. To our surprise we found little if any difference between the spatial properties (or indeed any other properties) of cells in the deprived and the non-deprived layers - certainly no more than the normal small difference in spatial resolution between layers with ipsilateral and contralateral input (Blakemore, Vital-Durand 1979).

STRIATE CORTEX

There now seems to be general agreement that the maturation of the primary visual cortex in kittens involves both innate and environmental influences (see Blakemore 1978; Movshon, Van Sluyters 1981). Without visual experience many cortical neurons in the kitten lack proper orientation selectivity and the full development of orientation columnar structure depends on visual stimulation. Nevertheless there is a substantial proportion of cells that have orientation selectivity without visual experience, and retain it even during binocular deprivation. This population of innately pre-specified, resistant neurons is distinguished by their 'simple' receptive field properties, their tendency to be monocularly driven and to be located in the deeper layers, and by their predominant preference for either vertical or horizontal orientations.

We have now examined this question in the monkey and are struck by certain similarities between kitten and monkey. We have recorded from 476 cortical neurons (collected in a manner designed to provide representative samples) from 13 normally experienced monkeys ranging in age from the day of birth to more than 5 years, and from 244 cells in 6 animals, between 5 days and 130 days of age, that had been deprived by lid suture of both eyes from within a few hours of birth. Even on the day of birth about 40% of cells were clearly orientation selective but, with normal vision, during the first week of life the percentage rose to the adult level and the average orientation 'tuning' became much narrower. With the same time course the proportion of binocularly driven cells rose from 50% to 80%. Without visual experience, however, the fraction of orientation selective cells remained quite constant for the first two weeks of life and then started to fall: binocularity also fell from 50% to 20%. Interestingly, those cells that were orientation selective without visual experience had a strong tendency to be 'tuned' to horizontal or vertical.

Now we have used techniques similar to those used for LGN neurons to measure spatial resolution and contrast sensitivity of cells in the foveal area of the primary visual cortex (Blakemore, Vital-Durand 1982, 1983). In normal monkeys, at each age the best cortical cells have spatial resolutions similar to those of the best LGN cells, and the peak contrast sensitivity (reciprocal of threshold contrast) improves, over the first year or more of life, by a factor of about 10. However, in contrast to the LGN, deprivation of vision leads to an almost complete failure of spatial maturation. Cortical neurons in deprived animals (up to 130 days of age) have spatial resolutions and contrast sensitivities similar to those seen on the day of birth.

These results imply that while the substantial maturation of spatial properties in the LGN does not require patterned visual stimulation, both the orientation selectivity of cortical neurons and their spatial characteristics depend on patterned visual experience. The difference in time course for the maturation of these two aspects of cortical neurons is striking: orientation selectivity and columnar structure are essentially mature at 1-2 weeks of age, whereas spatial resolution and contrast sensitivity take 1-2 years to mature fully. Work presently under way should reveal the mechanism by which cortical cells normally follow LGN neurons in their spatial resolution but fail to do so in the absence of adequate visual stimulation.

Baker FH, Grigg P, von Noorden GK (1974) Effects of visual deprivation and strabismus on the response of neurons in the visual cortex of the monkey, including studies on the striate and prestriate cortex in the normal animal. Brain Res 66: 185.

Blakemore C (1978) Maturation and modification in the developing visual system. In Held R, Leibowitz HW,

Teuber H-L (eds): "Handbook of Sensory Physiology", Vol VIII, "Perception", Berlin: Springer, p 377.

Blakemore C, Vital-Durand F (1979) Development of the neural basis of visual acuity in monkeys: speculation on the origin of deprivation amblyopia. Trans Ophthalmol Soc UK 99: 363.

Blakemore C, Vital-Durand F (1981) In "The Fetus and Independent Life", (Ciba Foundation Symposium 86), London: Pitman, p 152.

Blakemore C, Vital-Durand F (1982) Development of contrast sensitivity by neurones in monkey striate cortex. J. Physiol 334: 18.

Blakemore C, Vital-Durand F (1983) Visual deprivation prevents the postnatal maturation of spatial resolution and contrast sensitivity for neurones of the monkey's striate cortex. J. Physiol (in the press).

Blakemore C, Garey LJ, Vital-Durand F (1978) The physiological effects of monocular deprivation and their reversal in the monkey's visual cortex. J Physiol 283: 223.

Blakemore C, Vital-Durand F, Garey LJ (1981) Recovery from monocular deprivation in the monkey. 1. Reversal of physiological effects in the visual cortex. Proc R Soc Lond B Biol Sci 213: 399.

LeVay S, Wiesel TN, Hubel DH (1980) The development of ocular dominance columns in normal and visually deprived monkeys. J comp Neurol 191: 1.

Movshon J, Van Sluyters RC (1981) Visual neural development. Ann Rev Psychol 32: 477.

von Noorden GK, Crawford MLJ (1978) Morphological and physiological changes in the monkey visual system after short-term lid suture. Invest Ophthalmol Visual Sci 17: 762.

Swindale NV, Vital-Durand F, Blakemore C (1981) Recovery from monocular deprivation in the monkey. 3. Reversal of anatomical effects in the visual cortex. Proc R Soc Lond B Biol Sci 213:435.

Teller DY, (1981) The development of visual acuity in human and monkey infants. Trends NeuroSci 4(1): 21.

Teller DY, Regal DM, Videen TO, Pulos E (1978) Development of visual acuity in infant monkeys (Macaca nemestrina) during the early postnatal weeks. Vision Res 18: 561.

Index